Monte Carlo Optimization, Simulation and Sensitivity of Queueing Networks

Probability and Mathematical Statistics (Continued)

PURI and SEN • Nonparametric Methods in Multivariate Analysis
RANDLES and WOLFE • Introduction to the Theory of Nonparametric Statistics
RAO • Linear Statistical Inference and Its Applications, *Second Edition*
RAO • Real and Stochastic Analysis
RAO and SEDRANSK • W.G. Cochran's Impact on Statistics
ROHATGI • An Introduction to Probability Theory and Mathematical Statistics
ROHATGI • Statistical Inference
ROSS • Stochastic Processes
RUBINSTEIN • Simulation and The Monte Carlo Method
SCHEFFE • The Analysis of Variance
SEBER • Linear Regression Analysis
SEBER • Multivariate Observations
SEN • Sequential Nonparametrics: Invariance Principles and Statistical Inference
SERFLING • Approximation Theorems of Mathematical Statistics
SHORACK and WELLNER • Empirical Processes with Applications to Statistics
TJUR • Probability Based on Radon Measures

Applied Probability and Statistics

ABRAHAM and LEDOLTER • Statistical Methods for Forecasting
AGRESTI • Analysis of Ordinal Categorical Data
AICKIN • Linear Statistical Analysis of Discrete Data
ANDERSON, AUQUIER, HAUCK, OAKES, VANDAELE, and WEISBERG • Statistical Methods for Comparative Studies
ARTHANARI and DODGE • Mathematical Programming in Statistics
BAILEY • The Elements of Stochastic Processes with Applications to the Natural Sciences
BAILEY • Mathematics, Statistics and Systems for Health
BARNETT • Interpreting Multivariate Data
BARNETT and LEWIS • Outliers in Statistical Data, *Second Edition*
BARTHOLOMEW • Stochastic Models for Social Processes, *Third Edition*
BARTHOLOMEW and FORBES • Statistical Techniques for Manpower Planning
BECK and ARNOLD • Parameter Estimation in Engineering and Science
BELSLEY, KUH, and WELSCH • Regression Diagnostics: Identifying Influential Data and Sources of Collinearity
BHAT • Elements of Applied Stochastic Processes, *Second Edition*
BLOOMFIELD • Fourier Analysis of Time Series: An Introduction
BOX • R. A. Fisher, The Life of a Scientist
BOX and DRAPER • Empirical Model-Building and Response Surfaces
BOX and DRAPER • Evolutionary Operation: A Statistical Method for Process Improvement
BOX, HUNTER, and HUNTER • Statistics for Experimenters: An Introduction to Design, Data Analysis, and Model Building
BROWN and HOLLANDER • Statistics: A Biomedical Introduction
BUNKE and BUNKE • Statistical Inference in Linear Models, Volume I
CHAMBERS • Computational Methods for Data Analysis
CHATTERJEE and PRICE • Regression Analysis by Example
CHOW • Econometric Analysis by Control Methods
CLARKE and DISNEY • Probability and Random Processes: A First Course with Applications, *Second Edition*
COCHRAN • Sampling Techniques,
COCHRAN and COX • Experiment
CONOVER • Practical Nonparametr
CONOVER and IMAN • Introducti
CORNELL • Experiments with Mixt
of Mixture Data

Applied Probability and Statistics (Continued)

COX • Planning of Experiments
COX • A Handbook of Introductory Statistical Methods
DANIEL • Biostatistics: A Foundation for Analysis in the Health Sciences, *Third Edition*
DANIEL • Applications of Statistics to Industrial Experimentation
DANIEL and WOOD • Fitting Equations to Data: Computer Analysis of Multifactor Data, *Second Edition*
DAVID • Order Statistics, *Second Edition*
DAVISON • Multidimensional Scaling
DEGROOT, FIENBERG and KADANE • Statistics and the Law
DEMING • Sample Design in Business Research
DILLON and GOLDSTEIN • Multivariate Analysis: Methods and Applications
DODGE • Analysis of Experiments with Missing Data
DODGE and ROMIG • Sampling Inspection Tables, *Second Edition*
DOWDY and WEARDEN • Statistics for Research
DRAPER and SMITH • Applied Regression Analysis, *Second Edition*
DUNN • Basic Statistics: A Primer for the Biomedical Sciences, *Second Edition*
DUNN and CLARK • Applied Statistics: Analysis of Variance and Regression
ELANDT-JOHNSON and JOHNSON • Survival Models and Data Analysis
FLEISS • Statistical Methods for Rates and Proportions, *Second Edition*
FLEISS • The Design and Analysis of Clinical Experiments
FOX • Linear Statistical Models and Related Methods
FRANKEN, KÖNIG, ARNDT, and SCHMIDT • Queues and Point Processes
GIBBONS, OLKIN, and SOBEL • Selecting and Ordering Populations: A New Statistical Methodology
GNANADESIKAN • Methods for Statistical Data Analysis of Multivariate Observations
GREENBERG and WEBSTER • Advanced Econometrics: A Bridge to the Literature
GROSS and HARRIS • Fundamentals of Queueing Theory, *Second Edition*
GUPTA and PANCHAPAKESAN • Multiple Decision Procedures: Theory and Methodology of Selecting and Ranking Populations
GUTTMAN, WILKS, and HUNTER • Introductory Engineering Statistics, *Third Edition*
HAHN and SHAPIRO • Statistical Models in Engineering
HALD • Statistical Tables and Formulas
HALD • Statistical Theory with Engineering Applications
HAND • Discrimination and Classification
HOAGLIN, MOSTELLER and TUKEY • Exploring Data Tables, Trends and Shapes
HOAGLIN, MOSTELLER, and TUKEY • Understanding Robust and Exploratory Data Analysis
HOEL • Elementary Statistics, *Fourth Edition*
HOEL and JESSEN • Basic Statistics for Business and Economics, *Third Edition*
HOGG and KLUGMAN • Loss Distributions
HOLLANDER and WOLFE • Nonparametric Statistical Methods
IMAN and CONOVER • Modern Business Statistics
JAGERS • Branching Processes with Biological Applications
JESSEN • Statistical Survey Techniques
JOHNSON and KOTZ • Distributions in Statistics
Discrete Distributions
Continuous Univariate Distributions—1
Continuous Univariate Distributions—2
Continuous Multivariate Distributions

(continued on back)

Monte Carlo Optimization, Simulation and Sensitivity of Queueing Networks

REUVEN Y. RUBINSTEIN
Technion-Israel Institute of Technology
and Harvard University

JOHN WILEY & SONS
New York • Chichester • Brisbane • Toronto • Singapore

To the memory of Pinhas Naor,
Oded Levin, and Jonathan Halpern—
Late members of the
Operations Research Department
of the Technion

Library of Congress Cataloging in Publication Data:

Rubinstein, Reuven Y.
Monte Carlo optimization, simulation, and sensitivity of queueing networks.

(Wiley series in probability and mathematical statistics. Applied probability and statistics)
1. Monte Carlo method. 2. Queueing theory. I. Title.
II. Series.
QA298.R79 1986 519.4 86-4040
ISBN 0-471-81929-8

Printed in the United States of America

10 9 8 7 6 5 4 3 2 1

Preface

In the past decade there has been considerable activity related to Monte Carlo simulation. There is a growing recognition that topics such as Monte Carlo optimization, sensitivity and optimization of queueing networks, and variance reduction techniques (especially antithetic and control random variables) deserve separate study. This monograph deals with these topics and is based mainly on my research since 1982. When I was almost finished writing this book, I was introduced to several interesting papers on perturbation analysis by Y. C. Ho, R. Suri, X. R. Cao, C. Cassandras, M. A. Eyler, G. W. Diehl, J. W. Dille, and M. A. Zazanis, all from the Division of Applied Sciences at Harvard University. With these authors' kind permission, I included some material from several of their recent papers as a separate chapter (see Chapter 4). I feel that this will enhance the value of the monograph.

Although this book is related to my earlier *Simulation and the Monte Carlo Method* (Wiley, 1981), it is not meant to replace it. The material is presented mainly as a research monograph. However, selected sections, such as Section 1.1 and Chapters 2 and 5, have been successfully used in my graduate course in simulation. The treatment is theoretical, but emphasis is on concepts rather than on mathematical completeness. It is assumed that the reader has a basic background in probability theory, mathematical statistics, and nonlinear programming.

Chapters 1 and 2 deal with the optimality and efficiency properties of antithetic and control variables used to obtain accurate estimators while simulating complex stochastic systems.

Chapter 1 proves several theorems concerning optimality of antithetic random variables in estimating the expected value of the response sum or response difference of a pair of functions of vector arguments with dependent components. One of these theorems is a modified version of Handscomb's antithetic variates theorem. The efficiency of antithetic random variables for variance reduction under different assumptions on the response function, with application to reliability, network, and queueing systems are also discussed.

Chapter 2 deals with some theoretical and practical aspects of the multidimensional control variables technique. The optimal size of the vector of control variables is derived under the normality assumption, and recommendations are also made for choosing the size of this vector in the nonnormal case.

Chapter 3 considers several stochastic optimization models (inventory, allocation, etc.) under different levels of uncertainty about the underlying distribution function. We show how stochastic approximation (adaptive approach) can be used in this situation for finding the optimal policy. We discuss robust adaptive algorithms and optimization of computer simulation models and give simulation results. Some of the material of this chapter is based on J. Kreimer's Ph.D. thesis and our joint work.

Section 3.6 is of special importance. Here we deal with sensitivity and optimization of computer simulation models. We show how the performance measures and their sensitivities can be estimated simultaneously from a single simulation. Applications to reliability, stochastic networks, and simple queueing models are given. The material in this section is related to that of Chapter 4.

Chapter 4 deals with the perturbation analysis approach recently introduced as an efficient means of estimating performance sensitivities for a rather broad class of complex queueing networks. The beauty of this approach is that it permits estimation of performance sensitivities (gradient, Hessians, etc.) using only a *single* sample path from the underlying system.

Chapter 5 discusses Monte Carlo optimization methods, which now deserve a lot of attention, especially in light of the introduction of the so-called simulated annealing technique for solving N P complete combinatorial optimization problems and problems of computer design and machine learning. This chapter deals with random search methods and several other techniques for finding the global extremum for a multiextremal function. Special attention is paid to the smoothed functional method, according to which we construct an auxiliary (smoothed) function that has better "behavior" than the original one. Operating only with the smoothed function, we find the global extremum for the original one.

The final version of this book was written during my 1985–1986 sabbatical year at Harvard University. I express my gratitude to the perturbation analysis team at Harvard University, and especially to

Professor Yu-Chi Ho, for their hospitality and for providing a rich intellectual environment.

I would also like to express my indebtedness to Xi-ren Cao of Harvard University, Joseph Kreimer of the University of Negev, Israel, and Rajan Suri of the University of Wisconsin for several valuable suggestions for improvement of the earlier draft of the manuscript. Many thanks to Beatrice Shube and Diana Cisek of John Wiley & Sons and to Eliezer Goldberg of Technion for their efficient editorial guidance. I am also most grateful to the U.S. Office of Naval Research Contracts N00014-79-C-0776 and N00014-84-K-0465, the National Science Foundation Grants ECS-82-13680 and ECDR-85-001-08, and a grant from C. S. Draper Laboratories for support of portions of this work.

Finally, I thank the following authors and publishers for granting permission for publication of the cited material:

1. Pages 177–183 based on R. Y. Rubinstein and I. Weisman, The Monte Carlo Method for Global Optimization, *Cah. Cen. Etud. Rech. Oper.*, **21**, No. 2, 143–149 (1979)
2. Pages 229–236 based on R. Y. Rubinstein, Generating Random Vectors Uniformly Distributed Inside and on the Surface of Different Regions, *Eur. J. OR*, **10**, 205–209 (1982).
3. Pages 206–227 based on R. Y. Rubinstein and G. Samorodnitsky, The Efficiency of the Random Search Method, *Math. Comp. Simulation*, **XXIV**, 257–268 (1982).
4. Pages 43–45, 49–68 based on R. Y. Rubinstein and R. Marcus, Efficiency of Multivariate Control Variates in Monte Carlo Simulation, *Oper. Res.* **33**, No. 3, 661–677 (1985).
5. Pages 12–23 based on R. Y. Rubinstein, G. Samorodnitsky, and M. Shaked, Antithetic Variates, Multivariate Dependence and Simulation of Complex Stochastic Systems, *Manage. Sci.*, **31**, No. 1, 66–77 (1985).
6. Pages 236–255 based on R. Y. Rubinstein and G. Samorodnitsky, Optimal Coverage of Complex Regions, *J. Optim. Theory Appl.* (*JOTA*), to appear (1986).
7. Pages 146–147, 151–168 based on Y. C. Ho, R. Suri, X. R. Cao, G. W. Diehl, J. W. Dille, and M. Zazanis, Optimization of Large Multi-Class (Non-Product-Form) Queueing Networks Using Perturbation Analysis, *Large Scale System J.*, **7**, 165–180 (1984).
8. Pages 137–138 based on E. G. Gladyshev, On Stochastic Approximation, *Theory Prob. Appl.* **10**, No. 2, 275–276 (1965).

9. Pages 29–40 based on R. Y. Rubinstein and G. Samorodnitsky, A modified version of Handscomb's Antithetic Variates Theorem, *SIAM J.* (to appear) 1986.

10. Pages 6–12, 23–29 based on R. Y. Rubinstein and G. Samorodnitsky, Antithetic and Common Random Numbers in Simulation Stochastic Systems, *J. Stat. Comp. Simul.*, **2**, 161–184, 1985.

REUVEN Y. RUBINSTEIN

Harvard University
Cambridge, Massachusetts
January 1986

Contents

CHAPTER 1

Antithetic and Common Random Variables in Simulation of Complex Stochastic Systems

1.1 ANTITHETIC VARIABLES, MULTIVARIATE DEPENDENCE, AND SIMULATION OF COMPLEX STOCHASTIC SYSTEMS

1.1.1 Introduction

The problem of obtaining accurate estimators while simulating complex stochastic systems is of importance in Monte Carlo simulation. Recently there has been a rapid growth of interest in use of variance reduction techniques in simulation analysis for the purpose of improving accuracy when estimating output parameters of complex systems.

These techniques comprise [see Rubinstein (1981)] common and antithetic random variables, control random variables, important and stratified sampling, conditional Monte Carlo and random quadrature methods. Variance reduction techniques essentially transform the underlying simulation models into related ones that permit more accurate estimation of the parameters of interest. For a good survey of the current state of variance reduction techniques see Wilson (1984). Recently Nelson and Schmeiser (1983) developed a new analytic framework for variance reduction and a set of concepts from which most variance reduction techniques are composed.

One of the widely used variance reduction techniques is the use of common and antithetic random numbers [see Andreasson (1971); Granovsky (1982); Grant and Solberg (1982); Hammersley and Handscomb (1964); Hammersley and Mauldon (1956); Hammersley and Morton (1956); Kleijnen (1974); Mitchel (1973); Page (1967); Rubinstein (1981); Schruben (1979); Schruben and Margolin (1978); Sullivan, Hayya, and Schaul (1982); Wilson (1979), (1983); Wright and Ramsey (1979) and references therein].

This chapter deals with theoretical and practical aspects of antithetic and common random numbers for variance reduction in simulation of stochastic systems. A proof of their optimality in estimating the expected value of the response sum or the response difference of a pair of functions of vector arguments with independent and dependent components is presented, and variance reduction under weaker conditions is derived. The efficiency of antithetic and common random numbers for variance reduction under different assumptions on the response functions and the type of dependence (e.g., totally positive of order 2, conditionally increasing in sequence, associated, etc.) is discussed. Applications to reliability, networks, and queueing systems are given. Further a modified version of Handscomb's theorem is presented, and some procedures for finding the optimal cumulative distribution function that minimizes the variance of the response sum or the response difference of a pair of functions are presented.

The material of this chapter is based on the following papers: Rubinstein and Samorodnitsky (1985), (1986); and Rubinstein, Shaked, and Samorodnitsky (1985); a large part of them being reproduced almost verbatim.

To motivate the definition of CRNs (common random numbers) and ARNs (antithetic random numbers), consider the following case: Let X and Y be r.v.'s (random variates) with known c.d.f.'s (cumulative distribution functions) F_X and F_Y. We seek an estimator of the expected value of their difference with a minimal variance. Since

$$\operatorname{Var}(X - Y) = \operatorname{Var}(X) + \operatorname{Var}(Y) - 2\operatorname{Cov}(X, Y) \tag{1.1.1}$$

and since the marginal c.d.f.'s of X and Y are given in advance, it follows that the variance of $X - Y$ is minimized by maximizing the covariance. Assume that both X and Y are generated by the inverse transform method

$$\begin{aligned} X &= F_X^{-1}(U_1) = \inf\{x\colon F_X(x) \ge U_1\}\,, \\ Y &= F_Y^{-1}(U_2) = \inf\{y\colon F_Y(y) \ge U_2\}\,, \end{aligned} \tag{1.1.2}$$

where U_1 and U_2 are uniformly distributed on $(0, 1)$. We will say that CRNs are used if $U_1 = U_2$. We will say that ARNs are used if $U_2 = 1 - U_1$. It is known [see, e.g., Whitt (1976)] that when CRNs are used, then

Cov(X, Y) is maximized, so that Var($X - Y$) is minimized. Similarly, Var($X + Y$) is minimized when ARNs are used.

Consider now estimation of $E[g(X) - h(Y)]$ with a minimal variance, where X and Y are univariate r.v.'s g and h are real-valued measurable functions, and the marginal c.d.f.'s F_X and F_Y are known. Mathematically, the problem is: within the set of all two-dimensional c.d.f.'s F_{XY}

$$\begin{aligned}&\text{Minimize } \operatorname{Var}[g(X) - h(Y)]\\&\text{subject to the given marginal c.d.f.'s } F_X \text{ and } F_Y\,.\end{aligned} \tag{1.1.3}$$

Defining

$$g^*(U_1) = g[F_X^{-1}(U_1)]\,, \qquad h^*(U_2) = h[F_Y^{-1}(U_2)]\,, \tag{1.1.4}$$

one can prove† that if g and h are monotonic in the same direction, then

$$\begin{aligned}\min_{F_{XY}} \operatorname{Var}[g(X) - h(Y)] &= \operatorname{Var}[g(F_X^{-1}(U)) - h(F_Y^{-1}(U))]\\ &= \operatorname{Var}[g^*(U) - h^*(U)]\end{aligned} \tag{1.1.5}$$

This means that use of CRNs that is, $U_1 = U_2 = U$, is optimal for problem (1.1.3). If g and h are monotonic in the opposite directions, then by analogy to (1.1.5) the use of ARNs, that is, $U_2 = 1 - U_1$, is optimal.

Problem (1.1.3) lends itself naturally to extension to the case where $\mathbf{X}$ and $\mathbf{Y}$ are n-dimensional vectors. This multidimensional version reads as follows: Find a $2n$-dimensional distribution function $F_{\mathbf{XY}}$ so as to

$$\begin{aligned}\text{Minimize } \operatorname{Var}\{g(\mathbf{X}) - h(\mathbf{Y})\} &= \operatorname{Var}\{g(F_X^{-1}(\mathbf{U}_1)) - h(F_X^{-1}(\mathbf{U}_2))\}\\ &= \operatorname{Var}\{g^*(\mathbf{U}_1) - h^*(\mathbf{U}_2)\}\end{aligned}$$

subject to the given n-dimensional distributions (1.1.6)

$$F_{\mathbf{X}}(x) = \prod_{i=1}^{n} F_{X_i}(x_i) \quad \text{and} \quad F_{\mathbf{Y}}(y) = \prod_{i=1}^{n} F_{Y_i}(y_i)\,,$$

where $\mathbf{X} = F_{\mathbf{X}}^{-1}(\mathbf{U}_1) = \{F_{x_1}^{-1}(U_1^{(1)}), \ldots, F_{x_n}^{-1}(U_1^{(n)})\}$.

Mitchell (1973) suggested using CRNs while estimating $E[g^*(\mathbf{U}_1) - h^*(\mathbf{U}_2)]$, $\mathbf{U}_1, \mathbf{U}_2 \in R^n$; he proved (see his Theorem 1) that if $g^*(\mathbf{u}_1)$ and $h^*(\mathbf{u}_2)$, $\mathbf{u}_1$, $\mathbf{u}_2 \in R^n$, are monotonic in the same direction with respect to each argument of the vector $\mathbf{u}$, then

†For the proof of (1.1.5) see Gal, Rubinstein, and Ziv (1984).

$$\mathrm{Cov}\{g^*(\mathbf{U}),\ h^*(\mathbf{U})\} \geq 0\ , \tag{1.1.7}$$

where $\mathbf{U} = (U^{(1)}, \ldots, U^{(n)})$ is a vector of independent random variables each uniformly distributed on $[0, 1]$. From (1.1.7), it follows that

$$\mathrm{Var}[g^*(\mathbf{U}_1) - h^*(\mathbf{U}_2)] \geq \mathrm{Var}[g^*(\mathbf{U}) - h^*(\mathbf{U})]\ .$$

Thus, while estimating $E[g^*(\mathbf{U}_1) - h^*(\mathbf{U}_2)]$, variance reduction is achieved by using CRNs, that is, $U_1 = U_2 = U$.

Mitchell also demonstrated the efficiency of CRNs while simulating a GI/G/1 queueing system. (For more examples of use of CRNs, see Section 1.1.4).

In Proposition 1 (Section 1.1.5) we shall prove that within the set of all c.d.f.'s $F_{\mathbf{XY}}$ on R^{2n} where dependence is permitted only between like components of $\mathbf{X}$ and $\mathbf{Y}$ (compontents with the same indices), CRNs not only lead to variance reduction but are also optimal. Subsequently, we shall call $\mathbf{U}_1 = \mathbf{U}_2 = \mathbf{U}$ the vector of common random numbers (VCRNs).

In this section we introduce also a new class of function pairs g and h for which the monotonicity assumption is not required. We call these functions pairs *corresponding*. (See Definition 1.) In Proposition 2 (Section 1.1.5) we find conditions under which the VCRNs $\mathbf{U}_1 = \mathbf{U}_2 = \mathbf{U}$ is optimal for problem (1.1.6) while assuming that g and h are corresponding functions.

In many practical problems the r.v.'s X_i (and Y_i), $i = 1, \ldots, n$, are not independent. The problem of obtaining accurate estimators while simulating complex stochastic systems with dependent elements is an important and a difficult one and has received very little attention in recent literature.

In Proposition 4 we prove that under some assumptions on the type of multivariate dependence between the components of $\mathbf{X}$ (and of $\mathbf{Y}$) the use of VCRNs or VARNs minimizes $\mathrm{Var}[g(\mathbf{X}) - h(\mathbf{Y})]$. As a direct consequence of these results, we can conclude that use of VCRNs or VARNs is superior to the crude Monte Carlo method while estimating output parameters of complex stochastic systems with dependent elements. An added advantage of the common or antithetic random numbers technique is that one needs *only half* as many random numbers as needed in the crude Monte Carlo method.

In order to formulate the extended version of equation (1.1.6), assume that $\mathbf{X} = (X_1, \ldots, X_n)$ and $\mathbf{Y} = (Y_1, \ldots, Y_n)$ are random vectors each of which have dependent components. Denote

$$F_{X_i|x_1,\ldots,x_{i-1}}(x_i) = P(X_i \leq x_i | X_1 = x_1, \ldots, X_{i-1} = x_{i-1}) \quad \text{for } i = 2, \ldots, n\ .$$

Define

$$F^{-1}_{X_i|x_1,\ldots,x_{i-1}}(u_i) = \inf\{x_i\colon F_{X_i|x_1,\ldots,x_{i-1}}(x_i) \geq u_i\}\ .$$

Similarly, define $F^{-1}_{Y_i|y_1,\dots,y_{i-1}}(u_i)$. Assume also that X and Y can be generated by the inverse transform method

$$X_1 = F^{-1}_{X_1}(U_1^{(1)}) \equiv \tilde{a}_1(U_1^{(1)}) ,$$

$$X_i = F^{-1}_{X_i|\tilde{a}_1(U_1^{(1)}),\dots,\tilde{a}_{i-1}(U_1^{(1)},\dots,U_{i-1}^{(1)})}(U_i^{(1)}) \equiv \tilde{a}_i(U_1^{(1)}, \dots, U_i^{(1)}), \tag{1.1.8}$$

$$i = 2, \dots, n ,$$

$$Y_1 = F^{-1}_{Y_1}(U_1^{(2)}) \equiv \tilde{b}_1(U_1^{(2)}) ,$$

$$Y_i = F^{-1}_{Y_i|\tilde{b}_1(U)_1^{(2)}),\dots,\tilde{b}_{i-1}(U_1^{(2)},\dots,U_{i-1}^{(2)})}(U_i^{(2)}) \equiv \tilde{b}_i(U_1^{(2)}, \dots, U_i^{(2)}) , \tag{1.1.9}$$

$$i = 2, \dots, n ,$$

where $\mathbf{U}^{(1)}$ and $\mathbf{U}^{(2)}$ are vectors of independent r.v.'s uniformly distributed on $(0, 1)$. Define $a_i(\mathbf{U}^{(1)}) \equiv \tilde{a}_i(U_1^{(1)}, \dots, U_i^{(1)})$, $b_i(\mathbf{U}^{(2)}) \equiv \tilde{b}_i(U_1^{(2)}, \dots, U_i^{(2)})$, $i = 1, \dots, n$. Then we can rewrite equations (1.1.8) and (1.1.9) as

$$X_i = a_i(\mathbf{U}^{(1)}) , \qquad i = 1, \dots, n , \tag{1.1.10}$$

$$Y_i = b_i(\mathbf{U}^{(2)}) , \qquad i = 1, \dots, n . \tag{1.1.11}$$

Now the extended version of problem (1.1.6) reads: Within the set Ξ of all c.d.f.'s on R^{2n} with marginals $F_{\mathbf{X}}$ and $F_{\mathbf{Y}}$, when the dependence is permitted only between like components of vectors $\mathbf{U}^{(1)}$ and $\mathbf{U}^{(2)}$ of equations (1.1.10) and (1.1.11), find a c.d.f. $F_{\mathbf{XY}}$ so as to

$$\begin{aligned} &\text{Minimize } \operatorname{Var}[g(\mathbf{X}) - h(\mathbf{Y})] \\ &\text{subject to } F_{\mathbf{XY}} \in \Xi . \end{aligned} \tag{1.1.12}$$

The layout of this chapter is as follows. In Section 1.1.2 some theoretical results related to problems (1.1.6) and (1.1.2) are considered. Problem (1.1.12) is solved by making some special assumptions on the type of the multivariate dependence. In Section 1.1.3 we present several counterexamples to show that the assumptions of multivariate dependence used in Section 1.1.2 cannot be weakened. Section 1.1.4 deals with simulation results while estimating output parameters of stochastic networks, reliability, and queueing systems with both independent and dependent elements. In Section 1.2 we present a modified version of Handscomb's (1958) antithetic variables theorem and two practical procedures for solving problem (1.1.3) when the monotonicity assumption on g and h is removed.

1.1.2 Theoretical Results

(a) Problem (1.1.6) Let us rewrite problem (1.1.6) as follows:

$$\underset{F_{U_1U_2}}{\text{Minimize}} \ \text{Var}\{g^*(\mathbf{U}_1) - h^*(\mathbf{U}_2)\}$$

where $\mathbf{U}_1$ and $\mathbf{U}_2$ are vectors of independent random variables, each uniformly distributed on [0, 1] and (1.1.6′)

$$g(\mathbf{x}) = g\{F_{X_1}^{-1}(u_1^{(1)}), \ldots, F_{X_n}^{-1}(u_1^{(n)})\} \overset{\Delta}{=} g^*(\mathbf{u}_1)$$

$$h(\mathbf{y}) = g\{F_{Y_1}^{-1}(u_2^{(1)}), \ldots, F_{Y_n}^{-1}(u_2^{(n)})\} \overset{\Delta}{=} h^*(\mathbf{u}_2)$$

We distinguish the following three cases of problem (1.1.6):

A. The functions $g(\mathbf{x})$ and $h(\mathbf{y})$, $\mathbf{x}, \mathbf{y} \in R^n$, are monotonic in each component of $\mathbf{x}$ and $\mathbf{y}$. In this case $g^*(\mathbf{u}_1)$ and $h^*(\mathbf{u}_2)$ are also monotonic in each component of $\mathbf{u}_1$ and $\mathbf{u}_2$, respectively, since all $F_{X_i}^{-1}$ and $F_{Y_i}^{-1}$, $i = 1, \ldots, n$, are monotonic nondecreasing functions. In Proposition 1 of section 1.1.5 we prove that within the set of all c.d.f.'s $F_{\mathbf{U}_1,\mathbf{U}_2}$ on R^{2n}, where dependence is permitted *only between like components* of $\mathbf{U}_1$ and $\mathbf{U}_2$ (components with the same indices), the following policies are optimal: (a) the VCRNs ($\mathbf{U}_1 = \mathbf{U}_2 = \mathbf{U}$) if g and h are monotonic in the same direction; (b) the VARNs with $\mathbf{U}_2 = \mathbf{1} = \mathbf{U}_1$, where $\mathbf{1}$ is the unity vector if g and h are monotonic in the opposite direction. The last statement may be reformulated as follows: Within the set of all c.d.f.'s $F_{\mathbf{XY}}$ on R^{2n}, where dependence is permitted only between like components of $\mathbf{X}, \mathbf{Y}$ (components with the same indices), the random vector $\{F_{\mathbf{X}}^{-1}(\mathbf{U}), F_{\mathbf{Y}}^{-1}(\mathbf{U})\}$ has a c.d.f. with maximum correlation (and the random vector $\{F_{\mathbf{X}}^{-1}(\mathbf{U}), F_{\mathbf{Y}}^{-1}(\mathbf{1} - \mathbf{U})\}$ has a c.d.f. with minimum correlation) and therefore solves problem (1.1.6).

It also follows from Proposition 1 that if $g(\mathbf{u}_1)$ and $h(\mathbf{u}_2)$ are monotonic increasing with respect to some components of $\mathbf{u}_1, \mathbf{u}_2$ and monotonic decreasing with respect to other components, then the following policy is optimal:

$$\bar{u}_2^{(k)} = \begin{cases} u_1^{(k)}, & \text{if } g \text{ and } h \text{ are monotonic in the same direction in the } k\text{th component;} \\ 1 - u_1^{(k)}, & \text{if } g \text{ and } h \text{ are monotonic in the opposite direction in the } k\text{th component.} \end{cases} \quad (1.1.13)$$

We shall call the vector $\bar{\mathbf{u}}_2 = (\bar{u}_2^{(1)}, \ldots, \bar{u}_2^{(n)})$ with components as per (1.1.13), the vector of common-antithetic random numbers (VCARNs).

For example, let $g(u_1^{(1)}, u_1^{(2)})$ be monotonic increasing in $u_1^{(1)}$ and monotonic decreasing in $u_1^{(2)}$ and $h(u_2^{(1)}, u_2^{(2)})$ be monotonic increasing in both $u_2^{(1)}$ and $u_2^{(2)}$. Then the vector $\bar{\mathbf{u}}_2 = (\bar{u}_2^{(1)}, \bar{u}_2^{(2)}) = (u_1^{(1)}, 1 - u_1^{(2)})$ is the optimal VCARNs.

In the particular case $\mathbf{X} \sim U(\mathbf{a}_1, \mathbf{b}_1)$ and $\mathbf{Y} \sim U(\mathbf{a}_2, \mathbf{b}_2)$, where $\mathbf{X}, \mathbf{Y}, \mathbf{a}, \mathbf{b} \in R^n$, the optimal $F_{\mathbf{XY}}$ can be computed directly instead of the optimal $F_{\mathbf{U}_1\mathbf{U}_2}$. Indeed, let $g(\mathbf{X})$ and $h(\mathbf{Y})$ be monotonic in the same direction; then, taking into account that

$$(F_{X_1}(x_1), \ldots, F_{X_n}(x_n)) = \frac{\mathbf{x} - \mathbf{a}_1}{\mathbf{b}_1 - \mathbf{a}_1} = \mathbf{u}_1 , \quad \text{where } \frac{\mathbf{a}}{\mathbf{b}} \overset{\Delta}{=} \left(\frac{a^{(1)}}{b^{(1)}}, \ldots, \frac{a^{(n)}}{b^{(n)}} \right),$$

we see that the optimal $\mathbf{Y}$ satisfies

$$\mathbf{Y} = \mathbf{a}_2 + \frac{\mathbf{X} - \mathbf{a}_1}{\mathbf{b}_1 - \mathbf{a}_1} (\mathbf{b}_2 - \mathbf{a}_2) ,$$

which means that the optimal bivariate distribution is uniform over the positive-slope hyperdiagonal of the hyperrectangle $\mathbf{a}_1 < \mathbf{x} < \mathbf{b}_1$, $\mathbf{a}_2 < \mathbf{y} < \mathbf{b}_2$.

If $g(\mathbf{x})$ and $h(\mathbf{y})$ are monotonic in opposite direction, then analogous considerations yield that the optimal Y satisfies

$$\mathbf{Y} = \mathbf{a}_2 + \frac{\mathbf{b}_1 - \mathbf{X}}{\mathbf{b}_1 - \mathbf{a}_1} (\mathbf{b}_2 - \mathbf{a}_2) ,$$

which means that the optimal distribution is uniform over the negative-slope hyperdiagonal of the hyperrectangle in question.

The above formulas can be summarized in the following procedure:

Procedure 1

1. Generate $\mathbf{X} \sim U(\mathbf{a}_1, \mathbf{b}_1)$.
2. Calculate $\mathbf{A} = \mathbf{a}_2 + \dfrac{\mathbf{X} - \mathbf{a}_1}{\mathbf{b}_1 - \mathbf{a}_1} (\mathbf{b}_2 - \mathbf{a}_2)$.
3. If g and h are monotonic in the same direction, then choose $\mathbf{Y} = \mathbf{A}$; if they are monotonic in opposite directions, then choose $\mathbf{Y} = \mathbf{a}_2 + \mathbf{b}_2 - \mathbf{A}$.
4. Take $g(\mathbf{X}) - h(\mathbf{Y})$ as the desired estimate.

It is worthwhile to note that the VCARN (1.1.13) is optimal if dependence is permitted between like components of vectors $\mathbf{u}_1$ and $\mathbf{u}_2$ only. If we waive this restriction, then the VCARN (1.1.13) is generally not optimal for problem (1.1.6), as will be shown in Example 1 below.

B. Now consider the situation in which the functions g and h are nonmonotonic. In this case Proposition 1 cannot be directly applied for variance reduction; there is no practical procedure known to the authors that can be used for solving problem (1.1.6) except where analytical (or numerical) expressions are available for the inverses of the c.d.f.'s of $g(\mathbf{X})$ and $h(\mathbf{Y})$, denoted by $F_{g(\mathbf{X})}^{-1}$ and $F_{h(\mathbf{Y})}^{-1}$. In this case, applying the inverse transform method to $W = g(\mathbf{X})$ and $Z = h(\mathbf{Y})$, we obtain

$$W = g(\mathbf{X}) \triangleq F_{g(\mathbf{X})}^{-1}(U_1) = \tilde{g}(U_1),$$

and

$$Z = h(\mathbf{Y}) \triangleq F_{h(\mathbf{Y})}^{-1}(U_2) = \tilde{h}(U_2).$$

Since F_W^{-1} and F_Z^{-1} are monotonic nondecreasing functions on $(0, 1)$, use of CRNs ($U_2 = U_1$) minimizes the variance of $\tilde{g}(U_1) - \tilde{h}(U_2)$. The application of this technique is limited because of the difficulty in obtaining analytical expressions (or numerical tables) for $\tilde{g}$ and $\tilde{h}$.

We now describe a procedure for reducing (rather than minimizing) the variance of $g(\mathbf{X}) - h(\mathbf{Y}) = g^*(\mathbf{U}_1) - h^*(\mathbf{U}_2)$. Whereas this procedure is not optimal, it does not require the availability of $F_{g(\mathbf{X})}^{-1}$ and $F_{h(\mathbf{Y})}^{-1}$ while requiring certain information on the behavior of g^* and h^*, namely identification of the subregions of monotonicity of g^* and h^*. The intersections of these regions in turn yield disjoint subregions $L_i^{(1)} \times L_j^{(2)}$, $i = 1, \ldots, m$, $j = 1, \ldots, l$,

$$L_i^{(1)} = \{\mathbf{u}\colon \mathbf{v}_i \le \mathbf{u} < \mathbf{v}_{i+1}\}, \qquad L_j^{(2)} = \{\mathbf{u}\colon \mathbf{w}_j \le \mathbf{u} < \mathbf{w}_{j+1}\}$$

covering the $2n$-dimensional unit hypercube and such that both $g^*(\mathbf{u})$ and $h^*(\mathbf{u})$ are monotonic in each of them.

The idea behind our procedure can be illustrated as follows: Let $\{\mathbf{U}_1, \mathbf{U}_2\}$ be a $2n$-dimensional vector such that $\mathbf{U}_1 = (U_1^{(1)}, \ldots, U_1^{(n)})$ and $\mathbf{U}_2 = (U_2^{(1)}, \ldots, U_2^{(n)})$ are a pair of n-dimensional vectors with independent components each uniformly distributed on $[0, 1]$. The vector $(\mathbf{u}_1, \mathbf{u}_2)$ falls within one of the ml $2n$-dimensional hyperrectangles $L_i^{(1)} \times L_j^{(2)}$, $i = 1, \ldots, m$, $j = 1, \ldots, l$. Let us present g^* and h^* as

$$g^*(\mathbf{u}_1) = \sum_{i=1}^{m} g^*(\mathbf{u}_1) I_i^{(1)}(\mathbf{u}_1), \qquad h^*(\mathbf{u}_2) = \sum_{j=1}^{l} h^*(\mathbf{u}_2) I_j^{(2)}(\mathbf{u}_2), \tag{1.1.13a}$$

where

$$I_k^{(i)}(u) = \begin{cases} 1, & \text{if } \mathbf{u} \in L_k^{(i)},\ k = 1, \ldots, m \text{ or } k = 1, \ldots, l; \\ 0, & \text{otherwise}. \end{cases}$$

Now

$$E[g(\mathbf{X}) - h(\mathbf{Y})] = E[g^*(\mathbf{U}_1) - h^*(\mathbf{U}_2)] \tag{1.1.13b}$$

$$= \sum_{i=1}^{m} \sum_{j=1}^{l} P_{ij} E[g^*(\mathbf{U}_1) I_i^{(1)}(\mathbf{U}_1) - h^*(\mathbf{U}_2) I_j^{(2)}(\mathbf{U}_2)] ,$$

where $P_{ij} = P((\mathbf{U}_1, \mathbf{U}_2) \in L_i^{(1)} \times L_j^{(2)})$.

Since both $g^*(\mathbf{u}_1) I_i^{(1)}$ and $h^*(\mathbf{u}_2) I_j^{(2)}$ are monotonic functions in $L_i^{(1)}$ and $L_j^{(2)}$, respectively, variance reduction can be obtained by using common (antithetic) random numbers in each subregion, (see Proposition 3). Formally our procedure can be written as:

Procedure 2

1. Divide the region of variation of $g(\mathbf{x})$ into m_1 disjoint subregions $(\mathbf{a}_{i,1}, \mathbf{a}_{i+1,1})$, $1 \le i \le m_1$, such that $g(\mathbf{x})$ is monotonic in each of them. Perform an analogous operation for $h(\mathbf{y})$ and obtain a set of disjoint subregions $(\mathbf{a}_{j,2}, \mathbf{a}_{j+1,2})$, $1 \le j \le m_2$.
2. Define the sets $S_1^{(l)} = \{F_{X_l}(a_{i,1}^{(l)}),\ 1 \le i \le m_{1,l}\}$ and $S_2^{(l)} = \{F_{Y_l}(a_{j,2}^{(l)}),\ 1 \le j \le m_{2,l}\}$, $l = 1, 2, \ldots, n$. Arrange the elements of $S_t^{(l)}$, $t = 1, 2$, so that $S_1^{(l)} = (w_1^{(l)}, \ldots, w_{m_{1,l}}^{(l)})$, $S_2^{(l)} = (v_1^{(l)}, \ldots, v_{m_{2,l}}^{(l)})$. Then $S_t^{(1)} \times S_t^{(2)} \times \cdots \times S_t^{(n)}$, $t = 1, 2$, define the subregions $L_i^{(1)}$ and $L_i^{(2)}$.
3. Generate a pair of independent random vectors $\mathbf{U}_1 = (U_1^{(1)}, \ldots, U_1^{(n)})$ and $\mathbf{U}_2 = (U_2^{(1)}, \ldots, U_2)$ each component of which is uniformly distributed on (0, 1), and determine the corresponding $L_i^{(1)} \times L_j^{(2)}$.
4. Calculate the vector $\mathbf{A} = \mathbf{v}_j + \dfrac{\mathbf{U}_1 - \mathbf{v}_i}{\mathbf{v}_{i+1} - \mathbf{v}_i} (\mathbf{v}_{j+1} - \mathbf{v}_j)$ (see Procedure 1).
5. If g^* and h^* are monotonic in the same directions in $L_i^{(1)} \times L_j^{(2)}$ with respect to the kth component, choose $\bar{U}_2^{(k)} = A^{(k)}$, where $A^{(k)}$ is the kth component of $\mathbf{A}$. If they are monotonic in opposite directions, choose $\bar{U}_2^{(k)} = v_j^{(k)} + v_{j+1}^{(k)} - A^{(k)}$.
6. Take $g^*(\mathbf{U}_1) - h^*(\bar{\mathbf{U}}_2)$ as the desired estimate.

The following examples serve as illustrations for the cases A and B respectively.

Example 1 Let g and h be monotonic in each component function, and let $\mathbf{U}_1 = (U_1^{(1)}, U_1^{(2)})$ and $\mathbf{U}_2 = (U_2^{(1)}, U_2^{(2)})$ be a pair of vectors with independent components, each of which is uniformly distributed on [0, 1]. Define

$$g(u_1^{(1)}, u_2^{(1)}) = u_1^{(1)} + e^{-u_2^{(1)}}, \qquad h(u_1^{(2)}, u_2^{(2)}) = 5u_1^{(2)}(u_2^{(2)})^2$$

and assume that the dependence is permitted only between like components. We have

$$Eg = 1.132, \qquad Eg^2 = 1.398,$$
$$\text{Var } g = Eg^2 - [Eg]^2 = 0.116,$$
$$Eh = 0.833, \qquad Eh^2 = 1.667, \qquad \text{Var } h = 0.972,$$

and therefore for independent sampling

$$\text{Var}(g - h) = \text{Var } g + \text{Var } h = 1.088.$$

Noting that g is monotonic increasing in $u_1^{(1)}$ and monotonic decreasing in $u_1^{(2)}$ and h is monotonic increasing in both $u_2^{(1)}$ and $u_2^{(2)}$, it follows from (1.1.13) that the vector $\bar{\mathbf{U}}_2 = (\bar{U}_2^{(1)}, \bar{U}_2^{(2)}) = (U_1^{(1)}, 1 - U_1^{(2)})$ is the optimal VCARN.

In this case

$$Egh = E[(U_1 + e^{-U_2})5U_1(1 - U_2)^2] = 1.216,$$
$$\text{Cov}(g, h) = Egh - EgEh = 0.273,$$

and

$$\text{var}(g - h) = 0.542,$$

which is 50.18% less than the variance for independent sampling.

If dependence is permitted between any component of $\mathbf{U}_1$ and $\mathbf{U}_2$ rather than between like components, then the vector

$$\bar{\mathbf{U}}_2 = (\bar{U}_2^{(1)}, \bar{U}_2^{(2)}) = (1 - U_1^{(2)}, U_1^{(1)})$$

minimizes $\text{Var}(g - h)$. In this case

$$E(gh) = E[(U_1 + e^{-U_2})5(1 - U_2)U_1^2] = 1.238,$$
$$\text{Cov}(g, h) = 0.295,$$
$$\text{Var}(g - h) = 0.498,$$

which is 54.23% less than $\text{Var}(g - h) = 1.088$ for independent sampling.

Example 2 Let g and h be piece-wise monotonic functions and $\mathbf{U}_1 = (U_1^{(1)}, U_1^{(2)})$ and $\mathbf{U}_2 = (U_2^{(1)}, U_2^{(2)})$ be a pair of vectors with independent components each of which is uniformly distributed on $[0, 1]$. Define

$$g(u_1^{(1)}, u_2^{(1)}) = u_1^{(1)} + u_2^{(1)} ,$$
$$h(u_2^{(1)}, u_2^{(2)}) = (u_2^{(1)} - \tfrac{1}{2})^2 + (u_2^{(2)} - \tfrac{1}{2})^2 .$$

For independent sampling we have $\operatorname{Var} g = 0.167$, $\operatorname{Var} h = 0.011$, and therefore $\operatorname{Var}(g - h) = 0.178$.

It is readily seen that there the following are the four regions of monotonicity of h:

$$L_1 = [0, \tfrac{1}{2}) \times [0, \tfrac{1}{2}) , \qquad L_2 = [0, \tfrac{1}{2}) \times [\tfrac{1}{2}, 1) ,$$
$$L_3 = [\tfrac{1}{2}, 1) \times [0, \tfrac{1}{2}) , \qquad L_4 = [\tfrac{1}{2}, 1) \times [\tfrac{1}{2}, 1) ,$$

and g is monotone on $I = [0, 1) \times [0, 1)$.

We have therefore that in

(a) $I \times L_1$, functions g and h are monotonic in opposite direction in both u_1 and u_2;
(b) $I \times L_2$, functions g and h are monotonic in opposite direction in u_1 and are monotonic in the same direction in u_2;
(c) $I \times L_3$, functions g and h are monotonic in the opposite direction in u_2 and are monotonic in the same direction in u_1;
(d) $I \times L_4$, functions g and h are monotonic in the same direction in both u_1 and u_2.

We therefore have for Procedure 2 (step 5) that in

(a) $I \times L_1$, $\bar{u}_2^{(1)} = 0.5 - 0.5u_1^{(1)}$, $\bar{u}_2^{(2)} = 0.5 - 0.5u_1^{(2)}$;
(b) $I \times L_2$, $\bar{u}_2^{(1)} = 0.5 - 0.5u_1^{(1)}$, $\bar{u}_2^{(2)} = 0.5 + 0.5u_1^{(2)}$;
(c) $I \times L_3$, $\bar{u}_2^{(1)} = 0.5 + 0.5u_1^{(1)}$, $\bar{u}_2^{(2)} = 0.5 - 0.5u_1^{(2)}$;
(d) $I \times L_4$, $\bar{u}_2^{(1)} = 0.5 + 0.5u_1^{(1)}$, $\bar{u}_2^{(2)} = 0.5 + 0.5u_1^{(2)}$.

In this case it is straightforward to show that

$$E(gh) = 0.186 , \qquad EgEh = 0.167 ,$$
$$\operatorname{Cov}(gh) = 0.186 - 0.167 = 0.019 ,$$

and therefore

$$\operatorname{Var}(g - h) = 0.178 - 0.038 = 0.140 ,$$

which reduces the variance relatively to the independent sampling by 21.34%.

C. We shall now introduce a new class of pairs of function where the concept of monotonicity refers to the pair $\{g, h\}$ as such instead of to the components $\{x_i, y_i, 1 \le i \le n\}$, and prove the optimality of VCRN.

Definition 1 We say that a pair of real-valued measurable functions $g(\mathbf{x})$ and $h(\mathbf{x})$, $\mathbf{x} \in R^n$ are *corresponding* in the same (opposite) direction if for any $\mathbf{x}_1$, $\mathbf{x}_2$

$$g(\mathbf{x}_1) \le g(\mathbf{x}_2) \leftrightarrow h(\mathbf{x}_1) \underset{(\ge)}{\le} h(\mathbf{x}_2) .$$

Assume that $\mathbf{X}$ and $\mathbf{Y}$ have the same cumulative distribution functions, that is, $\mathbf{X} \overset{d}{=} \mathbf{Y}$, and g and h are corresponding functions. Since $\mathbf{X} \overset{d}{=} \mathbf{Y}$, correspondence of g and h leads to correspondence of g^* and h^*. In Proposition 2, we prove that if g^* and h^* are corresponding in the same direction, then within the set of all c.d.f.'s $F_{\mathbf{U}_1\mathbf{U}_2}$ on R^{2n} the VCRN ($\mathbf{U}_2 = \mathbf{U}_1 = \mathbf{U}$) is optimal† for problem (1.1.6).

Example 3 Let $\mathbf{U}_1 = (U_1^{(1)}, U_1^{(2)})$ and $\mathbf{U}_2 = (U_2^{(1)}, U_2^{(2)})$ be a pair of vectors with independent components each of which is uniformly distributed on [0, 1]. Define

$$g(\mathbf{u}_1^{(1)}, \mathbf{u}_1^{(2)}) = \begin{cases} \dfrac{1}{u_1^{(1)} + u_1^{(2)}} , & u_1^{(1)} + u_1^{(2)} \ge 1 , \\ (u_1^{(1)} u_1^{(2)})^2 + 1 , & u_1^{(1)} + u_1^{(2)} < 1 ; \end{cases}$$

$$h(\mathbf{u}_2^{(1)}, \mathbf{u}_2^{(2)}) = \begin{cases} -1 - u_2^{(1)} - u_2^{(2)} , & u_2^{(1)} + u_2^{(2)} \ge 1 , \\ u_2^{(1)} u_2^{(2)} , & u_2^{(1)} + u_2^{(2)} < 1 . \end{cases}$$

We have $Eg = 0.829$, $Eh = -1.125$, $\text{Var}\, g = 0.022$, $\text{Var}\, h = 1.490$, and therefore for independent sampling $\text{Var}(g - h) = 1.512$. Noting that g and h are corresponding in the same direction, we conclude that the VCRNs [i.e., $\mathbf{U}_2 = (U_2^{(1)}, U_2^{(2)}) = (U_1^{(1)}, U_1^{(2)})$] is optimal for problem (1.1.8). In this case it is readily found that $\text{Cov}(f, h) = 0.119$ and therefore $\text{Var}(g - h) = 1.512 - 2 \times 0.119 = 1.274$, which is 15.75% less than the variance for independent sampling.

†Note that we do not require the restriction that dependence is permitted only between like components of $\mathbf{U}_1$ and $\mathbf{U}_2$ (see Proposition 2).

(b) Problem (1.1.12) We shall show that if g and h are monotone in the same direction, then the use of VCRNs yields on optimal solution for problem (1.1.12) whenever each of the random vectors $\mathbf{X}$ and $\mathbf{Y}$ satisfy some conditions of positive dependence given in Definitions 1 and 2. Similarly, it will be shown that if g and h are monotone in opposite directions, then the use of VARNs yields an optimal solution for (1.1.12).

Definition 2 The r.v.'s $Z_1, \ldots, Z_n$ [or the random vector $\mathbf{Z} = (Z_1, \ldots, Z_n)$] are said to be CIS (conditionally increasing in sequence) if $P(Z_i \leq z_i \mid Z_1 = z_1, \ldots, Z_{i-1} = z_{i-1})$ is nonincreasing in $z_1, \ldots, z_{i-1}$ for every z_i, $i = 2, \ldots, n$.

Definition 3 Let $\mathbf{Z} = (Z_1, \ldots, Z_n)$ be a random vector with a probability density function (or a probability mass function) $f(z_1, \ldots, z_n)$. The r.v.'s $Z_1, \ldots, Z_n$ (or the random vector $\mathbf{Z}$) are said to be TP_2 (totally positive of order 2) in pairs if for every i and j $(1 \leq i < j \leq n)$ and every $z_1, \ldots, z_{i-1}, z_{i+1}, \ldots, z_{j-1}, z_{j+1}, \ldots, z_n$,

$$\begin{vmatrix} f(z_1, \ldots, z_i, \ldots, z_j, \ldots, z_n) & f(z_1, \ldots, z_i, \ldots, z_j', \ldots, z_n) \\ f(z_1, \ldots, z_i', \ldots, z_j, \ldots, z_n) & f(z_1, \ldots, z_i', \ldots, z_j', \ldots, z_n) \end{vmatrix} \geq 0$$

whenever $z_i < z_i'$ and $z_j < z_j'$.

It is well known [see, e.g., Barlow and Proschan (1975), Chapter 5] that:

FACT 1 If $Z_1, \ldots, Z_n$ are independent, then $\mathbf{Z}$ is TP_2 in pairs.
FACT 2 If $\mathbf{Z}$ is TP_2 in pairs, then $\mathbf{Z}$ is CIS.

A detailed list of c.d.f.'s of r.v.'s that are TP_2 in pairs is given in Karlin and Rinott (1980). Karlin and Rinott (1981) and Bolviken (1982) showed that some multinormal r.v.'s as well as their absolute values are TP_2 in pairs and hence CIS. Other examples are r.v.'s, which have multivariate logistic, multivariate gamma, multivariate F, and negative multinomial c.d.f.'s. Barlow and Proschan (1975, p. 151) showed that each pair of order statistics is CIS. Dykstra and Hewett (1978) observed that the characteristic roots of Wishart matrices are TP_2 in pairs. Karlin and Rinott (1981) noted that the absolute values of a multivariate Cauchy random vector are TP_2 in pairs. Many other examples of CIS and TP_2 in pairs random vectors are given in Barlow and Proschan (1975, Chapter 5).

The following lemma is the key step in obtaining the optimal solution for problem (1.1.12). The proof of the lemma is reminiscent of the proof of Lemma 4.8, p. 147, of Barlow and Proschan (1975).

Lemma 1 If $Z_1, \ldots, Z_n$ are CIS, then there exist nondecreasing functions c_i, $i = 1, \ldots, n$, from R^n to R such that

$$(Z_1, \ldots, Z_n) \stackrel{d}{=} (c_1(\mathbf{U}), \ldots, c_n(\mathbf{U})),$$

where $\mathbf{U} = (U_1, \ldots, U_n)$ is a vector of independent r.v.'s uniformly distributed on $(0, 1)$ and $\stackrel{d}{=}$ denotes equality in distribution.

Proof Let

$$c_1(\mathbf{u}) = F_{Z_1}^{-1}(u_1),$$

$$c_i(\mathbf{u}) = F_{Z_i|c_1(\mathbf{u}), \ldots, c_{i-1}(\mathbf{u})}^{-1}(u_i), \qquad i = 2, \ldots, n.$$

It is easy to see that

$$(Z_1, \ldots, Z_n) \stackrel{d}{=} (c_1(\mathbf{U}), \ldots, c_n(\mathbf{U})).$$

Thus we only have to demonstrate the monotonicity of c_i, $i = 1, \ldots, n$. The proof is by induction on i.

By definition, $c_1(\mathbf{u})$ is nondecreasing in $\mathbf{u}$. Assume now that $c_1(\mathbf{u}), \ldots, c_{i-1}(\mathbf{u})$ are nondecreasing in $\mathbf{u}$. Note that since $c_j(\mathbf{u})$, $j = 1, \ldots, i-1$, are independent of u_i, it follows from the definition of c_i that c_i is nondecreasing in u_i. To show the monotonicity of c_i in $u_1, \ldots, u_{i-1}$, let $u_j^{(1)} \le u_j^{(2)}$, $j = 1, \ldots, i-1$ and let $u_j^{(1)}$, $u_j^{(2)}$, $j = i, \ldots, n$, be any numbers such that $u_i^{(1)} = u_i^{(2)}$ [$= u_i$, say]. Since $c_j(\mathbf{u})$ depends only on $u_1, \ldots, u_j$, it follows from the induction hypothesis that $c_j(\mathbf{u}^{(1)}) \le c_j(\mathbf{u}^{(2)})$, $j = 1, \ldots, i-1$. Thus, since $\mathbf{Z}$ is CIS, it follows that

$$F_{Z_i|c_1(\mathbf{u}^{(1)}), \ldots, c_{i-1}(\mathbf{u}^{(1)})}(z_i) \ge F_{Z_i|c_1(\mathbf{u}^{(2)}), \ldots, c_{i-1}(\mathbf{u}^{(2)})}(z_i) \quad \text{for all } z_i. \tag{1.1.14}$$

But (1.1.14) is equivalent to

$$F_{Z_i|c_1(\mathbf{u}^{(1)}), \ldots, c_{i-1}(\mathbf{u}^{(1)})}^{-1}(u_i) \le F_{Z_i|c_1(\mathbf{u}^{(2)}), \ldots, c_{i-1}(\mathbf{u}^{(2)})}^{-1}(u_i) \quad \text{for all } u_i \in (0, 1).$$

Thus $c_i(\mathbf{u}^{(1)}) \le c_i(\mathbf{u}^{(2)})$, that is, $c_i(\mathbf{u})$ is nondecreasing in $\mathbf{u}$. Q.E.D.

In Proposition 4 of section 1.1.5 we prove that if X and Y are CIS and g and h are monotone in the same (opposite) direction, then the use of VCRNs (VARNs) yields an optimal solution to problem (1.1.12).

In Proposition 5 of Section 1.1.5 we prove the optimality of VCRNs for problem (1.1.12) assuming that g and h are corresponding and X and Y are CIS and have the same distribution.

1.1.3 Counterexamples

In Section 1.1.2 was shown that if X and Y are CIS and if g and h are monotone, then the optimal solution to problem (1.1.12) is obtained by using VCRNs or VARNs. One might conjecture that the conclusion of Proposition 4 is still true even if the positive dependence assumptions are weakened. In this section we show that this is not the case.

The positive dependence condition (iii) given in Definition 4 is weaker than CIS [see, e.g., Barlow and Proschan (1975), p. 149]. Condition (i) is weaker than TP_2 in pairs [see, e.g., Alam and Wallenius (1976) or Block and Ting (1981)]. Using their arguments, it can be shown also that condition (ii) is weaker than TP_2 in pairs. Other conditions that are weaker than those of Definition 4 can be found, for example, in Block and Ting (1981).

Definition 4 The r.v.'s $Z_1, \ldots, Z_n$ [or the random vector $\mathbf{Z} = (Z_1, \ldots, Z_n)$] are called

(i) RCSI (right corner set increasing) if $P(Z_i > z'_i, i = 1, \ldots, n \mid Z_i > z_i, i = 1, \ldots, n)$ is nondecreasing in $z_1, \ldots, z_n$ for all $z'_i, \ldots, z'_n$;

(ii) LCSD (left corner set decreasing) if $P(Z_i \geq z'_i, i = 1, \ldots, n \mid Z_i \leq z_i, i = 1, \ldots, n)$ is nonincreasing in $z_1, \ldots, z_n$ for all $z'_1, \ldots, z'_n$ [Brindley and Thompson (1972)]; and

(iii) associated if $\mathrm{Cov}(\psi(\mathbf{Z}), \phi(\mathbf{Z})) \leq 0$ for all nondecreasing functions ψ and ϕ such that the covariance exists.

In the following counterexamples it is shown that none of the conditions (i)–(iii) is sufficient for the conclusion of Proposition 4 to hold.

Counterexample 1 (i) and (iii) are not sufficient for the conclusion of Proposition 4. Let X_1 and X_2 have the following joint probabilities:

x_1 \ x_2	0	1	M	
0	0.2	0.2	0	0.4
1	0.2	0	0	0.2
M	0	0	0.4	0.4
	0.4	0.2	0.4	1

where $M > 1$ is some constant. The associated survival function $\bar{F}(x_1, x_2) \equiv P(X_1 > x_1, X_2 > x_2)$ is TP_2 [that is, it satisfies $\bar{F}(x_1, x_2)\bar{F}(x'_1, x'_2) \geq \bar{F}(x'_1, x_2)\bar{F}(x_1, x'_2)$ whenever $x_1 < x'_1$, $x_2 < x'_2$]. Hence

$(X_1, X_2)$is RCSI [see, e.g., Shaked (1977)]. To show that X_1 and X_2 are associated, we have to show [see Shaked (1982)] that

$$P(X \in A \cap B)P(X \in A^c \cap B^c) \geq P(X \in A \cap B^c)P(X \in A^c \cap B) \tag{1.1.15}$$

for all increasing sets A and B [a set C is increasing if $(x_1, x_2) \in C$ and $y_i \geq x_i$, $i = 1, 2$, imply that $(y_1, y_2) \in C$] where A^c and B^c denote the complements of A and B. Now, inequality (1.1.15) is not trivially satisfied only when $A \cap B = (M, M)$ and $A^c \cap B^c = (0, 0)$. But it is easy to check that even then (1.1.15) holds. Hence X_1 and X_2 are associated.

Using the notation of (1.1.10), we have

$$a_1(u_1, u_2) = \begin{cases} 0 & \text{if } 0 \leq u_1 < 0.4\,, \\ 1 & \text{if } 0.4 \leq u_1 < 0.6\,, \\ M & \text{if } 0.6 \leq u_1 < 1\,, \end{cases} \tag{1.1.16}$$

$$a_2(u_1, u_2) = \begin{cases} 0 & \text{if } 0 \leq u_1 < 0.4,\ 0 \leq u_2 < 0.5\,, \\ 1 & \text{if } 0 \leq u_1 < 0.4,\ 0.5 \leq u_2 < 1\,, \\ 0 & \text{if } 0.6 \leq u_1 < 0.6,\ 0 \leq u_2 < 1\,, \\ M & \text{if } 0.6 \leq u_1 < 1,\quad 0 \leq u_2 < 1\,. \end{cases} \tag{1.1.17}$$

Let (Y_1, Y_2)be a random vector such that $(Y_1, Y_2) \stackrel{d}{=} (X_1, X_2)$. Then, using the notation of (1.1.10)–(1.1.11), we have $b_1 = a_1$ and $b_2 = a_2$.

Let $g(\mathbf{x}) = x_2$ and $h(\mathbf{y}) = -y_2$ and consider $\operatorname{Var}(g(\mathbf{X}) - h(\mathbf{Y}))$. Define $\tilde{g}(u_1, u_2) = a_2(u_1, u_2)$ and $\tilde{h}(u_1, u_2) = -b_2(u_1, u_2)$. If the conclusion of Proposition 4 were true then, using VARN, we would have had $\operatorname{Cov}(\tilde{g}(U_1, U_2), \tilde{h}(1 - U_1, 1 - U_2)) \geq 0$, where U_1 and U_2 are independent r.v.'s uniformly distributed on (0, 1). However, here,

$$E[\tilde{g}(U_1, U_2)] = E[a_2(U_1, U_2)] = 0.2(1 + 2M)\,,$$
$$E[\tilde{h}(1 - U_1, 1 - U_2)] = -0.2(1 + 2M)\,,$$

and

$$E[\tilde{g}(U_1, U_2)\tilde{h}(1 - U_1, 1 - U_2)] = -0.4M\,.$$

So

$$\operatorname{Cov}(\tilde{g}(U_1, U_2), \tilde{h}(1 - U_1, 1 - U_2)) = 0.04(4M^2 - 6M + 1)$$

which is negative, for example, when $M = 1.1$.

Counterexample 2 [(ii) and (iii) are not sufficient for the conclusion of Proposition 4.] Let X_1 and X_2 have the following joint probabilities:

x_1 \ x_2	0	1	M	
0	0.4	0	0	0.4
1	0	0	0.2	0.2
M	0	0.2	0.2	0.4
	0.4	0.2	0.4	1

where $M > 1$ is some constant. The associated c.d.f. $F(x_1, x_2) = P(X_1 \le x_1, X_2 \le x_2)$ is TP_2, hence [using arguments such as in Brindley and Thompson (1972) or in Shaked (1977)] (X_1, X_2) in LCSD. The proof that X_1 and X_2 are associated is similar to the proof of the parallel result in Counterexample 1.

For this example

$$a_1(u_1, u_2) = \begin{cases} 0 & \text{if } 0 \le u_1 < 0.4 , \\ 1 & \text{if } 0.4 \le u_1 < 0.6 , \\ M & \text{if } 0.6 \le u_1 < 1 , \end{cases}$$

$$a_2(u_1, u_2) = \begin{cases} 0 & \text{if } 0 \le u_1 \le 0.4,\ 0 < u_2 < 1 , \\ M & \text{if } 0.4 \le u_1 < 0.6,\ 0 \le u_2 < 1 , \\ 1 & \text{if } 0.6 \le u_1 < 1,\ 0 \le u_2 < 0.5 , \\ M & \text{if } 0.6 \le u_1 < 1,\ 0.5 \le u_2 < 1 . \end{cases}$$

Let (Y_1, Y_2) be a random vector such that $(Y_1, Y_2) \stackrel{d}{=} (X_1, X_2)$. Then $b_1 = a_1$ and $b_2 = a_2$. Let $g(\mathbf{x}) = x_2$ and $H(\mathbf{y}) = -y_2$ and consider $\mathrm{Var}(g(\mathbf{X}) - h(\mathbf{Y}))$. Define $\tilde{g} = a_2$ and $\tilde{h} = -b_2$. If the conclusion of Proposition 4 were true, then using VARN, we would have had $\mathrm{Cov}(\tilde{g}(U_1, U_2), \tilde{h}(1 - U_1, 1 - U_2)) \ge 0$, where U_1 and U_2 are independent r.v.'s uniformly distributed on $(0, 1)$. However, here,

$$E[\tilde{g}(U_1, U_2)] = 0.2(1 + 2M) ,$$
$$E[\tilde{h}(1 - U_1, 1 - U_2)] = -0.2(1 + 2M) ,$$

and

$$E[g(U_1, U_2)h(1 - U_1, 1 - U_2)] = -0.2M^2 .$$

So

$$\mathrm{Cov}(\tilde{g}(U_1, U_2), \tilde{h}(1 - U_1, 1 - U_2)) = -0.04(M^2 - 4M - 1),$$

which is negative, for example, when $M \geq 5$.

1.1.4 Applications

In this section we present several applications related to Propositions 4 and 5.

(a) Introduction Suppose we seek to estimate $\theta^{(1)} = E[g(\mathbf{X})]$ where $\mathbf{X}$ is CIS and g is a real-valued measurable monotonic function.

An unbiased estimator for $\theta^{(1)}$ is the crude Monte Carlo estimator that can be written as

$$\bar{\theta}^{(1)} = \frac{1}{k} \sum_{i=1}^{k} g(\mathbf{X}^{(i)}). \tag{1.1.18}$$

Here $\mathbf{X}^{(1)}, \ldots, \mathbf{X}^{(k)}$ is a sample from the c.d.f. $F_{\mathbf{X}}(\mathbf{x})$. The r.v.'s $\mathbf{X}^{(i)}$ can be generated by applying the inverse transform method (1.1.10).

Another unbiased estimator for $\theta^{(1)}$ is

$$\bar{\theta}_a^{(1)} = \frac{1}{k} \sum_{i=1}^{k/2} [g(\mathbf{X}^{(i)}) + g(\mathbf{X}_a^{(i)})] \qquad (k \text{ even}), \tag{1.1.19}$$

where $\mathbf{X}^{(i)}$ is generated by (1.1.10) using $\mathbf{U}^{(i)}$ and $\mathbf{X}_a^{(i)}$ is generated by (1.1.10) using $\mathbf{1} - \mathbf{U}^{(i)}$. We call $\bar{\theta}_a^{(1)}$ the *antithetic estimator* of $\theta^{(1)}$.

Since $g(\mathbf{X}) + g(\mathbf{X}_a)$ is a particular case of $g(\mathbf{X}) - h(\mathbf{Y})$ [with $h(\mathbf{Y})$ replaced by $-g(\mathbf{X}_a)$], we obtain directly from Proposition 4 that $\mathrm{Var}[\bar{\theta}_a^{(1)}] \leq \mathrm{Var}[\bar{\theta}^{(1)}]$. That is, the antithetic estimator $\bar{\theta}_a^{(1)}$ is more accurate than the crude Monte Carlo estimator $\bar{\theta}^{(1)}$.

Assume that $\mathbf{X}$ is associated. If $\mathbf{X} \stackrel{d}{=} -\mathbf{X}$, then taking $h = -g$ and $\mathbf{X}_a^{(i)} = -\mathbf{X}^{(i)}$, $i = 1, \ldots, \frac{1}{2}k$ [see equation (1.1.19)], we obtain again $\mathrm{Var}[\bar{\theta}_a^{(1)}] \leq \mathrm{Var}[\bar{\theta}^{(1)}]$.

Assume now that we seek to estimate $\theta^{(2)} = E[g(\mathbf{X}) - h(\mathbf{Y})]$, $\mathbf{X}, \mathbf{Y} \in R^n$, that is, the expected value of the response difference of a pair of functions $g(\mathbf{X})$ and $h(\mathbf{Y})$. Assume that both $\mathbf{X}$ and $\mathbf{Y}$ are CIS and that both g and h are monotone functions. The crude Monte Carlo estimator for $\theta^{(2)}$ is

$$\bar{\theta}^{(2)} = \frac{1}{k} \sum_{i=1}^{k} [g(\mathbf{X}^{(i)}) - h(\mathbf{Y}^{(i)})], \tag{1.1.20}$$

where $\mathbf{X}^{(i)}$ and $\mathbf{Y}^{(i)}$ are generated by equations (1.1.10) and (1.1.11), respectively.

The antithetic estimator for $\theta^{(2)}$ is

$$\bar{\theta}_a^{(2)} = \frac{1}{k} \sum_{i=1}^{k} [g(\mathbf{X}^{(i)}) - h(\mathbf{Y}_a^{(i)})] \tag{1.1.21}$$

where $\mathbf{X}$ and $\mathbf{Y}_a$ are generated by using VCRN; that is, $\mathbf{X}^{(i)}$ is generated by (1.1.10) and $\mathbf{Y}_a^{(i)}$ is generated by (1.1.11) with $\mathbf{U}^{(2)}$ replaced by $\mathbf{U}^{(1)}$.

It follows from Proposition 4 that $\text{Var}[\bar{\theta}_a^{(2)}] \le \text{Var}[\bar{\theta}^{(2)}]$; that is, the antithetic estimator is more accurate than the crude Monte Carlo estimator.

(b) Network and Reliability Application The monotonicity assumption in Proposition 4 actually holds for a broad class of complex stochastic systems. Consider, for example, the following two problems:

(i) Estimate the shortest path in a stochastic network. Mathematically the problem reads: Estimate $\theta^{(1)} = E[g(\mathbf{X})] = E[\hat{g}(\mathbf{V})]$, where

$$\hat{g}(\mathbf{V}) = \min_{j=1,\ldots,m} V_j , \qquad V_j = \sum_{i \in L_j} X_i ; \tag{1.1.22}$$

L_j is the jth complete path from the source, say S, to the sink, say C, of the network; X_i, $i = 1, \ldots, n$, are the durations of the activities (links); and m is the number of complete paths in the network.

(ii) Estimate the expected value of a coherent life function. [For an exact definition of a coherent life function (system) $g(\mathbf{X})$ see Esary and Marshall (1970). Note that a coherent life function is nondecreasing in each component.] Mathematically the problem reads: Estimate $\theta^{(1)} = E[g(\mathbf{X})] = E[\hat{g}(\mathbf{V})]$, where

$$\hat{g}(\mathbf{V}) = \max_{j=1,\ldots,m} V_j , \qquad V_j = \min_{i \in L_j} X_i . \tag{1.1.23}$$

As a particular example, consider the Wheatstone bridge (see Fig. 1).

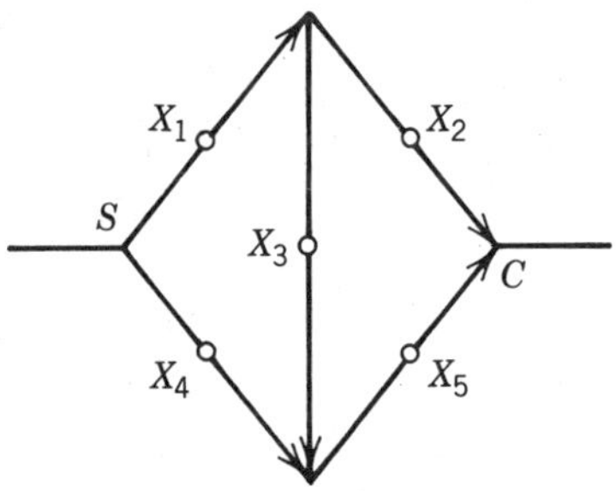

Fig. 1 Wheatstone Bridge.

In this case (1.1.22) reduces to $V_1 = X_1 + X_2$, $V_2 = X_1 + X_3 + X_5$, and $V_3 = X_4 + X_5$ or equivalently $g(\mathbf{X}) = \min\{X_1 + X_2, X_1 + X_3 + X_5, X_4 + X_5\}$ and (1.1.23) reduces to $V_1 = \min(X_1, X_2)$, $V_2 = \min(X_1, X_3, X_5)$, and $V_3 = \min(X_4, X_5)$ or equivalently $g(\mathbf{X}) = \max\{\min(X_1, X_2), \min(X_1, X_3, X_5), \min(X_4, X_5)\}$. Since $g(\mathbf{X})$ is monotone nondecreasing in each component of $\mathbf{X}$, we have from Proposition 4 that if $\mathbf{X}$ is CIS, then the antithetic estimator $\bar{\theta}_a^{(1)}$ of $Eg(\mathbf{X})$ is more accurate than the crude Monte Carlo estimator $\bar{\theta}^{(1)}$. Similarly, if g and h are two shortest paths of two stochastic networks or are two coherent life functions and if $\mathbf{X}$ and $\mathbf{Y}$ are CIS, then the estimator $\bar{\theta}_a^{(2)}$ of $E[g(\mathbf{X}) - h(\mathbf{Y})]$ is more accurate than $\bar{\theta}^{(2)}$.

We estimated the parameters $\theta^{(1)} = E[g(\mathbf{X})]$ and $\theta^{(2)} = E[g(\mathbf{X}) - h(\mathbf{Y})]$ for both stochastic networks and reliability systems considering the Wheatstone, double Wheatstone, and multiple Wheatstone bridges with different c.d.f.'s $F_{\mathbf{X}}(\mathbf{x})$ and $F_{\mathbf{Y}}(\mathbf{y})$, where $\mathbf{X}$ and $\mathbf{Y}$ were CIS. We generated $\mathbf{X}$ and $\mathbf{Y}$ as follows.

Let $\mathbf{X}^* = (X_1^*, \ldots, X_n^*)$ and $\mathbf{Y}^* = (Y_1^*, \ldots, Y_n^*)$ have the following c.d.f.'s:

$$F_{\mathbf{X}^*}(\mathbf{x}) = \prod_{i=1}^{n} F_{X_i^*}(x_i)\,, \qquad F_{\mathbf{Y}^*}(\mathbf{y}) = \prod_{i=1}^{n} F_{Y_i^*}(y_i)\,,$$

where $F_{X_i^*}(x_i)$ and $F_{Y_i^*}(y_i)$, $i = 1, \ldots, n$, are arbitrary univariate c.d.f.'s generated by the inverse transform method. Set

$$X_i = \sum_{j=1}^{i} X_j^*, \qquad Y_i = \sum_{j=1}^{i} Y_j^*\,, \qquad i = 1, \ldots, n\,.$$

Since $P(X_i \leq x_i \mid X_1 = x_1, \ldots, X_{i-1} = x_{i-1}) = P(X_i^* \leq x_i - x_{i-1})$, which is nonincreasing in x_{i-1}, it follows that $\mathbf{X} = (X_1, \ldots, X_n)$ is CIS. Similarly, $\mathbf{Y} = (Y_1, \ldots, Y_n)$ is CIS.

A sensible performance measure is $(t_{\bar{\theta}} \operatorname{Var}(\bar{\theta}))/(t_{\bar{\theta}_a} \operatorname{Var}(\bar{\theta}_a))$, where $t_{\bar{\theta}}$ and $t_{\bar{\theta}}$ are the central processing unit times for the crude Monte Carlo and the antithetic variates, respectively. Comparing equations (1.1.18) and (1.1.19) [or (1.1.20) and (1.1.21)], we see that $t_{\bar{\theta}_a} \leq t_{\bar{\theta}}$ because the antithetic estimator $\bar{\theta}_a$ needs only half as many random numbers as is needed by its counterpart $\bar{\theta}$. In the following we shall neglect this advantage of $\bar{\theta}_a$ and take the performance measure to be

$$\varepsilon = \frac{\operatorname{Var}(\bar{\theta})}{\operatorname{Var}(\bar{\theta}_a)}\,. \tag{1.1.24}$$

Table 1 Efficiency ε of the Antithetic Estimator $\theta_a^{(2)}$ as a Function of λ_X and λ_Y

λ_X \ λ_Y	1	2	3	4	5
1	3.81	1.34	1.01	1.19	1.21
2	1.93	2.95	1.72	1.59	1.26
3	1.20	3.42	4.29	1.64	1.39
4	1.06	1.89	5.01	2.90	2.28
5	1.01	1.52	2.62	5.57	3.24

The estimators $\bar{\theta}$ and $\bar{\theta}_a$ were calculated according to equations (1.1.18) and (1.1.19) while estimating $\theta^{(1)}$ and according to (1.1.20) and (1.1.21) while estimating $\theta^{(2)}$.

Table 1 gives the values of ε while estimating $\theta^{(2)}$ for the stochastic network for the Wheatstone bridge where we took $F_{X_i^*}(x) = 1 - \exp\{-(\lambda_X x), x \geq 0$, and $F_{Y_i^*}(y) = 1 - \exp\{-(\lambda_Y y)\}$, $y \geq 0$, $i = 1, \ldots, 5$, for various choices of λ_X and λ_Y. The number of replications was 500. Note that in general the efficiency ε is a complicated function of the topology (structure) of the underlying system, the number of its components, the c.d.f.'s, and the type of multivariate dependence (CIS, TP_2, etc.). For the specific example at hand ε is a function of λ_X and λ_Y.

It is worthwhile to note that our numerical results are similar to those obtained earlier by Sullivan et al. (1982) for stochastic networks.

(c) Queueing Applications Consider a GI/G/1 queue. The waiting time of the ith customer in the queue [see Gross and Harris (1985)] is $W^{(i)} = \max(0, W^{(i-1)} - X_1^{(i)} + X_2^{(i)})$, $W^{(0)} = 0$. Here $X_1^{(i)}$ and $X_2^{(i)}$ are, respectively, the interarrival and the service time of the ith customer. Denote $\mathbf{X} = (\mathbf{X}^{(1)}, \ldots, \mathbf{X}^{(n)})$, where $\mathbf{X}^{(i)} = (-X_1^{(i)}, X_2^{(i)})$, $i = 1, \ldots, n$. Let $g(\mathbf{X}) = W^{(n)}$. Note that $g(\mathbf{X})$ is a nondecreasing function of $\mathbf{X}$.

We want to estimate $\theta^{(1)} = E[g(\mathbf{X})]$, that is, the mean waiting time of the nth customer. We assume that $\mathbf{X}^{(1)}, \ldots, \mathbf{X}^{(n)}$ are mutually independent, but we allow the service time $X_2^{(i)}$ to depend on the interarrival time $X_1^{(i)}$. In particular, we assume that for each i, $\mathbf{X}^{(i)} = (-X_1^{(i)}, X_2^{(i)})$ is CIS, $i = 1, \ldots, n$. [This is a reasonable assumption when a long queue (which may result from a sequence of short interarrival times) reduces the effieciency of the server and causes the service times to increase stochastically.] Note that when $\mathbf{X}^{(i)}$, $i = 1, \ldots, n$, are CIS and mutually independent, then $\mathbf{X} = (\mathbf{X}^{(1)}, \ldots, \mathbf{X}^{(n)})$ is CIS.

We generated the CIS vector $\mathbf{X}^{(i)}$ as follows: Let $\mathbf{Z}^{(i)} = (Z_1^{(i)}, Z_2^{(i)})$ be a nonnegative random vector having the c.d.f. $F(\mathbf{z}^{(i)}) = F_1(z_1^{(i)})F_2(z_2^{(i)})$, where $F_j(z_j)$, $j=1,2$, are arbitrary c.d.f.'s whose supports are bounded from the right (i.e., the values of $Z_j^{(i)}$ have an upper bound, say $M>0$). Set

$$X_1^{(i)} = Z_1^{(i)}\,, \qquad X_2^{(i)} = M - c_1 Z_1^{(i)} - c_2 Z_2^{(i)}\,, \tag{1.1.25}$$

where $c_1 + c_2 = 1$ and $c_1, c_2 \geq 0$. Note that both $X_1^{(i)}$ and $X_2^{(i)}$ are nonnegative as required. Since $P(X_2^{(i)} \leq x_2 \mid X_1^{(i)} = x_1) = P(c_2 Z_2^{(i)} \geq M - c_1 x_1 - x_2)$ is nondecreasing in x_1, it follows that $\mathbf{X}^{(i)} = (-X_1^{(i)}, X_2^{(i)})$ is CIS.

Table 2 presents results of one of our runs. In this case we let the $Z_1^{(i)}$ and $Z_2^{(i)}$ of equation (1.1.25) be uniform r.v.'s on the intervals (0, 1) and (0, b), respectively, for various values of b. We set $M=1$, let c_1 vary, and set $c_2 = 1 - c_1$. We choose $\lambda \equiv EX_1^{(i)} = \frac{1}{2}$, $\mu \equiv EX_2^{(i)} = M - \frac{1}{2}c_1 - \frac{1}{2}[b(1-c_1)]$ so that the traffic intensity is

$$\rho = \frac{\lambda}{\mu} = [2M - c_1 - b(1-c_1)]^{-1}\,.$$

The correlation between $X_1^{(i)}$ and $X_2^{(i)}$ is easily found to be

$$\gamma = \frac{-c_1}{[c_1^2 + (1-c_1)^2 b^2]^{1/2}}\,.$$

Thus, any choice of c_1 and b determines a pair of values ρ and γ.

In Table 2 we give the values of ε [see equation (1.1.24)], which we found for various choices of c_1 and b. The corresponding values of ρ and γ are given in Table 3. In each case we made $k=400$ replications, and in every replication (sample path) we simulated $n=40$ customers. In the antithetic procedure there are $\frac{1}{2}k = 200$ pairs of replications [see equation (1.1.19)], and $\text{Var}(\bar{\theta}_a)$ of equation (1.1.20) was estimated by $(\frac{1}{2}k)^{-1}s_a^2$, where s_a^2 is the sample variance of the $\frac{1}{2}k = 200$ values, $\frac{1}{2}(g(\mathbf{X}^{(i)}) + g(\mathbf{X}_a^{(i)}))$, $i = 1, \ldots, \frac{1}{2}k$. To estimate $\text{Var}(\bar{\theta})$ of (1.1.20), we divided the

Table 2 Efficiency ε of the Antithetic Estimator $\bar{\theta}_a^{(1)}$ as a Function of c_1 and b

b \ c_1	0	0.25	0.50	0.75	1
0.1	1852.26	17.53	3.63	2.39	3.22
0.5	5.10	2.71	2.11	2.74	2.87
0.9	2.84	2.22	1.97	3.00	3.01

Table 3 Values of ρ (Top of Each Entry) and γ (Bottom of Each Entry) as Functions of b and c_1

b \ c_1	0	0.25	0.50	0.75	1
0.1	0.526	0.597	0.690	0.816	1.000
	0.000	−0.958	−0.995	−0.999	−1.000
0.5	0.667	0.727	0.800	0.889	1.000
	0.000	−0.555	−0.894	−0.986	−1.000
0.9	0.909	0.930	0.952	0.976	1.000
	0.000	−0.347	−0.743	−0.958	−1.000

$k = 400$ replications into $\frac{1}{2}k = 200$ pairs of replications and estimated $\text{Var}(\bar{\theta})$ by $(\frac{1}{2}k)^{-1}s^2$ where s^2 is the sample variance of the $\frac{1}{2}k = 200$ averages $\frac{1}{2}(g(\mathbf{X}^{(i)}) + g(\mathbf{X}^{(i+1)}))$, $i = 1, 3, \ldots, 399$. Other choices of c.d.f.'s for $Z_1^{(i)}$ and $Z_2^{(i)}$ (e.g., binary, triangular, and beta) yielded similar results.

1.1.5 Proof of the Propositions (Appendix)

Before proving Proposition 1 let us first show that problem (1.1.6) is equivalent to

$$\max_{F_{\mathbf{U}_1\mathbf{U}_2}} E\{g^*(\mathbf{U}_1)h^*(\mathbf{U}_2)\}\,, \tag{A.1}$$

where U_1 and U_2 are vectors of independent random variables, each uniformly distributed on [0, 1].

Indeed, we have

$$\text{Var}[g(\mathbf{X}) - h(\mathbf{Y})] = -2Eg(\mathbf{X})h(\mathbf{Y}) + Eg^2(\mathbf{X}) + Eh^2(\mathbf{Y}) - [Eg(\mathbf{X})]^2$$
$$- [Eh(\mathbf{Y})]^2 + 2Eg(\mathbf{X})Eh(\mathbf{Y})\,. \tag{A.2}$$

Since the distributions $F_{\mathbf{X}}$ and $F_{\mathbf{Y}}$ are given, it follows that only the first term on the right side of (A.2) depends on the specific form of $F_{\mathbf{XY}}$. Thus, problem (1.1.6) can be written as

$$\max_{F_{\mathbf{XY}}} E\{g(\mathbf{X})h(\mathbf{Y})\} \tag{A.3}$$

subject to given c.d.f.'s $F_{\mathbf{X}} = \Pi_{i=1}^n F_{X_i}(x_i)$ and $F_{\mathbf{Y}} = \Pi_{i=1}^n F_{Y_i}(y_i)$. Clearly that by the definition (1.1.6′) formula (A.3) is equivalent to (A.1).

Proposition 1 Let $g^*(\mathbf{u}_1)$ and $h^*(\mathbf{u}_2)$, $\mathbf{u}_1, \mathbf{u}_2 \in R^n$ be a pair of real-valued measurable functions monotonic in the same direction in each component of $\mathbf{u}_1$ and $\mathbf{u}_2$, and let $\mathbf{U}_1 = \{U_1^{(1)}, \ldots, U_1^{(n)}\}$, $\mathbf{U}_2 = \{U_2^{(1)}, \ldots, U_2^{(n)}\}$ be a pair of vectors with independent components, each uniformly distributed on $[0, 1]$; then within the set Ξ of all c.d.f.'s $F_{\mathbf{U}_1\mathbf{U}_2}$ on R^{2n}, where dependence is permitted only between like components of $\mathbf{U}_1$ and $\mathbf{U}_2$ (components with the same indices),

$$\max_{F_{\mathbf{U}_1\mathbf{U}_2} \in \Xi} E\{g^*(\mathbf{U}_1)h^*(\mathbf{U}_2)\} = E\{g^*(\mathbf{U})h^*(\mathbf{U})\} \tag{A.4}$$

and therefore the VCRN $(U_2 = U_1 = U)$ is an optimal policy for problem (1.1.6).

The proposition is proved by induction.

Proof For $n = 1$ our proposition is correct [see Proposition 3 of Gal, Rubinstein, and Ziv (1984).] Assuming that for some $1 \leq n_0$, VCRN $(\mathbf{U}_1 = \mathbf{U}_2 = \mathbf{U}, \mathbf{U} \in R^{n_0})$ is an optimal choice, then for $n_0 + 1$ we have

$$\begin{aligned} E[g^*(\mathbf{U}_1)h^*(\mathbf{U}_2)] &= E_{U_1^{(1)},\ldots,U_1^{(n_0+1)},U_2^{(1)},\ldots,U_2^{(n_0+1)}}\{g^*(\mathbf{U}_1)h^*(\mathbf{U}_2)\} \\ &= E_{U_1^{(n_0+1)},U_2^{(n_0+1)}}\{E_{U_1^{(1)},\ldots,U_1^{(n_0)},U_2^{(1)},\ldots,U_2^{(n_0)}}[g^*(\mathbf{U}_1)h^*(\mathbf{U}_2)]\} \\ &\leq E_{U_1^{(n_0+1)},U_2^{(n_0+1)}}\{E_U[g^*(\mathbf{U}, \mathbf{U}_1^{(n_0+1)})h^*(\mathbf{U}, \mathbf{U}_2^{(n_0+1)})]\} \end{aligned} \tag{A.5}$$

where $\mathbf{U} \in R^{n_0}$.

The above inequality follows from the induction assumption. So we conclude that for given index $n_0 + 1$, the other components must be common. The last statement is obviously also true for any other index $(n = n_0 + 1 \geq 2)$, which means that $U_1^{(n_0+1)}$ and $U_2^{(n_0+1)}$ must be CRN, that is, $\mathbf{U}_1^{(n_0+1)} = \mathbf{U}_2^{(n_0+1)} = \mathbf{U}^{(n_0+1)}$. This completes the proof. Q.E.D.

It is also obvious that if g^* and h^* are monotonic in opposite direction functions, then the vector of antithetic random numbers (VARN), that is, $\mathbf{U}_2 = \mathbf{1} = \mathbf{U}_1$, where $\mathbf{1}$ is an n-dimensional unit vector, is optimal for problem (1.1.6).

Proposition 2 Assume that **X** and **Y** have the same c.d.f.'s $\mathbf{X} \overset{d}{=} \mathbf{Y}$. Assume also that g and h are real valued *corresponding* in the same direction. Then the use of VCRN yields an optimal solution to problem (1.1.6).

Proof Since $\mathbf{X} \overset{d}{=} \mathbf{Y}$ and g and h are corresponding in the same direction, then $g^*(\mathbf{U}_1)$ and $h^*(\mathbf{U}_2)$ are also corresponding in the same direction. Clearly Proposition 2 will be proved if we prove that within the set L of all c.d.f.'s $F_{\mathbf{U}_1\mathbf{U}_2}$ on R^{2n}, where $\mathbf{U}_i = \{U_i^{(1)}, \ldots, U_i^{(n)}\}$, $i = 1, 2$, are vectors with independent components each uniformly distributed on [0, 1]:

$$\max_{F_{\mathbf{U}_1\mathbf{U}_2} \in L} \{E[g^*(\mathbf{U}_1)h^*(\mathbf{U}_2)]\} = E\{g^*(\mathbf{U})h^*(\mathbf{U})\} . \tag{A.6}$$

The proof is analogous to that of Proposition 2 in Gal, Rubinstein, and Ziv (1984).

Discrete case. Let us assume that the ith component of the vector $\mathbf{U}$ has a uniform discrete distribution with a p.m.f. (probability mass function) equal to $1/K_i$ at points $(x_1^{(i)}, x_2^{(i)}, \ldots, x_{K_i}^{(i)})$. Since the components of $\mathbf{U} = (U^{(1)}, \ldots, U^{(n)})$ are independent, the joint p.m.f. of the vector $\mathbf{U}$ equals $\Pi_{i=1}^{n} 1/K_i$ at points $(x_{\alpha_1}^{(1)}, \ldots, x_{\alpha_n}^{(n)})$ with $\alpha_i = 1, \ldots, K_i$, $i = 1, \ldots, n$. We denote the values of g and h at point $\{x_{\alpha_1}^{(1)}, \ldots, x_{\alpha_n}^{(n)}\}$ by $g_{\alpha_1,\ldots,\alpha_n} = g_\alpha$ and $h_{\alpha_1,\ldots,\alpha_n} = h_\alpha$ respectively.

We now introduce a $2n$-dimensional nonnegative matrix $P_{\alpha,\beta} = (P_{\alpha_1,\ldots,\alpha_n,\beta_1,\ldots,\beta_n})$ that satisfies

$$\begin{aligned} \sum_{\beta} P_{\alpha,\beta} &= \left(\prod_{i=i}^{n} K_i\right)^{-1} \qquad \forall \alpha = (\alpha_1, \ldots, \alpha_n) , \\ \sum_{\alpha} P_{\alpha,\beta} &= \left(\prod_{i=1}^{n} K_i\right)^{-1} \qquad \forall \beta = (\beta_1, \ldots, \beta_n) . \end{aligned} \tag{A.7}$$

It is clear that the proposition will be proven if the inequality

$$\sum_{\alpha} \sum_{\beta} g_\alpha^* h_\beta^* P_{\alpha,\beta} \le \left(\prod_{i=1}^{n} K_i\right)^{-1} g_\alpha^* h_\beta^* \tag{A.8}$$

holds.

To prove (A.8) (and thereby the proposition), it suffices to show that the solution of the linear programming problem

$$\begin{aligned} &\max \sum_{\alpha} \sum_{\beta} g_\alpha^* h_\beta^* P_{\alpha,\beta} \\ &\text{subject to (A.7)} \end{aligned} \tag{A.9}$$

is given by

$$P_{\alpha,\beta} = \begin{cases} \left(\prod_{i=1}^{n} K_i\right)^{-1}, & \text{for } \alpha_i = \beta_i,\ i = 1, \ldots, n\,, \\ 0\,, & \text{otherwise}\,. \end{cases} \tag{A.10}$$

To prove (A.10), we note that there are $\Pi_{i=1}^{n} K_i$ points at which the vector **u** has a nonzero probability. Now choose one of them, say the one where g attains maximum. (If there are several such points, any of them may be chosen arbitrarily.) Since g and h are corresponding in the same direction functions, their maxima coincide at a point denoted by $\alpha_0 = (\alpha_{01}, \ldots, \alpha_{0n})$.

We first prove that for the optimal solution

$$P_{\alpha_0,\alpha_0} = \left(\prod_{i=1}^{n} K_i\right)^{-1}$$

For this purpose we assume that $P_{\alpha_0,\alpha_0} < 1/\Pi_{i=1}^{n} K_i$. Then, by (A.7), there are indices $\alpha^+ = (\alpha_1^+, \ldots, \alpha_n^+)$ and $\beta^+ = (\beta_1^+, \ldots, \beta_n^+)$ such that $P_{\alpha^+\alpha_0} = \Delta_1 > 0$ and $P_{\alpha_0\beta^+} = \Delta_2 > 0$. We denote $\Delta = \min(\Delta_1, \Delta_2)$ and consider the nonnegative matrix $\hat{P}_{\alpha,\beta} = (\hat{P}_{\alpha_1,\ldots,\alpha_n,\beta_1,\ldots,\beta_n})$ with elements

$$\hat{P}_{\alpha_0\alpha_0} = P_{\alpha_0\alpha_0} + \Delta \qquad \hat{P}_{\alpha^+\alpha_0} = P_{\alpha^+\alpha_0} - \Delta$$
$$\hat{P}_{\alpha_0\beta^+} = P_{\alpha_0\beta^+} - \Delta \qquad \hat{P}_{\alpha^+\beta^+} = P_{\alpha^+\beta^+} + \Delta$$

all other elements of $\hat{P}$ being equal to P. Obviously, $\hat{P}$ satisfies (A.7) and therefore is a feasible solution for (A.9). Now,

$$\sum_{\alpha}\sum_{\beta} g_\alpha^* h_\beta^* \hat{P}_{\alpha,\beta} = \sum_{\alpha}\sum_{\beta} g_\alpha^* h_\beta^* P_{\alpha,\beta}$$
$$+ \Delta(g_{\alpha_0}^* h_{\alpha_0}^* + g_{\alpha^+}^* h_{\beta^+}^* - g_{\alpha_0}^* h_{\beta^+}^* - g_{\alpha^+}^* h_{\alpha_0}^*) \tag{A.11}$$

Because of the correspondence between g and h, we have

$$g_{\alpha_0}^* h_{\alpha_0}^* + g_{\alpha^+}^* h_{\beta^+}^* - g_{\alpha_0}^* h_{\beta^+}^* - g_{\alpha^+}^* h_{\alpha_0}^* = (g_{\alpha_0}^* - g_{\alpha^+}^*)(h_{\alpha_0}^* - h_{\beta^+}^*) \geq 0\,,$$

and it follows that $\hat{P}$ is superior to P. However, since either $\hat{P}_{\alpha^+\alpha_0} = 0$ or $\hat{P}_{\alpha_0\beta^+} = 0$, and since the number of elements in $\hat{P}$ is finite, it follows that after a finite number of such steps† (at each of which an element of $\hat{P}$ is reduced to zero) we obtain a better solution with $P_{\alpha_0\alpha_0} = 1/\Pi_{i=1}^{n} K_i$.

†Readily shown to equal $2\Pi_{i=1}^{n} K_i - 3$.

Next, deleting all elements with index α_0 from the solution of (A.9) and repeating the procedure for successive points $\{x_1^{(\alpha_1)}, \ldots, x_n^{(\alpha_n)}\}$ from the set of $\Pi_{i=1}^n K_i$, we see that the proof of (A.8)—and therefore of the proposition—is complete.

Continuous Case Let $F_{\mathbf{U}_1\mathbf{U}_2}$ be a nonnegative measure on the unit $2n$-dimensional hypersquare $[0, 1]^n \times [0, 1]^n$, which satisfies

$$\int\int_{A\times[0,1]^n} dF_{\mathbf{U}_1\mathbf{U}_2}(\mathbf{u}_1, \mathbf{u}_2) = \text{mes } A \quad \text{for all measurable } A \subset [0, 1]^n \tag{A.12}$$

and

$$\int\int_{[0,1]^n\times B} dF_{\mathbf{U}_1\mathbf{U}_2}(\mathbf{u}_1, \mathbf{u}_2) = \text{mes } B \quad \text{for all measurable } B \subset [0, 1]^n . \tag{A.13}$$

We divide the interval $[0, 1]$ into k equal subintervals $\{((i-1)/k, i/k], i = 1, \ldots, k\}$ (nonequal subintervals may be considered as well).

Denote

$$C_\alpha = C_{\alpha_1,\ldots,\alpha_n} = \left(\frac{\alpha_1 - 1}{k}, \frac{\alpha_1}{k}\right] \times \cdots \times \left(\frac{\alpha_n - 1}{k}, \frac{\alpha_n}{k}\right],$$

$\alpha_i = 1, \ldots, k$, $i = 1, \ldots, n$. Obviously the volume of C_α equals $1/k^n$, and therefore $P(\mathbf{u} \in C_\alpha) = \int_{C_\alpha} dF(\mathbf{u}) = 1/k^n$.

We next introduce a $2n$-dimensional matrix with elements

$$P_{\alpha,\beta} = P_{\alpha_1,\ldots,\alpha_n,\beta_1,\ldots,\beta_n} = \int_{C_\alpha}\int_{C_\beta} dF(u_1, u_2) . \tag{A.14}$$

It is evident that $P_{\alpha,\beta}$ satisfies condition (A.7) with $k_1 = \cdots = k_n = k$. Defining

$$g^*_{\alpha_1,\ldots,\alpha_n} = g^*\left(\frac{\alpha_1}{k}, \ldots, \frac{\alpha_n}{k}\right)$$

and

$$h^*_{\alpha_1,\ldots,\alpha_n} = h^*\left(\frac{\alpha_1}{k}, \ldots, \frac{\alpha_n}{k}\right),$$

we obtain, from (A.8),

$$\sum_{\alpha}\sum_{\beta} g_\alpha^* h_\beta^* P_{\alpha,\beta} \leq \frac{1}{k^n} \sum_{\alpha} g_\alpha^* h_\alpha^* . \tag{A.15}$$

But due to integrability of g^* and h^* as $k \to \infty$, the left and right sides of (A.15) approach, respectively,

$$\int\int_{[0,1]^n \times [0,1]^n} g^*(\mathbf{u}_1) h^*(\mathbf{u}_2)\, dF(\mathbf{u}_1, \mathbf{u}_2) = E[g(\mathbf{U}_1) h(\mathbf{U}_2)] \tag{A.16}$$

and

$$\int_{[0,1]^n} g^*(\mathbf{u}) h^*(\mathbf{u})\, d\mathbf{u} = E[g^*(\mathbf{U}) h^*(\mathbf{U})] . \tag{A.17}$$

Combining (A.16), (A.17), and (A.15), we finally have

$$E[g^*(\mathbf{U}_1) h^*(\mathbf{U}_2)] \leq E[g^*(\mathbf{U}) h^*(\mathbf{U})] . \qquad \text{Q.E.D.}$$

It is readily seen that if g^* and h^* correspond in the opposite direction, then VCRN ($\mathbf{U}_2 = \mathbf{U}_1$) is optimal for estimating $E[g^*(\mathbf{U}_1) + h^*(\mathbf{U}_2)]$.

Proposition 3 Procedure 2 permits variance reduction.

Proof Taking into account equation (1.1.13a), we obtain in analogy to (1.1.13b)

$$E[g(\mathbf{X}) h(\mathbf{Y})] = \sum_{i=1}^{m} \sum_{j=1}^{m} P_{ij} E[g^*(\mathbf{U}_1) h^*(\mathbf{U}_2) \mid (\mathbf{U}_1, \mathbf{U}_2) \in L_i \times L_j]$$

By Proposition 1 every conditional expectation in the last formula is less for dependent sampling with CRN than for independent one. Q.E.D.

Proposition 4 If **X** and **Y** are CIS and g and h are monotone in the same (the opposite) direction, then the use of VCRN (VARN) yields an optimal solution to problem (1.1.12).

Proof Since **X** and **Y** are CIS, it follows from Lemma 1 that there exist nondecreasing functions $a_1, \ldots, a_n$ and $b_1, \ldots, b_n$ such that

$$(X_1, \ldots, X_n) \stackrel{d}{=} (a_1(\mathbf{U}^{(1)}), \ldots, a_n(\mathbf{U}^{(1)})) ,$$
$$(Y_1, \ldots, Y_n) \stackrel{d}{=} (b_1(\mathbf{U}^{(2)}), \ldots, b_n(\mathbf{U}^{(2)})) ,$$

where $\mathbf{U}^{(1)}$ and $\mathbf{U}^{(2)}$ are vectors of independent r.v.'s uniformly distributed on $(0, 1)$. Define

$$\tilde{g}(\mathbf{u}) = g(a_1(\mathbf{u}), \ldots, a_n(\mathbf{u})) ,$$
$$\tilde{h}(\mathbf{u}) = h(b_1(\mathbf{u}), \ldots, b_n(\mathbf{u})) .$$

Then

$$E[g(\mathbf{X}) - h(\mathbf{Y})] = E[\tilde{g}(\mathbf{U}^{(1)}) - \tilde{h}(\mathbf{U}^{(2)})] . \tag{A.18}$$

Since the a_i and the b_i are nondecreasing, it follows that $\tilde{g}$ and $\tilde{h}$ have the same monotonicity properties as g and h. But $\tilde{g}(\mathbf{U}^{(1)})$ and $\tilde{h}(\mathbf{U}^{(2)})$ are functions of independent r.v.'s. Hence the use of VCRN (VARN) yields an optimal solution for (1.1.12). Q.E.D.

Using Fact 2 we obtain the following corollary from Proposition 1.

Corollary 1 If $\mathbf{X}$ and $\mathbf{Y}$ are TP_2 in pairs and g and h are monotonic in the same (the opposite) direction, then the use of VARN (VCRN) yields an optimal solution for (1.1.12).

Proposition 5 Let $\mathbf{X}$ and $\mathbf{Y}$ be CIS and let $\mathbf{Y}$ and $\mathbf{X}$ have the same c.d.f.'s, i.e., $\mathbf{Y} \overset{d}{=} \mathbf{X}$. Assume also that g and h are corresponding in the same direction. Then the use of VCRN yields an optimal solution to problem (1.1.12).

Proof The proof of this proposition is based on Proposition 2. Since $\mathbf{Y} \overset{d}{=} \mathbf{X}$, it is readily seen that correspondence of g and h leads to correspondence of $\tilde{g}$ and $\tilde{h}$ [see equation (A.18)]. But $\tilde{g}(\mathbf{U}^{(1)})$ and $\tilde{h}(\mathbf{U}^{(2)})$ are functions of independent r.v.'s $\mathbf{U}^{(1)}$ and $\mathbf{U}^{(2)}$, respectively. Now the proof of our proposition follows directly from Proposition 2 and therefore the use of VCRN yields an optimal solution to (1.1.12). Q.E.D.

It is readily seen that if g and h are corresponding in the opposite direction, then the VCRN is optimal for (1.1.12) with $g(\mathbf{X}) - h(\mathbf{Y})$ replaced by $g(\mathbf{X}) + h(\mathbf{Y})$.

1.2 ANTITHETIC VARIABLES FOR NONMONOTONIC FUNCTIONS

1.2.1 Introduction

In this section we present some new results related to the theory of antithetic random variables extending, in some sense, Handscomb's (1958) antithetic variables method. We also present two practical procedures for solving problem (1.1.3) when the monotonicity assumption about g and h is removed.

Suppose that we need to estimate

$$\theta = E\left[\sum_{i=1}^{n} \psi_i(X_i)\right] \tag{1.2.1}$$

with a minimal variance, where ψ_i are given functions, and X_i are r.v.'s with known c.d.f.'s $F_{X_i}(x)$, $i = 1, 2, \ldots, n$. Generally speaking the antithetic variable method consists of inducing an appropriate dependence between X_i in such a way that the variance of the estimator

$$\hat{\theta} = \sum_{i=1}^{n} \psi_i(X_i) \tag{1.2.2}$$

is reduced while preserving the marginal distributions of X_i. In other words, for given marginal c.d.f.'s $F_{X_i}(x)$, $i = 1, 2, \ldots, n$, the use of the antithetic variable method generates a joint c.d.f. $F_{X_1,\ldots,X_n}(\cdot)$ for which the variance of equation (1.2.2) is less than the variance in the case when the r.v.'s $X_1, \ldots, X_n$ are independent.

Assume that each r.v. X_i can be generated by using the inverse transform method, that is (see (1.1.2)),

$$X_i = F_{X_i}^{-1}(U_i) = \inf\{x\colon F_{X_i}(x) \geq U_i\}\,, \tag{1.2.3}$$

where U_i, $i = 1, 2, \ldots, n$, are r.v.'s uniformly distributed on $(0, 1)$. Then (1.2.1) can be rewritten as

$$\theta = E\left\{\sum_{i=1}^{n} \psi_i[F_{X_i}^{-1}(U_i)]\right\} = E\left[\sum_{i=1}^{n} g_i(U_i)\right] \tag{1.2.4}$$

where $g_i(U_i) = \psi_i[F_{X_i}^{-1}(U_i)]$.

Now we can restate the variance reduction problem as: Given $g_1, \ldots, g_n$, choose a joint c.d.f. of $U_1, \ldots, U_n$, denoted $F_{U_1,\ldots,U_n}(\cdot)$, to minimize Var $\theta = \text{Var}[\Sigma_{i=1}^{n} g_i(U_i)]$.

Handscomb's (1958) antithetic random variables theorem reads as follows:

Theorem If I denotes the infimum of $\text{Var}(\Sigma_{j=1}^{n} g_j(U_j))$ when all possible stochastic or functional dependences between the U_j are considered, then, provided that the g_j are bounded for all $j = 1, \ldots, n$,

$$\inf_{\substack{x_j \in \mathscr{X} \\ j=1,\ldots,n}} \operatorname{Var} \sum_{j=1}^{n} g_j[x_j(U)] = I , \tag{1.2.5}$$

where $\mathscr{X}$ denotes the class of all functions $x(U)$ such that

(i) $x(U)$ is a (one-to-one) mapping of the unit interval on itself and
(ii) dx/dU exists and equals 1, except perhaps at a finite set of points U.

Wilson (1979) has extended this theorem to unbounded functions $g_1, \ldots, g_n$ with finite variances. Unfortunately, Handscomb's and Wilson's theorems say nothing about (a) existence of the minimal value of Var θ exactly and (b) whether or not the minimal value of Var θ can be achieved while using *only a single random number U* in spite of the fact that the U appears in each $x_j(U)$ in (1.2.5).

In Section 1.2.2 we present a modified version (see Theorem 1) of Handscomb's theorem, giving positive answers on issues (a) and (b) while removing condition (ii).

In Section 1.2.2 we show that Theorem 1 can be readily extended to a rather general case (see Theorem 2 below). We chose to prove Theorem 1 rather than its generalized version (Theorem 2) only in order to provide an analogy with the well-known Handscomb theorem. In Section 1.2.3 we show how our theory can be used to construct practical procedures for generating an *optimal* two-dimensional c.d.f. $F_{U_1,U_2}(\cdot)$ that minimizes

$$\operatorname{Var}[g_1(U_1) + g_2(U_2)] . \tag{1.2.6}$$

1.2.2 Modified Version of Handscomb's Antithetic Variables Theorem

Theorem 1 Let $g_1(U_1), g_2(U_2), \ldots, g_n(U_n)$ be bounded measurable functions on $[0, 1]$ and Ξ_n be the set of all possible joint c.d.f.'s on $[0, 1]^n$ with uniform marginals. Then there exist transformations $h_1(U_1), \ldots, h_n(U_n)$ such that

$$\begin{aligned} \min_{F\in\Xi_n} & \operatorname{Var}[g_1(U_1) + g_2(U_2) + \cdots + g_n(U_n)] \\ &= \operatorname{Var}[g_1(h_1(U)) + \cdots + g_n(h_n(U))], \end{aligned}$$

where $h_1(U) \sim U(0,1), h_2(U) \sim U(0,1), \ldots, h_n(U) \sim U(0,1)$.

Note that the same U appears in all $h_k(\cdot)$, $k = 1, \ldots, n$.

We denote by F^* the optimal c.d.f. generated by the r.v.'s $h_1(U), \ldots, h_n(U)$.

The proof of Theorem 1 and its generalization Theorem 2 (see below) are given in Rubinstein and Samorodnitsky (1986)

Theorem 2 Let $\varphi[g_1(\mathbf{U}_1), \ldots, g_n(\mathbf{U}_n)]: R^n \to R^1$ be a continuous function defined on a closed bounded set, where $\mathbf{U}_i = (U_i^1, \ldots, U_i^m)$, $i = 1, \ldots, n$, is an m-dimensional vector with independent components each distributed $U(0, 1)$ and $g_1(\cdot), \ldots, g_n(\cdot)$ be bounded measurable functions on $[0, 1]^m$. Let Ξ_n^m be the set of all possible joint distributions on $[0, 1]^{n \times m}$ with uniform m-dimensional marginals. Then there exist transformations $h_i^j(U)$, $i = 1, \ldots, n$, $j = 1, \ldots, m$, that

$$\min_{F \in \Xi_n^m} E\varphi[g_1(U_1^1, \ldots, U_1^m), \ldots, g_n(U_n^1, \ldots, U_n^m)]$$
$$= E\varphi[g_1(h_1^1(U^1), \ldots, h_1^m(U^m)), \ldots, g_n(h_n^1(U^1), \ldots, h_n^m(U^m))], \tag{1.2.7}$$

where

$$h_i^j(U^j) \sim U(0, 1), \qquad i = 1, \ldots, n, j = 1, \ldots, m, \tag{1.2.8}$$

$$h_i^j(U^j) \text{ does not depend on } h_i^k(U^k),$$
$$k \neq j, k = 1, \ldots, m, i = 1, \ldots, n. \tag{1.2.9}$$

Corollary Suppose that we want to estimate

$$\theta = E\left[\sum_{i=1}^n g_i(\mathbf{U}_i)\right]$$

with minimal variance. Denote

$$\varphi[g_1(\mathbf{U}_1), \ldots, g_n(\mathbf{U}_n)] = \sum_{i=1}^n [g_i(\mathbf{U}_i) - \theta]^2.$$

Then it follows directly from Theorem 2 that there exist transformations $h_i(U)$ satisfying (1.2.8) and (1.2.9) such that

$$\min_{F \in \Xi_n^m} \operatorname{Var}[g_1(\mathbf{U}_1) + \cdots + g_n(\mathbf{U}_n)] = \operatorname{Var}[g_1(\mathbf{h}_1(\mathbf{U})) + \cdots + g_n(\mathbf{h}_n(\mathbf{U}))], \tag{1.2.10}$$

where $\mathbf{h}_i(\mathbf{U}) = (h_i^1(U^1), \ldots, h_i^m(U^m))$, $i = 1, \ldots, n$.

This corollary presents the multidimensional version of Theorem 1. Obviusly, in the particular case where $\mathbf{U}_i$ is a one-dimensional random variable, we obtain Theorem 1.

1.2.3 Optimal Procedures for $n = 2$

Theorem 1 is purely an existence theorem. It says nothing about constructing the optimal transformation $h_1(U), \ldots, h_n(U)$ that minimizes $\text{Var} \sum_{i=1}^{n} g_i(U_i)$.

In this section we present *two practical* procedures (see Procedures 3 and 4) for constructing $h_1(U)$ and $h_2(U)$ for the particular case $n = 2$ and therefore for generating the optimal c.d.f. F_{U_1,U_2}, which minimizes $\text{Var}[g_1(U_1) + g_2(U_2)]$ for nonmonotonic g_1 and g_2.

It is interesting to note that the application of antithetic variables to the case $n = 2$ has received much attention in the recent literature on simulation [see Page (1967); Andreasson (1971); Mitchel (1973); Kleijnen (1974); Schruben and Margolin (1978); Schruben (1979); Fishman and Huang (1981); Sullivan et al. (1982); Granovsky (1982); Rubinstein and Samorodnitsky (1985); and Rubinstein, Samorodnitsky, and Shaked (1985). Most of these applications and in particular our section 1.1 are related, however, to the case where both g_1 and g_2 are monotone or corresponding functions.

(a) Theoretical Results Let $g(u)$ be any measurable function on $[0, 1]$. let $F_g(x) = P[g(U) \le x]$ be the c.d.f. of the r.v. $g(U)$. By the inverse transform method $F_g^{-1}(U)$ is distributed as $g(U)$. It is clear that $F_g^{-1}(1-U)$ is also distributed as $g(U)$. Here $F_g^{-1}(U)$ and $F_g^{-1}(1-U)$ are, respectively, nondecreasing and nonincreasing functions of U. We denote $g^*(u) = F_g^{-1}(u)$ and $g_*(u) = F_g^{-1}(1-u)$ and shall call them accordingly nondecreasing and nonincreasing functions of U distributed as $g(U)$.

Proposition 6. For every measurable function $g(u)$ on $[0, 1]$ there exists a pair of transformations $\{\bar{h}_g(u), \underline{h}_g(u)\}$ such that

1. $\bar{h}_g(U) \sim U(0, 1)$ and $\underline{h}_g(U) \sim U(0, 1)$ and
2. $g^*[\bar{h}_g(U)] = g(U)$ a.s. (almost surely) on $[0, 1]$ and $g_*[\underline{h}_g(U)] = g(U)$ a.s.

Proof The proof of Proposition 6 is given in Rubinstein and Samorodnitsky (1986).

Example 4 Let (see Fig. 2)

$$g(u) = \begin{cases} 4u\,, & 0 < u \le \frac{1}{4}\,, \\ 0.5\,, & \frac{1}{4} < u \le \frac{1}{2}\,, \\ 4u - 2\,, & \frac{1}{2} < u \le \frac{3}{4}\,, \\ 0.5\,, & \frac{3}{4} < u \le 1\,. \end{cases}$$

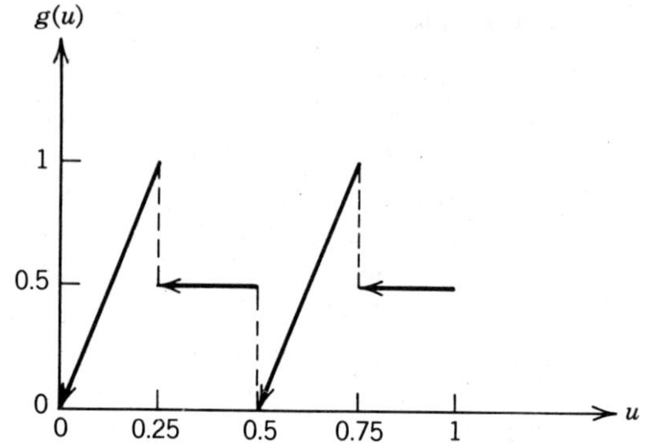

Fig. 2.

Then (see Fig. 3)

$$F_g(x) = \begin{cases} \frac{1}{2}x\,, & 0 < x < 0.5\,, \\ \frac{1}{2}x + 0.5\,, & 0.5 \le x \le 1\,. \end{cases}$$

Hence

$$g^*(u) = F_g^{-1}(u) = \begin{cases} 2u\,, & 0 < u < \frac{1}{4}\,, \\ 0.5\,, & \frac{1}{4} \le u \le \frac{3}{4}\,, \\ 2u - 1\,, & \frac{3}{4} < u \le 1\,, \end{cases}$$

and consequently (see Fig. 4)

$$g_*(u) = F_g^{-1}(1-u) = \begin{cases} 1 - 2u\,, & 0 < u < \frac{1}{4}\,, \\ 0.5\,, & \frac{1}{4} \le u \le \frac{3}{4}\,, \\ 2 - 2u\,, & \frac{3}{4} < u \le 1\,. \end{cases}$$

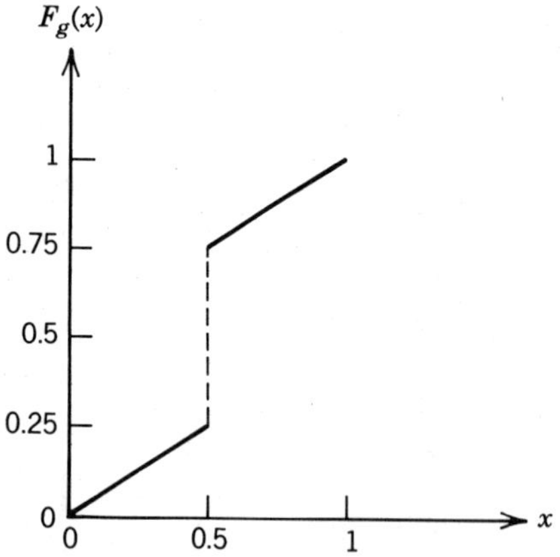

Fig. 3.

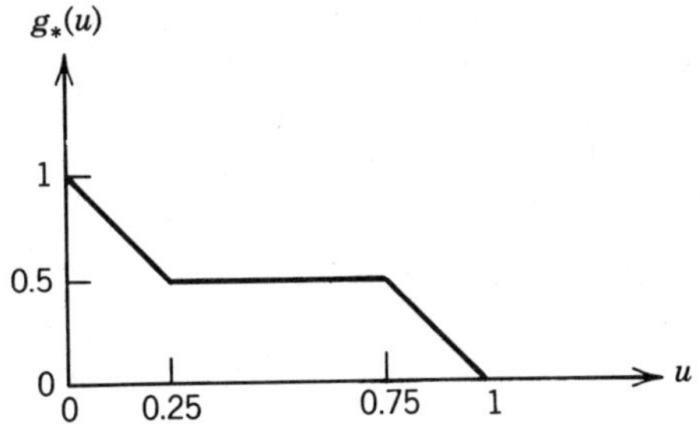

Fig. 4.

By Proposition 6 there exists a pair of transformations $\{\bar{h}_g(u), \underline{h}_g(u)\}$ [regardless of the continuity of $g(u)$] that is a function of a single r.v. U only. For example (see Fig. 5), one of the possible functions $\underline{h}_g(u)$ is

$$\underline{h}_g(u) = \begin{cases} 1 - 2u, & 0 < u < \frac{1}{8}, \\ 0.5 - 2u, & \frac{1}{8} \le u < \frac{1}{4}, \\ 1 - u, & \frac{1}{4} \le u < \frac{1}{2}, \\ 2 - 2u, & \frac{1}{2} \le u < \frac{5}{8}, \\ 1.5 - 2u, & \frac{5}{8} \le u < \frac{3}{4}, \\ \frac{5}{4} - u, & \frac{3}{4} \le u \le 1. \end{cases}$$

(b) Optimal Procedures and Examples Now, we shall use Proposition 6 to construct the pair of transformations $\{h_1(u), h_2(u)\}$ that generates the optimal c.d.f. F^* and therefore minimizes

$$\mathrm{Var}[g_1(U_1) + g_2(U_2)]. \tag{1.2.11}$$

We start with the case where one of the functions, say g_2, is monotone.

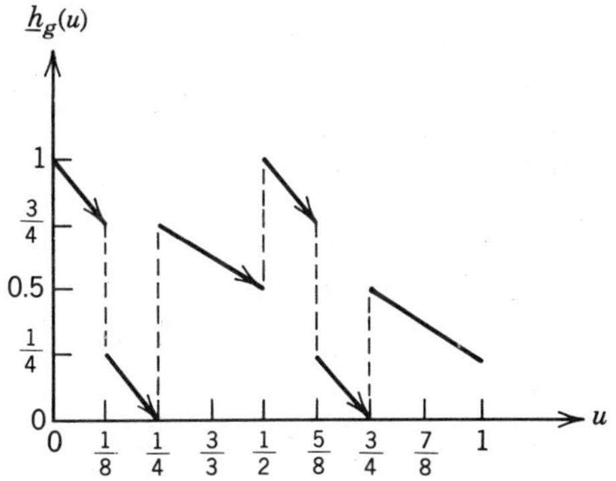

Fig. 5.

Lemma 2 If $g_1(u_1)$ and $g_2(u_2)$ are two measurable functions on [0, 1] and $g_2(u_2)$ is monotone, then there exists a transformation $h(u)$ such that

1. $h(U) \sim U(0, 1)$ and
2. $\min_{F\mathbf{u}_1\mathbf{u}_2 \in \Xi_2} \{\text{Var}[g_1(U_1) + g_2(U_2)]\} = \text{Var}\{g_1(U) + g_2[h(U)]\}$.

Proof Assume that $g_2(u)$ is nonincreasing in u. By Proposition 6 there exists a measure-preserving transformation $\bar{h}_{g_1}(u)$ such that

$$g_1^*[\bar{h}_{g_1}(u)] = g_1(u) \text{ a.s.}$$

Set $h(u) = \bar{h}_{g_1}(u)$. Then 1 holds. To show that 2 holds, let $F(u_1, u_2)$ be any c.d.f. from the set Ξ_2 of all possible c.d.f.'s. Then

$$\min_{F\mathbf{u}_1\mathbf{u}_2 \in \Xi_2} \text{Var}[g_1(U_1) + g_2(U_2)] = \min_{Fu_1u_2 \in \Xi_2} \text{Var}[g_1^*(\bar{h}_{g_1}(U_1)) + g_2(U_2)]$$

$$\geq \text{Var}[g_1^*(U) + g_2(U)] = \text{Var}[g_1^*(\bar{h}_{g_1}(U)) + g_2(\bar{h}_{g_1}(U))]$$

$$= \text{Var}[g_1(U) + g_2(\bar{h}_{g_1}(U))] = \text{Var}[g_1(U) + g_2(h(U))] . \tag{1.2.12}$$

The inequality in (1.2.12) holds since both $g_1^*(u) = F_{g_1}^{-1}(u)$ and $g_2(u)$ are monotone in opposite directions [see, e.g., Gal et al. (1984)].

The first equality after the inequality holds since $\bar{h}_{g_1}(U) \sim U(0, 1)$. The last two equalities in equation (1.2.12) are obvious. This completes the proof. Q.E.D.

We present now two examples demonstrating how Lemma 2 can be applied to minimize the variance (1.2.11) when g_2 is monotone.

Example 5 Let $g_2(u_2)$ be any monotone increasing function on [0, 1] and $g_1(u_1)$ be a symmetric unimodal function, that is, $g_1(0.5 - \theta) = g_1(0.5 + \theta)$, $\forall \theta \in [0, 0.5]$ and $g_1(u)$ is monotone on [0, 0.5], for example, nonincreasing. Then it is readily obtainable that

$$\bar{h}_{g_1}(u) = \begin{cases} 1 - 2u , & 0 < u \leq 0.5 , \\ 2u - 1 , & 0.5 < u \leq 1 , \end{cases}$$

and it follows from Lemma 2 (property 2) that

$$\min_{F\mathbf{u}_1\mathbf{u}_2 \in \Xi_2} \text{Var}[g_1(U_1) + g_2(U_2)] = \text{Var}\{g_1(U) + g_2[\bar{h}_{g_1}(U)]\} . \tag{1.2.13}$$

As a numerical example, let $g_1(u_1) = \sin \pi u_1$ and $g_2(u_2) = u_2^2$. Denote

$\theta_1 = g_1(U_1) + g_2(U_2)$ and $\theta_2 = g_1(U) + g_2[\bar{h}_{g_1}(U)]$. Then, for the crude Monte Carlo method and for our optimal procedure, we obtain Var $\theta_1 =$ 0.1836 and Var $\theta_2 = 0.0189$, respectively. The gain in variance reduction is

$$\epsilon = \frac{\text{Var } \theta_2}{\text{Var } \theta_1} = 9.71 .$$

Example 6 Let $g_1(u_1)$ and $g_2(u_2)$ be a piece-wise continuous function and a monotone function, respectively, and let the c.d.f. of the r.v. $g_1(U)$ be continuous. Then the following procedure can be used for minimizing the variance in (1.2.11):

Procedure 3

1. Generate $U_1 \sim U(0, 1)$
2. Find the set $\{u \in [0, 1]\colon g_1(u) \le g_1(u_1)\}$ that presents a union of some intervals.
3. Compute $l =$ sum of the lengths of these intervals.
4. Deliver $\theta_2 = g_1(u_1) + g_2(l)$ as a desired estimator.

To see that the Procedure 3 minimizes (1.1.12), we present $g_2(l)$ as

$$g_2(l) = g_2\{P[g_1(U)] \le g_1(u_1)]\} = g_2[F_{g_1}(g_1(u_1))] = g_2[\bar{h}_{g_1}(u_1)]$$

and the proof is completed.

As a numerical example, take

$$g_1(u_1) = \begin{cases} e^{10u_1^2} , & 0 < u_1 \le 0.3 , \\ 3u_1 - 0.9 , & 0.3 < u_1 \le 1 ; \end{cases}$$

$$g_2(u_2) = 1 - u_2 , \qquad 0 < u_2 \le 1 .$$

For example, if $0 < u_1 \le 0.272$, then (see Fig. 6) $l = u_1 + \frac{1}{3}e^{10u_1^2}$. Proceeding further, we obtain

$$l = \begin{cases} u_1 + {}_3e \quad , & 0 < u_1 \le 0.272 , \\ 0.7 + u_1 , & 0.272 < u_1 \le 0.3 , \\ u_1 - 0.3 , & 0.3 < u_1 \le 0.633 , \\ u_1 - 0.3 + \sqrt{\frac{1}{10}[\ln(3u_1 - 0.9)]} , & 0.633 < u_1 \le 1 . \end{cases}$$

By using the crude Monte Carlo method and by the optimal procedure,

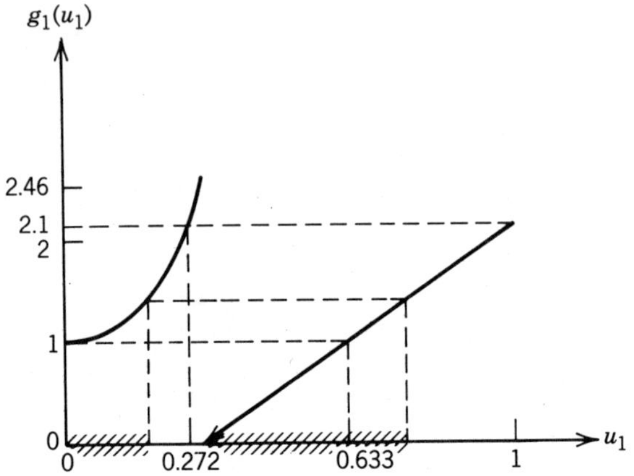

Fig. 6.

we obtain, accordingly,

$$\text{Var}\ \theta_1 = 0.717 \quad \text{and} \quad \text{Var}\ \theta_2 = 0.533\ .$$

The gain is 25.66%.

Consider now the case where g_1 and g_2 are arbitrary measurable functions. Let us show how Proposition 6 can be used to construct a pair of transformations $\{h_1(U), h_2(U)\}$ (to generate an optimal c.d.f.) that minimizes (1.2.11).

Consider first a pair of random vectors $\{u, \bar{h}_{g_1}(u)\}$ and $\{u, \underline{h}_{g_2}(u)\}$ that defines a pair of distributions on $[0, 1] \times [0, 1]$, with uniform marginals denoted $G_1(U^{(1)}, U^{(2)})$ and $G_2(U^{(1)}, U^{(2)})$ accordingly. Then the procedure for minimizing (1.2.11) is as follows:

Procedure 4

1. Generate $U \sim U(0, 1)$.
2. Find the conditional distributions $G_1(U^{(1)} \mid U^{(2)} = U)$ and $G_2(U^{(1)} \mid U^{(2)} = U)$.
3. Generate $\tilde{U}_1$ and $\tilde{U}_2$ from $G_1(\cdot \mid U^{(1)} = U)$ and $G_2(\cdot \mid U^{(2)} = U)$, respectively.
4. Deliver $g_1(\tilde{U}_1) + g_2(\tilde{U}_2)$ as a desired estimator.

To see that Procedure 4 indeed is optimal [minimizes (1.2.11)], note first that the c.d.f. $F_{\tilde{U}_1, \tilde{U}_2}$ has uniform marginals since the c.d.f.'s G_1, G_2 do. Further from construction of the procedure, it follows that $\bar{h}_{g_1}(\tilde{U}_1) = U$ and $\underline{h}_{g_2}(\tilde{U}_2) = U$. Hence, $g_1(\tilde{U}_1) = g_1^*(U)$ and $g_2(\tilde{U}_2) = g_{2*}(U)$. Now arguing similar to (1.2.12), we obtain

$$\min_{F_{U_1U_2}\in\Xi_2} \mathrm{Var}[g_1(U_1)+g_2(U_2)]$$

$$= \min_{F_{U_1U_2}\in\Xi_2} \mathrm{Var}[g_1^*[\bar{h}_{g_1}(U_1)] + g_{2*}[\underline{h}_{g_2}(U_2)]]$$

$$\geq \mathrm{Var}[g_1^*(U)+g_{2*}(U)] = \mathrm{Var}\{g_1^*[\bar{h}_{g_1}(\tilde{U}_1)] + g_{2*}[\underline{h}_{g_2}(\tilde{U}_2)]\}$$

$$= \mathrm{Var}[g_1(\tilde{U}_1)+g_2(\tilde{U}_2)] .$$

Thus, within the set Ξ_2 of all c.d.f.'s, the c.d.f. $F_{\tilde{U}_1\tilde{U}_2}$ minimizes (1.2.11).

Remark If the c.d.f.'s of g_1 and g_2 are continuous, then it is readily seen that Procedure 4 provides a one-to-one correspondence between the values of g_1 and g_2.

Example 7 Assume that g_1 and g_2 are given as (see Figs. 7 and 8, respectively)

$$g_1(u_1) = \sin \pi u_1 , \qquad 0 < u_1 \leq 1 ,$$

$$g_2(u_2) = \begin{cases} 1 - 3u_2 , & 0 < u_2 \leq \frac{1}{3} , \\ \frac{3}{4}u_2 - \frac{1}{4} , & \frac{1}{3} < u_2 \leq 1 . \end{cases}$$

Let us show how to construct h_1 and h_2 in order to minimize $\mathrm{Var}[g_1(U_1) + g_2(U_2)]$. We have (see Figs. 9 and 10, respectively)

$$g_1^*(u_1) = \sin(\tfrac{1}{2}\pi u_1)$$

$$g_2^*(u_2) = \begin{cases} \frac{3}{5}u_2 , & u_2 \leq \frac{5}{6} , \\ 3u_2 - 2 , & \frac{5}{6} < u_2 \leq 1 . \end{cases}$$

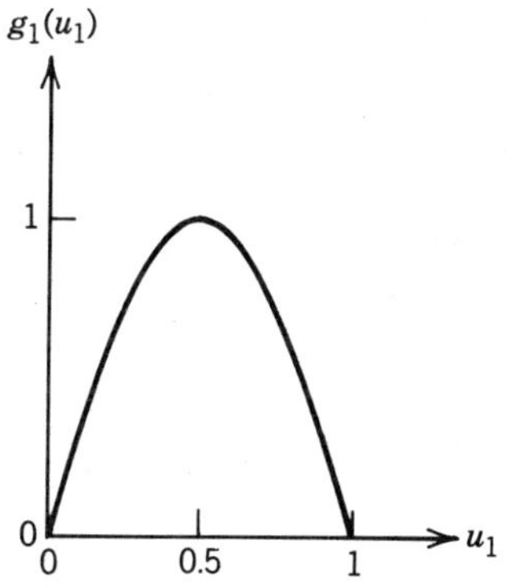

Fig. 7.

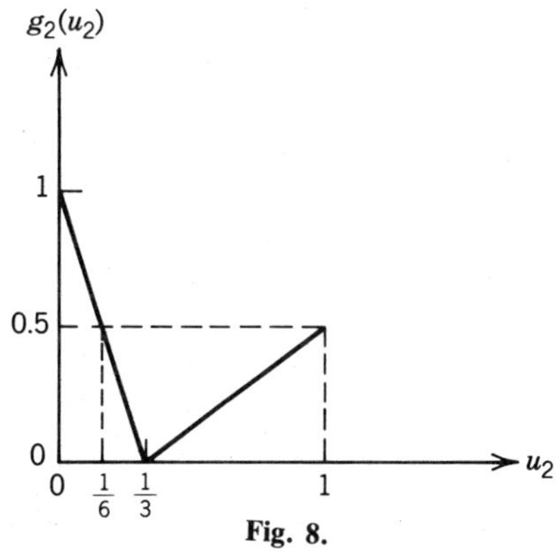

Fig. 8.

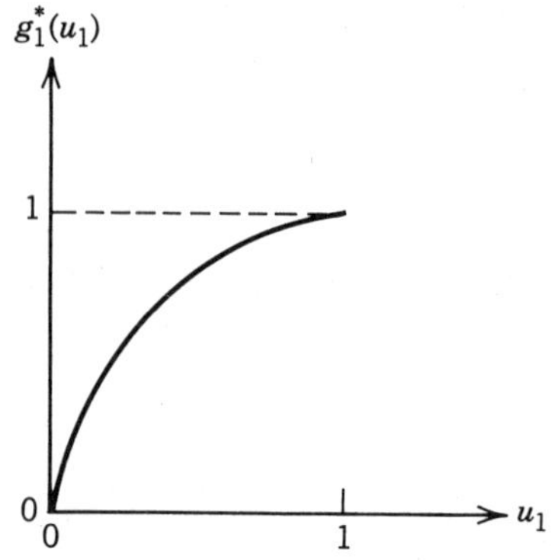

Fig. 9.

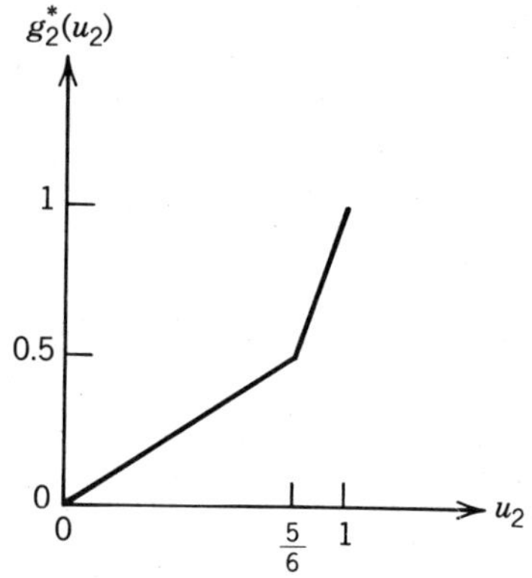

Fig. 10.

The optimal transformations $h_1(u)$ and $h_2(u)$ can be readily constructed according to Procedure 4. They are,

$$h_1(u) = \begin{cases} 5u\,, & 0 < u \le 0.1\,, \\ \frac{5}{4}u - \frac{1}{8}\,, & 0.1 < u \le 0.5\,, \\ 3.5 - 5u\,, & 0.5 < u \le 0.6\,, \\ \frac{7}{4} - \frac{5}{4}u\,, & 0.6 < u \le 1\,, \end{cases}$$

and

$$h_2(u) = \begin{cases} \frac{1}{3} - 2u\,, & 0 < u \le \frac{1}{12}\,, \\ 1 - 10u\,, & \frac{1}{12} < u \le \frac{1}{10}\,, \\ 2u + \frac{2}{15}\,, & \frac{1}{10} < u \le \frac{13}{30}\,, \\ \frac{5}{4} - \frac{5}{2}u\,, & \frac{13}{30} < u \le \frac{1}{2}\,, \\ \frac{4}{3} - 2u\,, & \frac{1}{2} < u \le \frac{7}{12}\,, \\ 6 - 10u\,, & \frac{7}{12} < u \le 0.6\,, \\ 2u - \frac{13}{15}\,, & 0.6 < u \le \frac{14}{15}\,, \\ \frac{5}{2} - \frac{5}{2}u\,, & \frac{14}{15} < u \le 1\,. \end{cases}$$

respectively.

By using the crude Monte Carlo method and by our optimal procedure, we obtain Var $\theta_1 = (1/2 - 4/\pi^2) + 7/72 = 0.1913$ and Var $\theta_2 = 0.0991$, respectively. The gain in variance reduction is 48.36%.

REFERENCES

Alam, K. and K. T. Wallenius (1976). Positive Dependence and Monotonicity in Conditional Distributions, *Comm. Stat.*, **5**, 525–534.

Andreasson, I. J. (1971). On the Generation of Negatively Correlated Random Numbers, Report A71,32, Department of Computer Science, Royal Institute of Technology, Stockholm.

Barlow, R. and F. Proschan (1975). *Statistical Theory of Reliability and Life Testing: Probability Models*, Holt, Reinhart and Winston, New York.

Barnett, V. (1980). Some Bivariate Uniform Distributions, *Comm. Stat.—Theor. Meth.*, **A9**, 453–461.

Block, H. W. and M. Ting (1981). Some Concepts of Multivariate Dependence, *Comm. Stat.—Theor. Meth.*, **A10**, 749–762.

Bolviken, E. (1982). Probability Inequalities for the Multivariate Normal with Non-negative Partial Correlations, *Scand. J. Stat.*, **9**, 49–58.

Brindley, E. C. and W. A. Thompson (1972). Dependence and Aging Aspects of Multivariate Survival, *J. Amer. Stat. Assoc.*, **67**, 822–830.

Dykstra, R. L. and J. E. Hewett (1978). Positive Dependence of the Roots of a Wishart Matrix, *Ann. Stat.*, **6**, 235–238.

Esary, J. D. and A. W. Marshall (1970). Coherent Life Functions, *SIAM J. Appl. Math.*, **18**, 810–814.

Fishman, G. S. and B. D. Huang (1981). Antithetic Variates Revised, Technical Report 80-4, Curriculum in Operations Research and System Analysis, University of North Carolina at Chapel Hill.

Gal, S., Y.R. Rubinstein, and A. Ziv (1984). On the Optimality and Efficiency of Common Random Numbers, *Math. Comp. Simulation*, **XXVI**, 502–512.

Granovsky, B. (1982). Optimal Variance Reduction Theorem in Simulation by the Monte Carlo Method, Technion Report Series No. MT-571.

Grant, F. H. and J. J. Solberg (1982). Variance Reduction Techniques in Stochastic Shortest Route Analysis, Pritsker & Associates, Inc., West Lafayette, Indiana 47906.

Gross, D. and C. M. Harris (1985) Fundamentals of queueing theory, Wiley, New York.

Hammersley, J. M. and D. C. Handscomb (1964). *Monte Carlo Methods*, Methuen, London.

Hammersley, J. M. and J. G. Mauldon (1956). General Principles of Antithetic Variates, *Math. Proc. Comb. Phil. Soc.*, **52**, 476–481.

Hammersley, J. M. and K. W. Morton (1956). A New Monte Carlo Technique: Antithetic Variates, *Proc. Cambridge Phil. Soc.*, **52**, 449–475.

Handscomb, D. C. (1958). Proof of the Antithetic Variates Theorem for $n > 2$, *Math. Proc. Comb. Phil. Soc.*, **1958**, 300–301.

Karlin, S. and Y. Rinott (1980). Classes of Orderings of Measures and Related Inequalities, *J. Multivar. Anal.*, **10**, 467–498.

Karlin, S. and Y. Rinott (1981). Total Positivity Properties of Absolute Value Multinormal Variables with Applications to Confidence Interval Estimates and Related Probability Inequalities, *Ann. Prob.*, **9**, 1035–1049.

Kleijnen, J. P. C. (1974). *Statistical Techniques in Simulation*, Part II, Marcel Dekker, New York.

Luenberger, D. G. (1969). *Optimization by Vector Space Methods*, Wiley, New York.

Mitchel, B. (1973). Variance Reduction by Antithetic Variates in GI/G/1 Queueing Simulation, *Oper. Res.*, **21**, 988–997.

Nelson, B. L. and Schmeiser, B. W. (1983). Variance Reduction: Basic Transformations, Proceedings of the 1983 Winther Simulation Conference, Institute of Electrical and Electronic Engineers, Washington D.C.

Page, E. S. (1967). On Monte Carlo Methods in Congestion Problems, II: Simulation of Queueing Systems, *Oper. Res.*, **13**, 300–305.

Rubinstein, Y. R. (1981). *Simulatin and the Monte Carlo Methods*, Wiley, New York.

Rubinstein, Y. R. and G. Samorodnitsky (1985). Antithetic and Common Random Numbers in Simulation Stochastic Systems, *J. Stat. Comp. Simul.*, **2**, 161–184.

Rubinstein, Y. R. and G. Samorodnitsky (1986). A Modified Version of Handscomb's Antithetic Variates Theorem, *SIAM J. Stat. Sci. Comp.* (to appear).

Rubinstein, Y. R., G. Samorodnitsky, and M. Shaked (1985). Antithetic Variates, Multivariate Dependence and Simulation of Complex Stochastic Systems, *Manage. Sci.*, **31**, 66–77.

Schruben, L. W. (1979). Designing Correlation Induction Strategies for Simulation Experiments, Current Issues in Computer Simulation, Academic Press, New York, pp. 235–256.

Schruben, L. W. and B. Margolin (1978). Pseudo-Random Number Assignment in Statistically Designed Simulation and Distribution Sampling Experiments, *J. Amer. Stat. Assoc.*, **73**, 504–525.

Shaked, M. (1977). A Family of Concepts of Positive Dependence for Bivariate Distributions, *J. Amer. Stat. Assoc.*, **72**, 642–650.

Shaked, M. (1982). A General Theory of Some Positive Dependence Notions, *J. Multivariate Anal.*, **12**, 199–218.

Sullivan, R. S., J. C. Hayya, and R. Schaul (1982). The Antithetic Variate Method for Simulating Stochastic Networks, *Manage. Sci.*, **28**, 563–572.

Whitt, W. (1976). Bivariate Distributions with Given Marginals, *Ann. Stat.*, **4**, 1280–1289.

Wilson, J. R. (1979). Proof of the Antithetic Variates Theorem for Unbounded Functions, *Math. Proc. Cambridge Phil. Soc.*, **86**, 477–479.

Wilson, J. R. (1983). Antithetic Sampling with Multivariate Inputs, *Amer. J. Math. Manage. Sci.*, **3**, 121–144.

Wilson, J. R. (1984). Variance Reduction Techniques for Digital Simulation, *Amer. J. Math. Manage. Sci.*, **4**, No. 3 and 4, 277–313.

Wright, R. D. and T. E. Ramsay, Jr. (1979). On the Effectiveness of Common Random Numbers, *Manage. Sci.*, **25**, 649–656.

CHAPTER 2

Multidimensional Control Variates in Monte Carlo Simulation

2.1 INTRODUCTION

In this chapter we shall deal with the control variable technique, which has recently received much attention in Monte Carlo simulation as a means of improving accuracy when estimating output parameters of complex stochastic systems [see, e.g., Glynn and Whitt (1985); Kleijnen (1974); Nozari, Arnold, and Pegden (1982); Lavenberg and Welch (1981); Lavenberg, Moeller, and Welch (1982); and Rubinstein and Marcus (1985)]. Applications of control variates for stochastic activity networks are given in Grant and Solberg (1983); those for queueing networks are given in Lavenberg, Moeller, and Welch (1982) and Wilson and Pritsker (1984a, b). Rothery (1982) used F-statistic as control variate for the one-way analysis of variance and Swain and Shmeiser (1984a, b) used control variates as linear approximations to nonlinear regression models.

We first explain the meaning of control variates.

Assume that Y is an unbiased estimator of μ, which has to be estimated from a simulation run. A random variable C is a control variate for Y if it is correlated with Y and its expectation r is known. The control variate C is used to construct an unbiased estimator for μ, say $Y(\alpha)$, with a variance smaller than var(Y). The new estimator $Y(\alpha)$ is defined as

$$Y(\alpha) = Y - \alpha(C - r) , \qquad (2.1.1)$$

where α is any real number.

Now

$$\text{var}[Y(\alpha)] = \text{var}[Y] - 2\alpha \,\text{cov}[Y, C] + \alpha^2 \,\text{var}[C] . \qquad (2.1.2)$$

The value of α that minimizes $\text{var}[Y(\alpha)]$ is

$$\beta = \frac{\text{cov}[Y, C]}{\text{var}[C]}, \tag{2.1.3}$$

and the minimum variance is equal to

$$\text{var}[Y(\beta)] = (1 - \rho_{YC}^2)\,\text{var}[Y], \tag{2.1.4}$$

where ρ_{YC} is the correlation coefficient between Y and C. Hence the more C is correlated with Y, the greater is the reduction in variance.

Procedure (2.1.1) can readily be extended to the case of more than one control variate [see Lavenberg et al. (1982)]. Indeed, let $\mathbf{C} = (C_1, \ldots, C_g)'$ be a vector of g control variates with known mean vector $\mathbf{r} = (r_1, \ldots, r_g)'$, where $r_i = E(C_i)$, and let $\boldsymbol{\alpha}$ be any $g \times 1$ vector. In analogy with (2.1.1),

$$Y(\boldsymbol{\alpha}) = Y - \boldsymbol{\alpha}'(\mathbf{C} - \mathbf{r}) \tag{2.1.5}$$

is an unbiased estimator of $\mu = E(Y)$, where $'$ denotes the transpose operator. The vector $\boldsymbol{\beta}$ that minimizes $\text{var}[Y(\boldsymbol{\alpha})]$ [see, e.g., Lavenberg et al. (1982)] is

$$\boldsymbol{\beta} = \boldsymbol{\Sigma}_C^{-1}\boldsymbol{\sigma}_{YC}, \tag{2.1.6}$$

where $\boldsymbol{\Sigma}_C$ is the covariance matrix of $\mathbf{C}$ and $\boldsymbol{\sigma}_{YC}$ is a g-dimensional vector whose components are the covariances between Y and C. Substituting (2.1.6) into (2.1.5), we obtain the resulting minimal variance

$$\text{var}[Y(\boldsymbol{\beta})] = (1 - R_{YC}^2)\,\text{var}[Y], \tag{2.1.7}$$

where

$$R_{YC}^2 = \frac{\boldsymbol{\sigma}_{YC}'\boldsymbol{\Sigma}_C^{-1}\boldsymbol{\sigma}_{YC}}{\text{var}[Y]}$$

is the square of the multiple correlation coefficient betwen Y and $\mathbf{C}$. As before, the larger the multiple correlation coefficient between $\mathbf{C}$ and Y, the greater the reduction in variance.

A rigorous treatment of the control variable procedure (2.1.5) with application in simulation of a wide class of closed queueing networks that allow multiple types of customers, priorities, and blocking is discussed in

Lavenberg et al. (1982). A survey article on control variables and their potential for increasing the efficiency of Monte Carlo simulations is presented in Lavenberg and Welch (1981). Both papers deal with the efficiency of control variates in *single-population single-response* simulation experiments. Our object in this work is to extend the results of Lavenberg et al. (1982) to *single-population multivariate response case.*

Nozari et al. (1982) extend the application of control variates to the *multipopulation single-response* case, assuming that the response and the control variates have a multinormal distribution. In contrast to Nozari et al. (1982), we assume that there is only one population but more than one response. We note that the above results are also an extension of those of Lavenberg et al. (1982) but, as mentioned, in a different context.

Let us now introduce our multidimensional version of (2.1.5) in which we seek to estimate a p-dimensional vector $\boldsymbol{\mu} = E(\mathbf{Y})$, $\boldsymbol{\mu} = (\mu_1, \ldots, \mu_p)'$, rather than a single parameter $\mu = E(Y)$. As before, let $\mathbf{C} = (C_1, \ldots, C_g)'$ be a vector of g control variates with known mean vector $\mathbf{r} = (r_1, \ldots, r_g)'$. Then for any $p \times g$ matrix $\mathbf{A}$ the multidimensional procedure of control variates can be defined as

$$\underset{(p\times 1)}{\mathbf{Y}(\mathbf{A})} = \underset{(p\times 1)}{\mathbf{Y}} - \underset{(p\times g)}{\mathbf{A}}\,(\underset{(g\times 1)}{\mathbf{C}} - \mathbf{r})\,. \tag{2.1.8}$$

In the particular case where $p = 1$, procedure (2.1.8) reduces to produce (2.1.5).

To motivate procedure (2.1.8), let us consider the following two examples.

(a) GI/G/1 Queue We assume that, if an arriving customer finds the server free, the service commences immediately, and he or she departs from the system after completion of the service. If the arriving customer finds the server busy, the customer enters the waiting room and waits to be served. Customers are served on a first-in–first-out (FIFO) basis.

Let S_i denote the service time of the ith customer who arrives at time t_i, and let $A_i = t_i - t_{i-1}$, $i \geq 1 (t_0 \equiv 0)$ denote the interarrival time [the time between the arrivals of the $(i-1)$th and ith customers]. Assume that the sequences $\{S_i, i \geq 1\}$ and $\{A_i, i \geq 1\}$ each consist of i.i.d. r.v.'s and are themselves independent. Let μ be the mean service rate, and let λ be the mean arrival rate, that is,

$$E(S_i) = \mu^{-1} \quad \text{and} \quad E(A_i) = \lambda^{-1}\,.$$

The parameter $\rho = \lambda/\mu$ is called the *traffic intensity* and measures the congestion of the queueing system. The necessary and sufficient conditions for the system to reach steady-state condition is $\rho < 1$.

As measures for the performance of the system (vector **Y**), we can use the mean waiting time of the ith customer (time from arrival to commencement of service), the mean number of customers in the system at time t, or the mean number of customers who have been served in the interval $[0, t]$. For the vector of control variates (vector **C**) we can use the service time, the interarrival time, the interdeparture time of the ith customer, or the busy fraction of the server (the number of customers in service).

(b) Closed Queueing Networks We shall describe now a class of closed queueing networks taken from Lavenberg et al. (1982). Their networks consist of a finite number, say S, of interconnected service centers. A service center is a single-server or multiple-server queue. There are D different types of customers and there is a finite number $N(d)$ of type d customers, $d = 1, \ldots, D$. These customers circulate indefinitely among the service centers.

In Fig. 1 an example taken from Lavenberg et al. (1982) of such a closed queueing network is presented. It is assumed that the number of servers at service center 1 is equal to the number of customers at the network and that the service centers $2, \ldots, S$ present single-server queues with customers served in FIFO discipline. The transition probability matrix for each type d customer is assumed to be

$$P(d) = \begin{vmatrix} 0 & 1 & 0 & \cdots & 0 \\ p_1(d) & 0 & p_3(d) & \cdots & p_s(d) \\ 0 & 1 & 0 & \cdots & 0 \\ & \vdots & \vdots & & \vdots \\ 0 & 1 & 0 & \cdots & 0 \end{vmatrix}$$

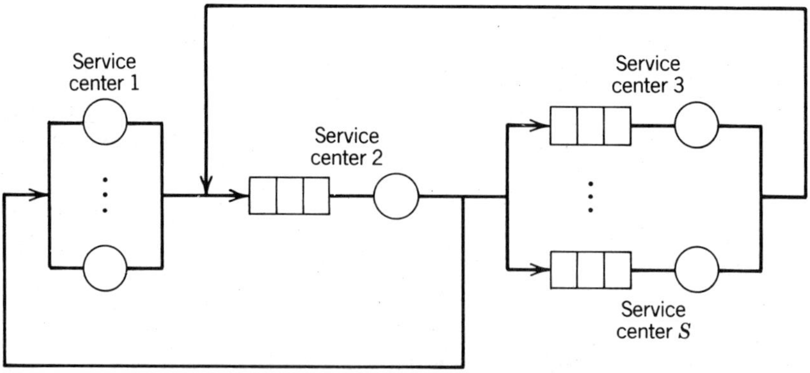

Fig. 1 A closed queueing network.

where $p_i(d) > 0$, $i = 1, 3, \ldots, S$. It follows from $P(d)$ that a type d customer that departs from service center 1 enters service center 2, a type d customer that departs from service center 2 enters service center i with probability $p_i(d)$, $i = 1, 3, \ldots, S$, and a type d customer that departs from service center i, $i = 3, \ldots, S$, enters service center 2.

Such a network can model, for example, an interactive multiprogrammed computer system, where customers represent users of the system, service center 1 represent the terminals, service center 2 represents the processor, and finally service centers $3, \ldots, S$ represent secondary storage devices (e.g., drums and disks). A service time at service center 2 represents an interval of processing time for a user's task until either that task is completed, in which case the user starts another think time, or until data from a secondary storage device are required. A service time at any of service centers $3, \ldots, S$ represents the time to access and transfer data from a secondary storage device to main memory.

Lavenberg et al. (1982) make the following assumptions about their networks:

- **(i)** The sequence of service centers visited by each type d customer, $d = 1, \ldots, D$, is an irreducible Markov chain with state space $I(d)$, which is a subset of $\{1, \ldots, S\}$, and transition probability matrix $\mathbf{P}(d)$.
- **(ii)** The sequence of service times for each type d customer, $d = 1, \ldots, D$, at service center $i \in I(d)$ is a sequence of independent and identically distributed nonnegative random variables with finite mean and variance. Service time random variables that are identically zero are allowed.
- **(iii)** The sequences of service centers visited (there is one for each customer) and the service time sequences (there is one for each customer and each service center that customers can visit) are mutually independent random sequences.

These networks also allow priorities and blocking, and therefore their class is quite broad. Moreover, no assumptions are made about the queueing disciplines or about the capacity of any service center or group of service centers.

The response characteristics for these networks (vector $E(\mathbf{Y}) = E(Y_1, \ldots, Y_p)'$ are:

$E[Y_1^{(i,d)}]$ The average waiting time (not including the service time) for type d customers at service center i, $i = 2, \ldots, S$, $d = 1, \ldots, D$;

$E[Y_2^{(d)}]$ The rate at which events (i.e., departures) occur for type d customers, $d = 1, \ldots, D$;

$E[Y_3^{(d)}]$ The average response time for type d customers, $d = 1, \ldots, D$; where a response time is the time between the departure of a customer from service center 1 and the subsequent arrival at service center 1.

The vector of control variables is associated with the service time random variables and the multinomial routing random variables that control the flow of customers in the networks. More specifically, define an event in the simulation of a queueing network to be the departure of a customer from a service center, and define a type d event to be the departure of a type d customer. Now, let $e_{d,i}(M)$ be number of type d events at service center i in the first M events, $W_{d,i}(M)$ the sum of the service times associated with the above $e_{d,i}(M)$ events, and $e_d(M)$ the number of type d events in the first M events $[=\Sigma_i^s e_{d,i}(M)]$. The quantity $W_{d,i}(M)/e_{d,i}(M)$ is the sample mean service time for type d customers at service center i. The quantity $e_{d,i}(M)/e_d(M)$ is the fraction of type d events at service center i. Finally, $W_{d,i}(m)/e_d(M)$, the product of these two quantities, is the sum of the service times for type d customers at service center i per type d event in the system. Lavenberg et al. (1982) call $W_{d,i}(M)/e_{d,i}(M)$ a *service time variable*, $e_{d,i}(M)/e_d(M)$ a *flow variable*, and $W_{d,i}(M)/e_d(M)$ a work variable, using them as a vector of control variables. They showed that a little computational effort is required to generate the vector of control variables and to compute its expected value; however, the effect of using the control variables for variance reduction is considerable.

Finally, it is worthwhile to note that in their procedure the vector of control variables was used separately for each component Y_i, $i = 1, \ldots, p$ [see procedure (2.1.5)], and therefore variance is reduced in the sense of reducing the width of the confidence interval for each $\mu_i = E(Y_i)$, $i = 1, \ldots, p$. However, by using procedure (2.1.8) and the results below, we can simultaneously obtain the confidence region for the vector $E(\mathbf{Y}) = \boldsymbol{\mu}$.

It is not the intention here to present simulation results based on the use of procedure (2.1.8) for any specific system, but as mentioned previously, our purpose is rather to extend the statistical results of procedure (2.1.5) to its multidimensional case (2.1.8) and to make some practical recommendations for choosing the optimal size of the vector of control variates $\mathbf{C}$.

We believe that our discussion will promote wider use of control variates in the simulation of complex stochastic systems.

In Section 2.2 we derive an expression for the optimal matrix $\mathbf{A}$ that minimizes the generalized variance of $\mathbf{Y}(\mathbf{A})$ and discuss the efficiency of procedure (2.1.8) under the assumption of the normality of $\mathbf{Y}$ and $\mathbf{C}$.

Section 2.3 deals with the choice of the optimal size of the vector of control variates where the sample size and the size of the vector **Y** are given. In the Appendix we present some theoretical results related to procedure (2.1.8).

2.2 STATISTICAL RESULTS ON CONTROL VARIABLES

Denote by $\mathbf{\Sigma}_{YY}$, $\mathbf{\Sigma}_{CC}$, $\mathbf{\Sigma}_{Y(A)}$ the respective covariance matrices of **Y**, **C** and $\mathbf{Y}(A)$ and by $\mathbf{\Sigma}_{YC}$ the covariance matrix between **Y** and **C**. Then by (2.1.8) we have

$$\mathbf{\Sigma}_{Y(A)} = \mathbf{\Sigma}_{YY} + \mathbf{A}\mathbf{\Sigma}_{CC}\mathbf{A}' - \mathbf{A}\mathbf{\Sigma}'_{YC} - \mathbf{\Sigma}_{YC}\mathbf{A}' . \tag{2.2.1}$$

We define the efficiency of the control variates (2.1.8) as

$$\delta_1^2 = \frac{\text{var}[\mathbf{Y(A)}]}{\text{var}[\mathbf{Y}]} = \frac{|\mathbf{\Sigma}_{Y(A)}|}{|\mathbf{\Sigma}_{YY}|} , \tag{2.2.2}$$

where $|\mathbf{\Sigma}_{XX}|$ is the determinant of $\mathbf{\Sigma}_{XX}$, called the generalized variance of **X** [see Anderson (1984)]. Accordingly, the efficiency of the control variates is measured by the reduction in the generalized variance when **Y(A)** is used instead of **Y**.

It is shown in Lemma A1 that the matrix **B** that minimizes the generalized variance $|\mathbf{\Sigma}_{Y(A)}|$ is equal to

$$\mathbf{B} = \mathbf{\Sigma}_{YC}\mathbf{\Sigma}_{CC}^{-1} . \tag{2.2.3}$$

The resulting minimum variance is

$$\begin{aligned} |\mathbf{\Sigma}_{Y(B)}| &= |\mathbf{\Sigma}_{YY} - \mathbf{\Sigma}_{YC}\mathbf{\Sigma}_{CC}^{-1}\mathbf{\Sigma}'_{YC}| \\ &= |\mathbf{I} - \mathbf{\Sigma}_{YC}\mathbf{\Sigma}_{CC}^{-1}\mathbf{\Sigma}'_{YC}\mathbf{\Sigma}_{YY}^{-1}|\,|\mathbf{\Sigma}_{YY}| , \end{aligned} \tag{2.2.4}$$

and the efficiency of the control variates is

$$\delta_1^2 = |\mathbf{I} - \mathbf{\Sigma}_{YC}\mathbf{\Sigma}_{CC}^{-1}\mathbf{\Sigma}'_{YC}\mathbf{\Sigma}_{YY}^{-1}| = \prod_{i=1}^{q} (1 - \rho_i^2) , \tag{2.2.5}$$

where $q = \min(p, g)$, and ρ_i^2, $i = 1, \ldots, q$, are the squares of the canonical correlations between **Y** and **C** that satisfy $\rho_1^2 \geq \cdots \geq \rho_q^2$ [this follows from equations (33)–(38) in Section 12.2 of Anderson (1984)]. In the particular case where $p = 1$, $\rho_1 = R_{YC}$ is the multiple correlation coefficient between Y and **C**, and we have (2.1.5).

The ratio δ_1 measures the relationship between vectors **Y** and **C**: the larger the canonical correlations ρ_i, $i = 1, \ldots, q$, between **C** and **Y**, the greater the reduction in the generalized variance.

Let

$$\mathbf{Z} = \begin{pmatrix} \mathbf{Y} \\ \mathbf{C} \end{pmatrix}$$

be a $(p + g)$-dimensional vector with mean

$$\boldsymbol{\mu}_Z = \begin{pmatrix} \boldsymbol{\mu} \\ \mathbf{r} \end{pmatrix}$$

and covariance matrix

$$\boldsymbol{\Sigma}_Z = \begin{pmatrix} \boldsymbol{\Sigma}_{YY} & \boldsymbol{\Sigma}_{YC} \\ \boldsymbol{\Sigma}'_{YC} & \boldsymbol{\Sigma}_{CC} \end{pmatrix} .$$

Let us take a random sample of k observations

$$\mathbf{Z}_i = \begin{pmatrix} \mathbf{Y}_i \\ \mathbf{C}_i \end{pmatrix}, \qquad i = 1, \ldots, k ,$$

of the random vector **Z**. We consider two cases:

(a) the matrix **B** is known and

(b) the matrix **B** is unknown and has to be estimated.

(a) For large k due to normality assumption

$$k(\bar{\mathbf{Y}} - \boldsymbol{\mu})' \boldsymbol{\Sigma}_{YY}^{-1} (\bar{\mathbf{Y}} - \boldsymbol{\mu})$$

has approximately a χ^2 distribution with p degrees of freedom, and the confidence ellipsoid for $\boldsymbol{\mu}$ is obtained from

$$\Pr\{k(\bar{\mathbf{Y}} - \boldsymbol{\mu})' \boldsymbol{\Sigma}_{YY}^{-1} (\bar{\mathbf{Y}} - \boldsymbol{\mu}) \leq \chi^2_{p;1-\alpha}\} = 1 - \alpha , \qquad (2.2.6)$$

where $\bar{\mathbf{Y}} = (1/k) \sum_{j=1}^{k} \mathbf{Y}_j$. The volume of this ellipsoid, which with probability $1 - \alpha$ covers the unknown vector $\boldsymbol{\mu}$, is [see Anderson (1984), p. 170]

$$V_1 = p^{-1} C(p) |\boldsymbol{\Sigma}_{YY}|^{1/2} \left[\frac{\chi^2_{p;1-\alpha}}{k} \right]^{p/2} , \qquad (2.2.7)$$

where

$$C(p) = \frac{2\pi^{p/2}}{\Gamma(p/2)} \tag{2.2.8}$$

and Γ denotes the gamma function. Another confidence ellipsoid for $\boldsymbol{\mu}$ is obtainable from

$$\Pr\{k[\bar{\mathbf{Y}}(\mathbf{B}) - \boldsymbol{\mu}]\boldsymbol{\Sigma}_{Y(B)}^{-1}[\bar{\mathbf{Y}}(\mathbf{B}) - \boldsymbol{\mu}] \leq \chi^2_{p;1-\alpha}\} = 1 - \alpha \tag{2.2.9}$$

by using control variates, where

$$\bar{\mathbf{Y}}(\mathbf{B}) = \bar{\mathbf{Y}} - \mathbf{B}(\bar{\mathbf{C}} - \mathbf{r}) .$$

In analogy to (2.2.7), the volume of the ellipsoid defined by (2.2.9) for a given $\bar{\mathbf{Y}}(\mathbf{B})$ is

$$V_2 = p^{-1}C(p)|\boldsymbol{\Sigma}_{Y(B)}|^{1/2}\left[\frac{\chi^2_{p;1-\alpha}}{k}\right]^{p/2} . \tag{2.2.10}$$

In this case the efficiency of the control variables can be measured by the ratio between the volumes V_2 and V_1, that is, by

$$\delta_2 = \frac{V_2}{V_1} ,$$

which, together with (2.2.7) and (2.2.10), yields

$$\delta_2^2 = \frac{V_2^2}{V_1^2} = \frac{|\boldsymbol{\Sigma}_{Y(B)}|}{|\boldsymbol{\Sigma}_{YY}|} = \prod_{i=1}^{q} (1 - \rho_i^2) . \tag{2.2.11}$$

Therefore, when $\mathbf{B}$ is known, so are $\delta_1 = \delta_2$, since the ratio of the generalized variances of $\mathbf{Y}(B)$ to $\mathbf{Y}$ is equal to the ratio of their squared volumes. As a result, the squared volume of the ellipsoid, which with prescribed probability $1 - \alpha$ covers the unknown $\boldsymbol{\mu}$, is reduced by

$$\prod_{i=1}^{q} (1 - \rho_i^2) .$$

(b) Assuming $\boldsymbol{\Sigma}_Z$ to be unknown, we denote by $\mathbf{S}_Z$ the sample covariance matrix equal to

$$\mathbf{S}_Z = \begin{pmatrix} \mathbf{S}_{YY} & \mathbf{S}_{YC} \\ \mathbf{S}'_{YC} & \mathbf{S}_{CC} \end{pmatrix} , \tag{2.2.12}$$

while the (u, v)th elements of $\mathbf{S}_{YY}$, $\mathbf{S}_{YC}$, and $\mathbf{S}_{CC}$ are

$$\frac{1}{k-1}\sum_{j=1}^{k}(Y_{uj}-\bar{Y}_u)(Y_{vj}-\bar{Y}_v)\,, \tag{2.2.13}$$

$$\frac{1}{k-1}\sum_{j=1}^{k}(Y_{uj}-\bar{Y}_u)(C_{vj}-\bar{C}_v)\,, \tag{2.2.14}$$

$$\frac{1}{k-1}\sum_{j=1}^{k}(C_{uj}-\bar{C}_u)(C_{vj}-\bar{C}_v)\,, \tag{2.2.15}$$

respectively, where

$$\bar{Y}_u=\frac{1}{k}\sum_{j=1}^{k}Y_{uj} \quad \text{and} \quad \bar{C}_v=\frac{1}{k}\sum_{j=1}^{k}C_{vj}\,.$$

Denoting

$$\mathbf{Y}_j(\hat{\mathbf{B}})=\mathbf{Y}_j-\hat{\mathbf{B}}(\mathbf{C}_j-\mathbf{r})\,, \qquad j=1,\ldots,k\,, \tag{2.2.16}$$

we estimate $\boldsymbol{\mu}$ by

$$\bar{\mathbf{Y}}(\hat{\mathbf{B}})=\frac{1}{k}\sum_{j=1}^{k}\mathbf{Y}_j(\hat{\mathbf{B}})=\bar{\mathbf{Y}}-\hat{\mathbf{B}}(\bar{\mathbf{C}}-\mathbf{r})\,. \tag{2.2.17}$$

In general, $\bar{\mathbf{Y}}(\hat{\mathbf{B}})$ is a biased estimator of $\boldsymbol{\mu}$, since $\hat{\mathbf{B}}$ and $\bar{\mathbf{C}}$ are interdependent. Also the $\mathbf{Y}_j(\hat{\mathbf{B}})=\mathbf{Y}_j-\hat{\mathbf{B}}(\mathbf{C}_j-\mathbf{r})$ are interdependent, and hence χ^2 cannot be directly used to obtain a confidence ellipsoid for $\boldsymbol{\mu}$.

To construct a confidence ellipsoid for $\boldsymbol{\mu}$, we assume $\mathbf{Z}\sim N(\boldsymbol{\mu}_Z, \boldsymbol{\Sigma}_Z)$. The maximum-likelihood estimator of the matrix $\mathbf{B}$ is given by the sample equivalent

$$\hat{\mathbf{B}}=\mathbf{S}_{YC}\mathbf{S}_{CC}^{-1}\,. \tag{2.2.18}$$

[see Section 4.3.1 of Anderson (1984)].

Under the assumption that $\mathbf{Y}$ is distributed $N(\boldsymbol{\mu}, \boldsymbol{\Sigma}_{YY})$, a confidence ellipsoid for $\boldsymbol{\mu}$ is [see Anderson (1984)] given by

$$\Pr\left\{k(\bar{\mathbf{Y}}-\boldsymbol{\mu})'\mathbf{S}_{YY}^{-1}(\bar{\mathbf{Y}}-\boldsymbol{\mu})\leq\frac{(k-1)p}{k-p}F_{p,k-p;1-\alpha}\right\}=1-\alpha\,, \tag{2.2.19}$$

where $F_{p,k-p;1-\alpha}$ denotes the $(1-\alpha)$-quantile of the F distribution with p and $k-p$ degrees of freedom.

Let

$$\hat{\boldsymbol{\Sigma}}_{Y\cdot C} = \frac{k-1}{k-g-1}(\mathbf{S}_{YY} - \mathbf{S}_{YC}\mathbf{S}_{CC}^{-1}\mathbf{S}_{CY})\,; \tag{2.2.20}$$

then it is known [see Anderson (1984)] that $\hat{\boldsymbol{\Sigma}}_{Y\cdot C}(\mathbf{d}'\mathbf{d})$ is an unbiased estimator of the covariance matrix of $\bar{\mathbf{Y}}(\hat{\mathbf{B}})$ given $\mathbf{C}_j = \mathbf{c}_j$, $j = 1, \ldots, k$, where

$$\mathbf{d}' = \frac{1}{k}\mathbf{e}' - (\bar{\mathbf{c}} - \mathbf{r})'(\mathbf{M}'\mathbf{M})^{-1}\mathbf{M}'\,, \tag{2.2.21}$$

$$\mathbf{M} = \begin{pmatrix} c_{11} - \bar{c}_1 & \cdots & c_{g1} - \bar{c}_g \\ \vdots & & \vdots \\ c_{1k} - \bar{c}_1 & \cdots & c_{gk} - \bar{c}_g \end{pmatrix}, \tag{2.2.22}$$

and $\mathbf{e}'$ is a row vector with all elements equal to unity.

It is proved in Theorem A1 that under the normality assumption on $\mathbf{Z}$ the confidence ellipsoid for $\boldsymbol{\mu}$ while using control variates is given by

$$\Pr\left\{(\bar{\mathbf{Y}}(\hat{\mathbf{B}}) - \boldsymbol{\mu})'\hat{\boldsymbol{\Sigma}}_{Y\cdot C}^{-1}(\bar{\mathbf{Y}}(\hat{\mathbf{B}}) - \boldsymbol{\mu}) \le (\mathbf{d}'\mathbf{d})\frac{(k-g-1)}{k-g-p}F_{p,k-g-p;1-\alpha}\right\} = 1-\alpha\,. \tag{2.2.23}$$

By analogy with (2.2.2), define the efficiency of control variates as the ratio of the expectation of the generalized sample variance of $\bar{\mathbf{Y}}(\hat{\mathbf{B}})$ to the expectation of the generalized sample variance of $\bar{\mathbf{Y}}$, that is, define the efficiency of the control variates to be

$$\varepsilon_1^2 = \frac{E_C E_Y\{|\hat{\boldsymbol{\Sigma}}_{Y\cdot C}|(\mathbf{d}'\mathbf{d})^p \,|\, \mathbf{C}_i = \mathbf{c}_i,\ i = 1, \ldots, k\}}{E_Y\left\{\frac{1}{k}\,p|\mathbf{S}_{YY}|\right\}}\,. \tag{2.2.24}$$

Another way of defining the efficiency of control variates is to take the ratio of the expectation of the squared volume of the confidence ellipsoid (2.2.23) (determined by using control variates) to the expectation of the squared volume of the confidence ellipsoid (2.2.19) (determined without using control variates). In this case the efficiency is

$$\varepsilon_2^2 = \frac{E_C E_Y(V_2^2)}{E_Y(V_1^2)}, \tag{2.2.25}$$

where by analogy with (2.2.7) and (2.2.10) we have, respectively,

$$V_1 = p^{-1}C(p)|\mathbf{S}_{YY}|^{1/2}\left\{\frac{p(k-1)}{k(k-p)} \cdot F_{p,k-p;1-\alpha}\right\}^{p/2}, \tag{2.2.26}$$

$$V_2 = p^{-1}C(p)|\hat{\mathbf{\Sigma}}_{Y\cdot C}(\mathbf{d}'\mathbf{d})|^{1/2}\left\{\frac{(k-g-1)p}{k-g-p} \cdot F_{p,k-g-p;1-\alpha}\right\}^{p/2}. \tag{2.2.27}$$

It is proved in Theorem A1 that, for $k-g-2p>0$,

$$\varepsilon_1^2 = C_1(k,g,p)C_2(k,g,p)\prod_{i=1}^{q}(1-\rho_i^2), \tag{2.2.28}$$

where ρ_i^2 are the squares of the canonical correlations between $\mathbf{Y}$ and $\mathbf{C}$, and

$$C_1(k,g,p) = \prod_{i=1}^{p}\frac{(k-g-i)(k-1)}{(k-g-1)(k-i)}, \tag{2.2.29}$$

$$C_2(k,g,p) = 1 + \sum_{j=1}^{p}\binom{p}{j}\frac{g(g+2)\cdots[g+2(j-1)]}{(k-g-2)\cdots(k-g-2j)}. \tag{2.2.30}$$

Substituting (2.2.26) and (2.2.27) in (2.2.25) and taking into account (2.2.24), we obtain

$$\varepsilon_2^2 = \frac{E_C E_Y(V_2^2)}{E_Y(V_1^2)} = \frac{\left[\dfrac{(k-g-1)p}{(k-g-p)}F_{p,k-g-p;1-\alpha}\right]^p}{\left[\dfrac{p(k-1)}{(k-p)}F_{p,k-p;1-\alpha}\right]^p}\varepsilon_1^2$$

$$= C_2(k,g,p)C_3(k,g,p)\prod_{i=1}^{q}(1-\rho_i^2), \tag{2.2.31}$$

where

$$C_3(k,g,p) = \prod_{i=1}^{p}\frac{(k-g-i)(k-p)}{(k-g-p)(k-i)}\left(\frac{F_{p,k-g-p;1-\alpha}}{F_{p,k-p;1-\alpha}}\right)^p. \tag{2.2.32}$$

For $p=1$ we have

$$\varepsilon_1^2 = \frac{k-2}{k-g-2}(1 - R_{YC}^2), \tag{2.2.33}$$

and

$$\varepsilon_2^2 = \frac{k-2}{k-g-2} \frac{F_{1,k-g-1;1-\alpha}}{F_{1,k-1;1-\alpha}}(1 - R_{YC}^2),$$

$$= \frac{k-2}{k-g-2} \frac{t^2_{k-g-1;1-\alpha/2}}{t^2_{k-1;1-\alpha/2}}(1 - R_{YC}^2). \tag{2.2.34}$$

Note that formulas (2.2.33) and (2.2.34) were obtained earlier by Lavenberg et al. (1982).

2.3 OPTIMIZATION AND SIMULATION RESULTS

Comparing (2.2.28) and (2.2.31) with (2.2.5) and (2.2.11), we see that our estimator $\bar{\mathbf{Y}}(\hat{\mathbf{B}})$ does not yield the full potential variance reduction $\delta_1^2 = \delta_2^2 = \Pi_{i=1}^q (1 - \rho_i^2)$. The actual variance reduction is

$$\varepsilon_1^2 = C_4(k, g, p) \prod_{j=1}^{q} (1 - \rho_i^2) = C_4(k, g, p)\delta_1^2, \tag{2.3.1}$$

where

$$C_4(k, g, p) = C_1(k, g, p)C_2(k, g, p).$$

It is readily seen that, for any p and g, $C_4(k, g, p) > 1$ and increases with g. Hence we conclude that in comparison with the potential variance reduction δ_2^2 (with $\mathbf{B}$ known) the actual variance reduction ε_1^2 (with $\mathbf{B}$ unknown) is reduced by the loss function $C_4(k, g, p)$. Note also that, instead of being reduced by $\delta_2 = [\Pi_{i=1}^q (1 - \rho_i^2)]^{1/2}$, the mean volume of the confidence ellipsoid is reduced by only

$$\left(C_2(k, g, p)C_3(k, g, p) \prod_{i=1}^{q} (1 - \rho_i^2)\right)^{1/2} > \left(\prod_{i=1}^{q} (1 - \rho_i^2)\right)^{1/2} = \delta_2 .$$

We can see from (2.3.1) that there exists a trade-off between $\Pi_{i=1}^q (1 - \rho_i^2)$ and $C_4(k, g, p)$. At one extreme, if k is not large relative to g (with k given and $k - g - 2p > 0$), the function $C_4(k, g, p)$ may nullify the potential variance reduction $\delta_2^2 = \Pi_{i=1}^q (1 - \rho_i^2)$. At the other extreme, for any p, ε_1^2

is a decreasing function of g. To find the optimal number of control variates as a function of k and p, we have to minimize (2.3.1) with respect to g. However, the minimization is, in general, difficult to implement analytically, and we shall restrict ourselves to the covariance matrix.

$$\mathbf{\Sigma}_Z = \begin{pmatrix} \mathbf{\Sigma}_{YY} & \mathbf{\Sigma}_{YC} \\ \mathbf{\Sigma}_{CY} & \mathbf{\Sigma}_{CC} \end{pmatrix} = \left(\begin{array}{cccc:cccc} 1 & \gamma & \cdots & \gamma & \gamma & \gamma & \cdots & \gamma \\ \gamma & 1 & \cdots & \gamma & \gamma & \gamma & \cdots & \gamma \\ \vdots & \vdots & & \vdots & \vdots & \vdots & & \vdots \\ \gamma & \gamma & \cdots & 1 & \gamma & \gamma & \cdots & \gamma \\ \hdashline \gamma & \gamma & \cdots & \gamma & 1 & \gamma & \cdots & \gamma \\ \gamma & \gamma & \cdots & \gamma & \gamma & 1 & \cdots & \gamma \\ \vdots & \vdots & & \vdots & \vdots & \vdots & & \vdots \\ \gamma & \gamma & \cdots & \gamma & \gamma & \gamma & \cdots & 1 \end{array}\right), \tag{2.3.2}$$

which is chosen for illustration purposes only.

Recalling that the generalized variance $|\mathbf{\Sigma}_Z|$ is the determinant of $\mathbf{\Sigma}_Z$, we obtain, for $p = 1$ after same simple algebra,

$$|\mathbf{\Sigma}_Z| = (1-\gamma)^g(1+g\gamma)\,,$$

and

$$\delta_1^2 = 1 - R_{YC}^2 = \frac{|\mathbf{\Sigma}_Z|}{|\mathbf{\Sigma}_C|\,\mathrm{var}(Y)} = \frac{(1-\gamma)^g(1+g\gamma)}{(1-\gamma)^{g-1}[1+(g-1)\gamma]}$$

$$= (1-\gamma)\,\frac{1+g\gamma}{1+(g-1)\gamma}\,. \tag{2.3.3}$$

The efficiency is

$$\varepsilon_1^2 = \frac{k-2}{k-g-2}\,(1-R_{YC}^2) = (1-\gamma)\,\frac{1+g\gamma}{1+(g-1)\gamma}\,\frac{k-2}{k-g-2}\,. \tag{2.3.4}$$

The optimal size g^* of the vector $\mathbf{C}$ can be found from (2.3.4) while noting that

$$\varepsilon_1^2(g+1) \geq \varepsilon_1^2(g) \Leftrightarrow Q(g) \geq 0\,,$$

where

$$Q(g) \equiv \gamma^2 g^2 + (2\gamma + \gamma^2)g + [1 - \gamma^2(k-2)]\,.$$

It follows immediately that

$$g^* = \min\{\text{nonnegative integers } g\text{: } Q(g) \geq 0\} .$$

After some calculations we obtain

$$g_0 = \frac{\{-(2\gamma + \gamma^2) + [(4k - 7)\gamma^4 + 4\gamma^3]^{1/2}\}}{2\gamma^2} , \qquad g^* = \max\{0, \lceil g_0 \rceil\} , \tag{2.3.5}$$

where $\lceil x \rceil$ denotes the ceiling of x. Table 1 presents g^* from (2.3.5) as a function of k and γ, from which it follows that g^* increases both with k and γ. Asymptotically,

$$g^* \to \max\{0, \lceil (k - \tfrac{3}{4})^{1/2} - \tfrac{3}{2} \rceil\} \quad \text{as } \gamma \to 1 ,$$

It follows from Table 1 that the optimal g has to be relatively small compared with k and sometimes (for small values of k and γ) must even be equal to zero. Figure 2 presents efficiencies δ_1^2 and ε_1^2 as functions of g for $\gamma = 0.5$ and $k = 10$, 50, and 100, respectively (δ_1^2 yields the full potential variance reduction: it does not depend on k; and for $\gamma > 0$, the optimal $g = \infty$).

Equating (2.3.4) to unity, we have

$$(1 - \gamma) \frac{1 + g\gamma}{1 + (g - 1)\gamma} \frac{k - 2}{k - g - 2} = 1 . \tag{2.3.6}$$

Solving (2.3.6) with respect to g, we obtain approximately (assuming $(1 + g\gamma)/[1 - (g - 1)\gamma] \approx 1$)

$$\hat{g} \approx \gamma(k - 2) . \tag{2.3.7}$$

Table 1 Optimal Size of Vector of Control Variates g^* as Function of k and γ for $p = 1$

k \ γ	0.1	0.2	0.4	0.5	0.8	0.9
10	0	0	1	1	2	2
50	0	2	5	5	6	6
100	0	5	8	8	9	9
1000	22	27	29	30	30	31

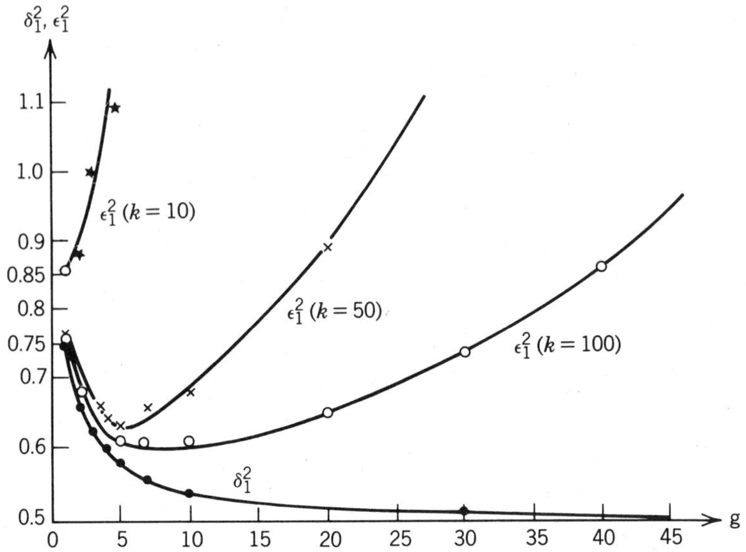

Fig. 2 The efficiencies δ_1^2 and ε_1^2 as functions of the size of the vector of control variates g for $\gamma = 0.5$ and for $k = 10$, 50, and 100, respectively.

It is readily seen that $\hat{g}$ defines the threshold, which can be written as

$$\text{var } \bar{Y}(\hat{\mathbf{B}}) \lesseqgtr \begin{cases} \text{var}(\bar{Y}) & \text{if } g \leq \hat{g} , \\ \text{var}(\bar{Y}) & \text{if } g > \hat{g} . \end{cases} \tag{2.3.8}$$

Formula (2.3.8) means that for $g > \hat{g}$ the method of control variates [estimator $\bar{Y}(\hat{B})$] becomes inferior to the usual sample mean Monte Carlo method (estimator $\bar{Y}$). The reason is that for g large enogh ($\hat{g}$ is always greater than g^*), the system becomes "overestimated", "overtrained" with the ineffective "help" of the vector of control variates, which itself is estimated from simulation. In other words, the conclusion is that too much imperfect information is worse than no such information at all.

In view of the impracticability of finding g^* analytically for $p > 1$, we resorted to simulation instead, and results for $p = 1, 3, 5$ and $\mathbf{\Sigma}_Z$ given in (2.3.2) are presented in Table 2, which readily shows that g has to be small relative to k.

The efficiency of control variates was defined as

$$\hat{\varepsilon}_1^2 = \frac{|\mathbf{S}_{Y(\hat{B})}|}{|\mathbf{S}_{YY}|} , \tag{2.3.9}$$

Table 2 Simulation Results for Optimal Size of Vector of Control Variates g^* for Different k, γ, and p under the Normality Assumption

	$p=1$			$p=3$			$p=5$		
k \ γ	0.1	0.5	0.9	0.1	0.5	0.9	0.1	0.5	0.9
50	2	6	8	7	8	12	10	14	16
100	6	8	10	8	13	15	12	14	18

where the (u, v)th elements of $\mathbf{S}_{YY}$ and $S_{Y(\hat{B})}$ are, respectively,

$$(\mathbf{S}_{YY})_{uv} = \frac{1}{k-1} \sum_{j=1}^{k} (Y_{uj} - \bar{Y}_u)(Y_{vj} - \bar{Y}_v), \tag{2.3.10}$$

$$(\mathbf{S}_{Y(\hat{B})})_{uv} = \frac{1}{k-g-1} \sum_{j=1}^{k} [Y_{uj}(\hat{\mathbf{B}}) - \bar{Y}_u(\hat{\mathbf{B}})][Y_{vj}(\hat{\mathbf{B}}) - \bar{Y}_v(\hat{\mathbf{B}})]. \tag{2.3.11}$$

Comparing the data of Tables 1 and 2 for $p = 1$, we see that both the theoretical results (Table 1) and the simulation results (Table 2) are fairly close. The curves ε_1^2 and δ_1^2, which we obtained from simulation for $p = 1, 3, 5$ (the points of minimum g^* are given in Table 2), are similar to those presented in Fig. 2.

The results of this section were addressed to the case where $\mathbf{Z} \sim N(\mathbf{0}, \mathbf{\Sigma})$ with the covariance matrix defined in (2.3.2). We also performed simulation of procedure (2.1.8) by removing the normality assumption on $\mathbf{Z}$. Before discussing these results, which are presented in Table 3, we describe a procedure for generating a random vector $\mathbf{Z}$ with any desirable covariance matrix $\mathbf{\Sigma}_Z$ [see Rubinstein (1981)]. The procedure can be written as follows.

(a) Generate $U_1, \ldots, U_{p+g}$ from $U(0, 1)$.

Table 3 Simulation Results for Optimal Size of Vector of Control Variates g^* for Different k, γ, and p under the Assumptions of Nonnormality

	$p=1$			$p=3$			$p=5$		
k \ γ	0.1	0.5	0.9	0.1	0.5	0.9	0.1	0.5	0.9
50	2	6	10	6	8	12	8	10	14
100	6	8	12	8	10	12	10	11	16

(b) From any desired distribution $F_X(x)$, generate $X_1, \ldots, X_{p+g}$ by using the inverse transform method, that is, by $X_j = F_X^{-1}(U_j)$, $j = 1, \ldots, p+g$.

(c) Deliver $\mathbf{Z} = \mathbf{TX}$, where $\mathbf{T}$ is a triangular matrix equal to

$$\mathbf{T} = \begin{vmatrix} t_{11} & 0 & \cdots & 0 \\ t_{21} & t_{22} & \cdots & 0 \\ \vdots & \vdots & & \vdots \\ t_{p+g,1} & t_{p+g,2} & \cdots & t_{p+g,p+g} \end{vmatrix},$$

and such that $\mathbf{\Sigma}_Z = \mathbf{TT}'$.

It should be noted that if $\mathbf{X} \sim N(\mathbf{0}, \mathbf{1})$, then $\mathbf{Z} \sim N(\mathbf{0}, \mathbf{\Sigma})$, since, as follows from $\mathbf{Z} = \mathbf{TX}$, the linear combination

$$Z_i = t_{i1}X_1 + t_{i2}X_2 + \cdots + t_{ii}X_i \tag{2.3.12}$$

is also distributed normally. If, however, $\mathbf{X}$ is a nonnormal r.v., then $\mathbf{Z} = \mathbf{TX}$ will generally have a distribution different from $\mathbf{X}$.

In our simulation we used $X_j \sim U(0, 1)$ and $X_j \sim \exp(1)$ with the covariance matrix defined in (2.3.2). Table 3 presents the same data as Table 2 for $X_j \sim U(0, 1)$ [for $X_j \sim \exp(1)$ the results were similar].

Comparison of the results of Tables 2 and 3 shows that for large k the optimal g^* is fairly insensitive to the distribution of $\mathbf{Z}$. We also obtained that the optimal size of the vector of control variates g^* is a nondecreasing function of k and p irrespective of the structure of $\mathbf{\Sigma}_Z$. For moderate k $(20 \leq k \leq 50)$ we observed, however, that g^* is rather sensitive to the structure of $\mathbf{\Sigma}_Z$. In this case good variance reduction was achieved when the number of control variates g was not large relative to p $(p \leq g \leq 7p)$.

Recently Brown and Rubinstein (1985) showed that the efficiency of control variate procedure (2.2.17) can be improved if we consider the *discount* control variate procedure, which can be written as

$$\bar{\mathbf{Y}}_\mu(\hat{\mathbf{B}}) = \bar{\mathbf{Y}} - \mu\hat{\mathbf{B}}(\bar{\mathbf{C}} - \mathbf{r}), \tag{2.3.13}$$

where $0 < \mu \leq 1$ is the discount factor.

2.4 CONCLUSIONS

We have considered some statistical aspects of the multidimensional control variate procedure to increase the efficiency (reduce the confidence region) in Monte Carlo simulation. The main results obtained so far follow.

(i) The estimator $\bar{\mathbf{Y}}(\hat{\mathbf{B}}) = \bar{\mathbf{Y}} - \hat{\mathbf{B}}(\bar{\mathbf{C}} - \mathbf{r})$ of the control variate procedure (2.1.8) does not yield its full potential variance reduction $\delta_1^2 = \Pi_{i=1}^{q}(1-\rho_i^2)$. The loss factor is equal to $C_4(k, g, p)$ [see equations (2.2.29)–(2.2.31) and (2.3.1)].

(ii) The estimator $\bar{\mathbf{Y}}(\hat{\mathbf{B}})$ yields substantial variance reduction only if the size of the vector of control variates is chosen carefully (close to the optimal size g^*). This is, however, difficult to find analytically for a general covariance matrix $\mathbf{\Sigma}_Z$. Nevertheless, the results obtained with the matrix $\mathbf{\Sigma}_Z$ [(2.3.2)] showed the following.

(a) The value of g^* must be small relative to the sample size k. For small values of k the value of g^* can even be chosen equal to zero (see Tables 1–3).

(b) g^* increases with k (the sample size), with p (the size of the vector $\mathbf{Y}$), and with γ [the correlation coefficient of the matrix $\mathbf{\Sigma}_Z$, (2.3.2)]. Since the efficiency of the control variates is very sensitive to the choice of the size g of the vector $\mathbf{Y}$, it is always preferable to take g too small rather than too large (see Fig. 2).

(c) g^* is practically independent of the marginal distributions of vector $\mathbf{Z}$ (see Tables 2 and 3). However, it depends on the structure of the covariance matrix $\mathbf{\Sigma}_Z$.

2.5 FUTURE RESEARCH

We suggest the following directions for future research:

(i) In this paper it has been assumed that $\hat{\mathbf{B}} = \mathbf{S}_{YC}\mathbf{S}_{CC}^{-1}$. In many practical situations the covariance matrix $\mathbf{\Sigma}_{CC}$ or some of its components are known. It would be interesting to find the efficiencies of control variates ε_1^2, ε_2^2, the loss factor, and the optimal size of the vector of control variates when B is approximated by $\hat{\hat{\mathbf{B}}} = \mathbf{S}_{YC}\mathbf{\Sigma}_{CC}^{-1}$.

(ii) It would be interesting to know how, given a set of components of size g available for forming the vector of control variates, the optimal subset of size g^* ($g^* \ll g$) could be constructed so as to maximize the efficiency of (2.1.8).

(iii) Consider the following particular case of (2.1.5). Let the expected values of the components $E(C_i) = r_i$, $i = 1, \ldots, g$, of the vector of control variates be unknown, but let all of them be equal to $E(Y) = \mu$, that is, $r_i = \mu$, $i = 1, \ldots, g$. As an unbiased estimator for μ consider

$$Y(\gamma) = \sum_{i=1}^{g} \gamma_i C_i + \gamma_{g+1} Y = Y + \sum_{i=1}^{g} \gamma_i (C_i - Y)\,, \qquad (2.5.1)$$

where

$$\sum_{i=1}^{g+1} \gamma_i = 1 . \tag{2.5.2}$$

The vector that minimizes the variance of $Y(\boldsymbol{\gamma})$ is [see Rubinstein (1981), p. 218] given by

$$\boldsymbol{\beta} = \frac{\mathbf{e}\boldsymbol{\Sigma}_Z^{-1}}{\mathbf{e}\boldsymbol{\Sigma}_Z^{-1}\mathbf{e}'} , \tag{2.5.3}$$

and

$$\operatorname{Var}[Y(\boldsymbol{\beta})] = \frac{1}{\mathbf{e}\boldsymbol{\Sigma}_Z^{-1}\mathbf{e}'} , \tag{2.5.4}$$

where $\mathbf{e}$ denotes the $(g+1)$-dimensional row vector each of whose components is 1 and $\boldsymbol{\Sigma}_Z$ is the covariance matrix of the vector $\binom{\mathbf{Y}}{\mathbf{C}}$.

It would be interesting to obtain statistical results for the procedure (2.5.1) and (2.5.2) with $\boldsymbol{\beta}$ given in (2.5.3), that is, to obtain the potential variance reduction, the loss factor, and the optimal size of the vector of control variates while sampling from $\mathbf{Y}$ and $\mathbf{C}$.

(iv) It would also be interesting to investigate the efficiency of, for example, jackknife and bootstrap estimators in reducing the bias of the multidimensional control variate estimator $\bar{\mathbf{Y}}(\hat{\mathbf{B}}) = \bar{\mathbf{Y}} - \hat{\mathbf{B}}(\bar{\mathbf{C}} - \mathbf{r})$.

(v) To extend the applications of control variates to multipopulation multivariate response case.

(vi) To apply procedure (2.1.8) (to achieve variance reduction) for the class of queueing networks considered in Lavenberg et al. (1982) while using the service time, flow, and work variable as control variables.

APPENDIX

Lemma A1 The matrix $\mathbf{B}$ that minimizes the generalized variance $|\boldsymbol{\Sigma}_{Y(A)}|$ in (2.2.1) is equal to

$$\mathbf{B} = \boldsymbol{\Sigma}_{YC}\boldsymbol{\Sigma}_{CC}^{-1} .$$

Proof Let $\mathbf{A}$ be an arbitrary $p \times g$ matrix. Define

$$\mathbf{D} = \mathbf{A} - \mathbf{B} ,$$

where

$$\mathbf{B} = \mathbf{\Sigma}_{YC}\mathbf{\Sigma}_{CC}^{-1} .$$

Therefore

$$\mathbf{A} = \mathbf{B} + \mathbf{D} = \mathbf{\Sigma}_{YC}\mathbf{\Sigma}_{CC}^{-1} + \mathbf{D} .$$

Now according to (2.2.1),

$$\begin{aligned}\mathbf{\Sigma}_{Y(A)} &= \mathbf{\Sigma}_{YY} + \mathbf{A}\mathbf{\Sigma}_{CC}^{-1}\mathbf{A}' - \mathbf{A}\mathbf{\Sigma}_{YC}' - \mathbf{\Sigma}_{YC}\mathbf{A}' \\ &= \mathbf{\Sigma}_{YY} + \mathbf{\Sigma}_{YC}\mathbf{\Sigma}_{CC}^{-1}\mathbf{\Sigma}_{YC}' + \mathbf{D}\mathbf{\Sigma}_{CC}^{-1}\mathbf{D}' + \mathbf{D}\mathbf{\Sigma}_{YC}' \\ &\quad + \mathbf{\Sigma}_{YC}\mathbf{D}' - 2\mathbf{\Sigma}_{YC}\mathbf{\Sigma}_{CC}\mathbf{\Sigma}_{YC}' - \mathbf{D}\mathbf{\Sigma}_{YC}' - \mathbf{\Sigma}_{YC}\mathbf{D}' \\ &= (\mathbf{\Sigma}_{YY} - \mathbf{\Sigma}_{YC}\mathbf{\Sigma}_{CC}^{-1}\mathbf{\Sigma}_{YC}') + \mathbf{D}\mathbf{\Sigma}_{CC}^{-1}\mathbf{D}' .\end{aligned}$$

Thus $\mathbf{\Sigma}_{Y(A)}$ is expressed as a sum of two nonnegative matrices,

$$\mathbf{\Sigma}_{Y(A)} = \mathbf{\Sigma}_{YY\cdot C} + \mathbf{D}\mathbf{\Sigma}_{CC}^{-1}\mathbf{D}' ,$$

where

$$\mathbf{\Sigma}_{YY\cdot C} = \mathbf{\Sigma}_{YY} - \mathbf{\Sigma}_{YC}\mathbf{\Sigma}_{CC}^{-1}\mathbf{\Sigma}_{YC}' .$$

Let $\mathbf{x}_1, \ldots, \mathbf{x}_p$ be the normalized eigenvectors of $\mathbf{\Sigma}_{Y(A)}$ and let $\lambda_1, \ldots, \lambda_p$ be its correspondent eigenvalues. We have

$$\lambda_i = \mathbf{x}_i'\mathbf{\Sigma}_{Y(A)}\mathbf{x}_i = \mathbf{x}_i'\mathbf{\Sigma}_{YY\cdot C}\mathbf{x}_i + \mathbf{x}_i'\mathbf{D}\mathbf{\Sigma}_{CC}^{-1}\mathbf{D}'\mathbf{x}_i , \qquad i = 1, \ldots, p .$$

Therefore λ_i is expressed as the sum of two nonnegative terms (both $\mathbf{\Sigma}_{YY\cdot C}$ and $\mathbf{D}\mathbf{\Sigma}_{CC}^{-1}\mathbf{D}'$ are nonnegative matrices).

Now since

$$|\mathbf{\Sigma}_{Y(A)}| = \prod_{i=1}^{p} \lambda_i$$

and since given $\mathbf{\Sigma}_{YY\cdot C}$ the product $\Pi_{i=1}^{p} \lambda_i$ will be minimized if $D = 0$, the proof of Lemma A1 follows immediately. Q.E.D.

Theorem A1 Under the assumption that **Z** is multivariate normally distributed, a confidence ellipsoid for $\boldsymbol{\mu}$ is given by

$$P\left\{(\bar{\mathbf{Y}}(\hat{\mathbf{B}}) - \boldsymbol{\mu})'\hat{\boldsymbol{\Sigma}}_{Y\cdot C}^{-1}(\bar{\mathbf{Y}}(\hat{\mathbf{B}}) - \boldsymbol{\mu}) \leq (\mathbf{d}'\mathbf{d}) \frac{(k-g-1)p}{k-g-p} F_{p,k-g-p;1-\alpha}\right\}$$
$$= 1 - \alpha \,, \qquad \text{(A.1)}$$

where $F_{v_1,v_2;1-\alpha}$ denote the upper α percentile of the F distribution with v_1 and v_2 degrees of freedom and d defined in (2.2.21).

Proof Since **Z** has a multivariate normal distribution, the conditional distribution of **Y** given $\mathbf{C} = \mathbf{c}$ is also multivariate normal with mean vector

$$E(\mathbf{Y} \mid \mathbf{C} = \mathbf{c}) = \boldsymbol{\mu} + \mathbf{B}(\mathbf{c} - \mathbf{r}) \,, \qquad \text{(A.2)}$$

where $\mathbf{B} = \boldsymbol{\Sigma}_{YC}\boldsymbol{\Sigma}_{CC}^{-1}$, and covariance matrix

$$\boldsymbol{\Sigma}_{Y\cdot C} = \boldsymbol{\Sigma}_{YY} - \boldsymbol{\Sigma}_{YC}\boldsymbol{\Sigma}_{CC}^{-1}\boldsymbol{\Sigma}_{CY} \,. \qquad \text{(A.3)}$$

It is well known from the theory of multivariate linear models that given $\mathbf{C}_i = \mathbf{c}_i$ $(i = 1, \ldots, k)$ an unbiased estimator for $\boldsymbol{\mu}$ is

$$\bar{\mathbf{Y}}(\hat{\mathbf{B}}) = \bar{\mathbf{Y}} - \hat{\mathbf{B}}(\bar{\mathbf{c}} - \mathbf{r}) \,, \qquad \text{(A.4)}$$

where $\hat{\mathbf{B}} = \mathbf{S}_{YC}\mathbf{S}_{CC}^{-1}$. That is,

$$E(\bar{\mathbf{Y}}(\hat{\mathbf{B}}) \mid \mathbf{C}_i = \mathbf{c}_i, \ i = 1, \ldots, k) = \boldsymbol{\mu} \,. \qquad \text{(A.5)}$$

Since the right side is constant, it is true unconditionally, that is,

$$E\bar{\mathbf{Y}}(\hat{\mathbf{B}}) = \boldsymbol{\mu} \,. \qquad \text{(A.6)}$$

Let $\hat{b}_j$ denote the jth row of $\hat{B}$. Then

$$\bar{Y}_j(\hat{\mathbf{b}}_j) = \bar{\mathbf{Y}}_j - \hat{\mathbf{b}}_j(\bar{\mathbf{c}} - \mathbf{r}) \equiv \mathbf{d}'\mathbf{Y}_j \qquad (j = 1, \ldots, p) \,, \qquad \text{(A.7)}$$

where $\mathbf{Y}_j' = (Y_{j1}, \ldots, Y_{jk})$ corresponds to the k independent observations on the jth variable, and $\mathbf{d}'$ is a row vector defined by

$$\mathbf{d}' = \frac{1}{k}\mathbf{e}' - (\bar{\mathbf{C}} - \mathbf{r})'(M'M)^{-1}M' \,. \qquad \text{(A.8)}$$

Hence, conditional on $\mathbf{C}_i = \mathbf{c}_i$ $(i = 1, \ldots, k)$, we have

$$\operatorname{var} \bar{Y}_j(\hat{\mathbf{b}}_j) = \sigma_{jj\cdot C}(\mathbf{d}'\mathbf{d}) , \tag{A.9}$$

and

$$\operatorname{cov}(\bar{Y}_j(\hat{\mathbf{b}}_j),\ \bar{Y}_{j'}(\hat{\mathbf{b}}_{j'})) = \sigma_{jj'\cdot C}(\mathbf{d}'\mathbf{d}) , \tag{A.10}$$

where $\sigma_{jj'\cdot C}$ denotes the jj'th element of $\boldsymbol{\Sigma}_{Y\cdot C}$. Thus, given $\mathbf{C}_i = \mathbf{c}_i$ $(i = 1, \ldots, k)$, the covariance matrix of $\bar{\mathbf{Y}}(\hat{\mathbf{B}})$ is

$$\operatorname{var}(\bar{\mathbf{Y}}(\hat{\mathbf{B}}) \mid \mathbf{C}_i = \mathbf{c}_i, i = 1, \ldots, k) = (\mathbf{d}'\mathbf{d})\boldsymbol{\Sigma}_{Y\cdot C} . \tag{A.11}$$

Now, an unbiased estimator of $\boldsymbol{\Sigma}_{Y\cdot C}$ is given by

$$\hat{\boldsymbol{\Sigma}}_{Y\cdot C} = \frac{k-1}{k-(g+1)} (\mathbf{S}_{YY} - \mathbf{S}_{YC}\mathbf{S}_{CC}^{-1}\mathbf{S}_{CY}) . \tag{A.12}$$

Conditional on $\mathbf{C}_i = \mathbf{c}_i$ $(i = 1, \ldots, k)$, we have

$$\bar{\mathbf{Y}}(\hat{\mathbf{B}}) \sim N(\boldsymbol{\mu}, (\mathbf{d}'\mathbf{d})\boldsymbol{\Sigma}_{Y\cdot C}) , \tag{A.13}$$

$$(k-g-1)\hat{\boldsymbol{\Sigma}}_{Y\cdot C} \sim \text{Wishart}_p(k-g-1, \boldsymbol{\Sigma}_{Y\cdot C}) , \tag{A.14}$$

and $\bar{\mathbf{Y}}(\hat{\mathbf{B}})$ and $\hat{\boldsymbol{\Sigma}}_{Y\cdot C}$ are independent. Thus, conditional on $\mathbf{C}_i = \mathbf{c}_i$, we have

$$T^2 = \frac{1}{(\mathbf{d}'\mathbf{d})} [\bar{\mathbf{Y}}(\hat{\mathbf{B}}) - \boldsymbol{\mu}]'\hat{\boldsymbol{\Sigma}}_{Y\cdot C}^{-1}[\bar{\mathbf{Y}}(\hat{\mathbf{B}}) - \boldsymbol{\mu}] \tag{A.15}$$

distributed as

$$\frac{p(k-g-1)}{k-g-p} F_{p,k-g-p} . \tag{A.16}$$

Conditional on $\mathbf{C}_i = \mathbf{c}_i$ $(i = 1, \ldots, k)$, we obtain the confidence ellipsoid for $\boldsymbol{\mu}$ as

$$P\left\{(\bar{\mathbf{Y}}(\hat{\mathbf{B}}) - \boldsymbol{\mu})'\hat{\boldsymbol{\Sigma}}_{Y\cdot C}^{-1}(\bar{\mathbf{Y}}(\hat{\mathbf{B}}) - \boldsymbol{\mu}) \le (\mathbf{d}'\mathbf{d}) \frac{(k-g-1)p}{k-g-p} F_{p,k-g-p;1-\alpha}\right\}$$

$$= 1 - \alpha . \tag{A.17}$$

Now, the left side of (A.17) is a conditional probability given $\mathbf{C}_i = \mathbf{c}_i$, $i = 1, \ldots, k$, and equals the constant $1 - \alpha$ independent of $\mathbf{c}_i$, $i = 1, \ldots, k$. The expectation over the joint distribution of $\mathbf{C}_i$, $i = 1, \ldots, k$, equals the same constant. Thus the confidence ellipsoid given by (A.17) holds unconditionally. Q.E.D.

Theorem A2 For $k-g-2p>0$,

$$\varepsilon_1^2 = C_1(k, g, p)C_2(k, g, p)\prod_{j=1}^{p}(1-\rho_j^2), \tag{A.18}$$

where ρ_j^2 are the squares of the canonical correlations, and C_1, C_2 are

$$C_1(k, g, p) = \sum_{i=1}^{p} \frac{(k-g-i)(k-1)}{(k-g-1)(k-i)}, \tag{A.19}$$

$$C_2(k, g, p) = 1 + \sum_{j=1}^{p} \binom{p}{j} \frac{g(g+2)\cdots[g+2(j-1)]}{(k-g-2)\cdots(k-g-2j)}. \tag{A.20}$$

Proof Since

$$(k-1)S_{YY} \sim W_p(k^{-1}, \mathbf{\Sigma}_{YY}), \tag{A.21}$$

where W denotes the Wishart distribution, we have from [Theorem 7.5.3 of Anderson (1984), p. 171] that

$$\frac{E|S_{YY}|}{k^p} = |\mathbf{\Sigma}_{YY}| \prod_{i=1}^{p} \frac{k-i}{(k-1)^p k^p} = |\mathbf{\Sigma}_{YY}| \prod_{i=1}^{p} \frac{k-i}{(k-1)k}. \tag{A.22}$$

Similarly, conditioning on $\mathbf{C}_j = \mathbf{c}_j$, $j = 1, \ldots, k$, $(k-g-1)\hat{\mathbf{\Sigma}}_{Y\cdot C} \sim W_p(k-g-1, \mathbf{\Sigma}_{Y\cdot C})$ implies

$$\begin{aligned}
E_C E_Y\{|\hat{\mathbf{\Sigma}}_{Y\cdot C}|(\mathbf{d}'\mathbf{d})^p\} &= E_C|\hat{\mathbf{\Sigma}}_{Y\cdot C}| \prod_{i=1}^{p} \frac{(k-g-i)(\mathbf{d}'\mathbf{d})^p}{(k-g-1)^p} \\
&= |\mathbf{\Sigma}_{Y\cdot C}| \prod_{i=1}^{p} \frac{(k-g-1)}{(k-g-1)^p E_C(\mathbf{d}'\mathbf{d})^p} \\
&= |\mathbf{\Sigma}_{Y\cdot C}| \frac{\prod_{i=1}^{p}(k-g-i)}{(k-g-1)^p E_C\left[\frac{1}{k} + (\bar{\mathbf{C}}-\mathbf{r})'(\mathbf{M}'\mathbf{M})^{-1}(\mathbf{C}-\mathbf{r})\right]^p} \\
&= |\mathbf{\Sigma}_{Y\cdot C}| \prod_{i=1}^{p}\left[\frac{k-g-i}{(k-g-1)k}\right] E_C[1 + k(\bar{\mathbf{C}}-\mathbf{r})'(\mathbf{M}'\mathbf{M})^{-1}(\bar{\mathbf{C}}-\mathbf{r})]^p \\
&= |\mathbf{\Sigma}_{Y\cdot C}| \prod_{i=1}^{p}\left[\frac{k-g-i}{(k-g-1)k}\right] \sum_{j=0}^{p}\binom{p}{j} E_C[k(\bar{\mathbf{C}}-\mathbf{r})'(\mathbf{M}'\mathbf{M})^{-1}(\bar{\mathbf{C}}-\mathbf{r})]^j.
\end{aligned} \tag{A.23}$$

However,

$$k(\bar{\mathbf{C}}-\mathbf{r})'(\mathbf{M}'\mathbf{M})^{-1}(\bar{\mathbf{C}}-\mathbf{r}) = \frac{k}{k-1}(\bar{\mathbf{C}}-\mathbf{r})'\mathbf{S}_{CC}^{-1}(\bar{\mathbf{C}}-\mathbf{r}) \sim \frac{g}{k-g} F_{g,k-g} . \tag{A.24}$$

Therefore (A.23) is equal to

$$|\mathbf{\Sigma}_{Y\cdot C}| \prod_{i=1}^{p} \left[\frac{k-g-i}{(k-g-1)k}\right] \sum_{j=0}^{p} \binom{p}{j} E\left[\frac{g}{k-g} F_{g,k-g}\right]^j$$

$$= |\mathbf{\Sigma}_{Y\cdot C}| \prod_{i=1}^{p} \frac{k-g-i}{(k-g-1)k} \sum_{j=0}^{p} \binom{p}{j} \frac{g^j}{(k-g)^j} E(F_{g,k-g})^j . \tag{A.25}$$

The δth moment of F_{r_1,r_2} about zero is [see Mood, Graybill, and Boes (1974)]

$$\mu'_\delta = \left(\frac{r_2}{r_1}\right)^\delta \cdot \frac{r_1(r_1+2)\cdots[r_1+2(\delta-1)]}{(r_2-2)(r_2-4)\cdots(r_2-2\delta)} \qquad (0<\delta<\tfrac{1}{2}r_2) \quad (\mu'_0 \equiv 1) .$$

Finally (A.23) becomes

$$|\mathbf{\Sigma}_{Y\cdot C}| \prod_{i=1}^{p} \frac{k-g-i}{(k-g-1)k} \sum_{j=0}^{p} \binom{p}{j} \frac{g^j}{(k-g)^j} \frac{k-g}{g}$$
$$\times \frac{g(g+2)\cdots[g+2(j-1)]}{(k-g-2)(k-g-4)\cdots(k-g-2j)} \qquad [p \le \tfrac{1}{2}(k-g)] ,$$

and the efficiency

$$\varepsilon_1^2 = \prod_{i=1}^{p} \frac{(k-g-i)(k-1)}{(k-g-1)(k-i)} \left(1 + \sum_{j=1}^{p} \binom{p}{j} \frac{g(g+2)\cdots[g+2(j-1)]}{(k-g-2)\cdots(k-g-2j)}\right)$$
$$\times \frac{|\mathbf{\Sigma}_{Y\cdot C}|}{|\mathbf{\Sigma}_{YY}|}$$
$$= C_1(k, g, p) C_2(k, g, p) \prod_{i=1}^{p} (1-\rho_i^2) ,$$

where

$$C_1(k, g, p) = \prod_{i=1}^{p} \frac{(k-g-i)(k-1)}{(k-g-1)(k-i)},$$

$$C_2(k, g, p) = 1 + \sum_{j=1}^{p} \binom{p}{j} \frac{g(g+2)\cdots[g+2(j-1)]}{(k-g-2)\cdots(k-g-2j)}. \quad \text{Q.E.D.}$$

REFERENCES

Anderson, T. W. (1984). *An Introduction to Multivariate Statistical Analysis*, John Wiley & Sons, New York.

Brown, D. and R. Rubinstein (1985). The discount version of multivariate control variates procedure (in preparation).

Glynn, P. and W. Whitt (1985). Indirect estimation via $L = \lambda W$. AT&T Bell Laboratories, Holmdel, NJ 07733.

Grant, F. H. and J. J. Solberg (1983). Variance Reduction Techniques in Stochastic Shortest Route Analysis: Application Procedures and Results, *Math. Comp. Simul.*, **XXV**, 366–375.

Kleijnen, J. P. C. (1974). *Statistical Techniques in Simulation*, *Part I.* Marcel Dekker, New York.

Nozari, A., S. F. Arnold, and C. D. Pegden (1984). Control Variates for Multipopulation Simulation Experiments, *IEEE Trans.*, **16**, 159–169.

Lavenberg, S. S., T. L. Moeller, and P. D. Welch (1982). Statistical Results on Control Variables with Application to Queueing Nework Simulation, *Oper. Res.*, **30**, 1, 182–202.

Lavenberg, S. S. and P. D. Welch (1981). A Perspective on the Use of Control Variables to Increase the Efficiency of Monte Carlo Simulation, *Manage. Sci.*, **27**, 3, 322–335.

Mood, A. M., F. A. Graybill, and D. C. Boes (1974). *Introduction to the Theory of Statistics*, 3rd ed., McGraw-Hill, New York.

Rothery, P. (1982). The Use of Control Variates in Monte Carlo Estimation of Power, *Appl. Stat.*, **31**, 125–129.

Rubinstein, R. Y. and R. Marcus (1985). Efficiency of Multivariate Control Variates in Monte Carlo Simulation, *Oper. Res.*, **33**, 3, 661–677.

Swain, J. J. and B. W. Schmeiser (1984a). Monte Carlo Analysis of Nonlinear Statistical Models, I: Theory, Technical Report J-84-11, School of Industrial and Systems Engineering, Georgia Institute of Technology, Atlanta, Georgia.

Swain, J. J. and B. W. Schmeiser (1984b). Monte Carlo Analysis of Nonlinear Statistical Models, II: Moments and Variances, Technical Report J-84-12, School of Industrial and Systems Engineering, Georgia Institute of Technology, Atlanta, Georgia.

Wilson, J. R. and A. A. B. Pritsker (1984a). Variances Reduction in Queueing Simulation using Generalized Concomitant Variables, *J. Stat. Comp. Simulation*, **19**, 129–153.

Wilson, J. R. and A. A. B. Pritsker (1984b). Experimental Evaluation of Variance Reduction Techniques for Queueing Simulations Using Generalized Concomitant Variables, *Manage. Sci.*, **30**, 1459–1472.

CHAPTER 3

Stochastic Optimization Via Stochastic Approximation

3.1 INTRODUCTION

The purpose of this chapter is to show how stochastic approximation can be used for finding the optimal solution for stochastic optimization models under uncertainty.

Let us consider the following problem:

$$
\text{(P1)} \begin{cases} \min\limits_x g_0(x) = \min\limits_x \int \psi_0(x, y) f_Y(y, x)\, dy = \min\limits_x E[\psi_0(x, Y)] \\ \text{subject to} \\ g_j(x) = \int \psi_j(x, y) f_Y(y, x)\, dy = E[\psi_j(x, Y)] \le 0\,, \quad j = 1, \ldots, M\,, \\ x \in X \subset R^n\,, \end{cases}
$$

where x and y are n- and m-dimensional vectors, respectively, and $f_Y(y, x)$ is a p.d.f. (probability density function) of an r.v. (random variable) Y that depends on a vector of parameters† x.

If the p.d.f. $f_Y(y, x)$ is *known* and the expected values $E[\psi_j(x, Y)]$, $j = 0, 1, \ldots, M$, can be found analytically for all x then (P1) represents the following nonlinear programming problem:

†We shall deal also with the particular case $f_Y(y, x) = f_Y(y)$, that is, when $f_Y(\cdot)$ does not depend on x.

$$
\text{(P2)} \begin{cases} \min g_0(x) \\ \text{subject to} \\ g_j(x) \le 0, \; j = 1, \ldots, M, \quad x \in X \subseteq R^n , \end{cases}
$$

which can be solved either analytically or numerically by methods of nonlinear programming [see Avriel (1976)].

If at the other extreme one of the expectations $E[\psi_j(x, Y)]$, $j = 0, 1, \ldots, M$, *cannot be found explicitly* (say they can be estimated only through Monte Carlo simulation) or the p.d.f. $f_Y(y, x)$ is *unknown*, then there is no way to solve problem (P1) analytically. In most cases we shall assume that $f_Y(y, x)$ is unknown. We shall always assume, however, that a sample $Y_1, \ldots, Y_N$ from $f_Y(y, x)$ is available. Many stochastic optimization problems are of this type.

We shall solve problem (P1) by using the *stochastic approximation method*, according to which the optimal solution of (P1) is approximated iteratively, that is, step by step.

Stochastic approximation originated with the pioneering paper of Robbins and Monro, who suggested in 1951 an iterative procedure for finding a root of a regression function in R^1 measured with noise. Kiefer and Wolfowitz (1952) suggested a procedure for finding the extremum in the unconstrained optimization problem (P1) in R^1. The procedures of Robbins and Monro and Kiefer and Wolfowitz were generalized by Dvoretsky (1956). During the past 20 years hundreds of papers and several books [see, e.g., Nevelson and Hasminskii (1973), Ermoliev (1983), Kushner and Clark (1978), and Wasan (1969)] were written on stochastic approximations, their convergence, and speed of convergence. Tsypkin (1970) was the first to show that many problems from diverse fields such as pattern recognition, filtering, identification, game theory, reliability, and economics can be fitted into the framework of stochastic approximation. More applications of stochastic approximation can be found in Goodwin, Ramadge, and Caines (1981); Ermoliev and Mirzoakhmedov (1976); Kushner and Clark (1978); and Wasan (1969). Now stochastic approximation is a recognized and widely used approach for many optimization problems under uncertainty.

The purpose of our study is to apply stochastic approximation to a variety of operations research models, optimizing in the situation under uncertainty. Moreover, for each particular problem we seek to find a *stochastic approximation algorithm* (SAA) and discuss its convergence conditions. In addition, we present simulation results related to convergence of stochastic approximation.

This chapter is organized as follows: In Section 3.2 we present some mathematical background related to the convergence of stochastic approximation. Here we also discuss robust stochastic approximation algorithms assuming some a priori knowledge about the p.d.f. $f_Y(y, x)$. In Section 3.3 we apply stochastic approximation to several well-known stochastic inventory models assuming that the p.d.f. of the demand is *unknown*, and in Section 3.4 we apply stochastic approximation to product mix and allocation models. Section 3.5 deals with sensitivity and opimization of computer simulation models. We assume that the p.d.f. $f_Y(y, x)$ is *known* but the underlying model is so complex that $E[\psi_j(x, Y)]_{j=0,1,\ldots,M}$ can be estimated only through Monte Carlo simulation. Section 3.6 deals with variance reduction, and in Section 3.7 simulation results are presented that indicate feasibility of SAA.

3.2 MATHEMATICAL BACKGROUND

3.2.1 Introduction

Let $g(x)$, $x \in R^1$, be an unknown function whose values can be computed (evaluated) at any point x and whose unique root θ is sought while solving the nonlinear equation

$$g(x) = \alpha \ .$$

If the error of computation of $g(x)$ is negligible, we can find θ by applying, say, the well-known Newton–Raphson iterative procedure

$$x_{i+1} = x_i - \frac{g(x_i) - \alpha}{g'(x_i)} \ ,$$

where g' is the first derivative of g. Under some mild condition on g and g' [see Avriel (1976)] x_i converges to θ.

It may happen, however, that $g(x)$ is computed with an error that cannot be neglected, or, more specifically, at each point x instead of $g(x)$ we observe $\psi(x, Y)$ such that†

$$g(x) = E[\psi(x, Y)] \ ,$$

where Y is a random variable, $\psi(x, Y) = g(x) + \Xi(x, Y)$, and $\Xi(x, Y)$ is likewise a random variable (called noise) defined on some probability

†The function $g(x) = E[\psi(x, Y)]$ is usually called the regression function.

space $(\Omega, \mathscr{F}, P)$. The c.d.f. (cumulative distribution function) of $\Xi(x, Y)$ is unknown for all x and $E[\Xi(x, Y)] = 0$.

Robbins and Monro (1951) suggested, for finding the root θ of $g(x) = \alpha$, the iterative procedure

$$x_{i+1} = x_i - a_i[\psi(x_1, Y_i) - \alpha]$$

and called it SA (stochastic approximation)

Note The original Robbins-Monro version was presented in a slightly different form. Namely, random variables $\psi(x, Y)$ depending on a real parameter x are considered, which may be defined on different probability spaces $(\Omega_x, \mathscr{F}_x, P_x)$ for different x, an observation $\psi_i(x_i, Y_i)$ of $\psi(x, Y)$ being possible at each instant of time i. It is also assumed that the conditional distribution of $\psi_i(x_i, Y_i)$, given $X_1 = x_1, X_2 = x_2, \ldots, X_{i-1} = x_{i-1}$, coincides with that of $\psi(x_i, Y)$. In this case the Robbins-Monro procedure can be written as

$$x_{i+1} = x_i - a_i[\psi_i(x_i, Y_i) - \alpha] .$$

This procedure is somewhat more general than the preceding one, since $\psi(x, Y)$ need not be defined on the same probability space. However, this generalization is mainly illusory, since the r.v.'s ψ_i can always be placed in a single probability space.

Taking $a_i > 0$, $\Sigma_{i=1}^{\infty} a_i = \infty$, and $\Sigma_{i=1}^{\infty} a_i^2 < \infty$, Robbins and Monro found conditions on g and ψ under which their procedure converges asymptotically in the mean square, that is,

$$E(x_i - \theta)^2 \rightarrow 0 \quad \text{as } i \rightarrow \infty .$$

Note that in order to perform the $(i + 1)$th iteration we need to know the current observations x_i and $\psi(x_i, Y_i)$ but not the earlier ones,† which makes their procedure extremely simple and attractive for applications.

In the following sections we present several algorithms of the stochastic approximation type and discuss their convergence for both constrained stochastic optimization problems and unconstrained stochastic optimization problem in which the constraints $g_j(x) \leq 0$, $j = 1, \ldots, M$, in (P1) are removed and with which we open our presentation.

† This actually means that the Robbins-Monro procedure represents a Markov chain.

3.2.2 Unconstrained Stochastic Optimization Problems

Let $\theta \in X \subseteq \mathbb{R}^n$ satisfy the system of equations $R(x) = \alpha$, where α is a given vector, $R(x)$ a real measurable function such that† $R(x) > \alpha$ for $x > \theta$, $R(x) < \alpha$ for $x < \theta$, and $\psi(x, Y)$ a real function depending on a random vector Y ($Y \in R^m$) and a vector of parameters x ($x \in R^n$) such that $E[\psi(x, Y)] = R(x)$.

Consider the random sequence $x_1, x_2, \ldots$ in which x_1 is a vector with real components and $x_2, x_3, \ldots$ are determined by the following recursive formula (SAA).

$$x_{i+1} = \pi_X[x_i - a_i(\psi_i - \alpha)] . \tag{3.2.1}$$

Here $\pi_X(z)$ denotes the projection operator of the point z onto the set X (i.e., for every $x \in R^n$, $\pi(x) \in X$, and $\|x - \pi(x)\| = \min_{y \in X} \|x - y\|$), a_i ($i = 1, 2, \ldots$) are a sequence of positive numbers such that

$$\sum_1^\infty a_i = \infty , \tag{3.2.2}$$

$$\sum_1^\infty a_i^2 < \infty , \tag{3.2.3}$$

and ψ_i is a random variable whose conditional distribution for given $x_1, \ldots, x_i$ coincides with that of the random variable $\psi(x_i, Y)$.

We suppose further that X is a "nice" set (region) that readily lends itself to projection (e.g., it is given by conventional linear programming constraints, $X = \{x; Ax \leq b\}$). We present now the one-dimensional version of Gladyshev's (1965) theorem,‡ which will be extensively used in the sequel.

Theorem 1 Let (3.2.2)–(3.2.3) hold and

$$\inf_{\varepsilon < |x - \theta| < \varepsilon^{-1}} (x - \theta)[R(x) - \alpha] > 0 \quad \text{for each } 1 > \varepsilon > 0 , \tag{3.2.4}$$

and let there exist a positive number d such that, for all x,

$$E[\psi^2(x, Y)] \leq d(1 + x^2) . \tag{3.2.5}$$

†All results correct if $R(x) < \alpha$ for $x > \theta$, $R(x) > \alpha$ for $x < \theta$ ($x > \theta$ signifies coordinate-wise inequality, that is, $x_j > \theta_j$, $j = 1, \ldots, n$).

‡In the original version of the theorem, the projection operator π_x in (3.2.1) was omitted.

Then the sequence x_i, defined by the SAA (3.2.1), converges to θ almost surely (a.s.) and in the mean square. The proof is given in Appendix 1.

The formulation and proof of Theorem 1 remain valid in the multi-dimensional case, that is, when α, θ, x, $\psi(x, Y)$, $R(x)$ are vectors.

Consider now the unconstrained problem (P1), assuming that $g_0(x)$ is a convex differentiable function. Then the optimal solution x^* is obtainable from the solution of

$$\nabla g_0(x) = \nabla E[\psi_0(x, Y)] = 0 , \tag{3.2.6}$$

where ∇ denotes the gradient. Assuming that the latter is *interchangeable* with the expectation, then for the particular case $f_Y(y, x) = f_Y(y)$ we have†

$$\nabla g_0(x) = E[\nabla \psi_0(x, Y)] = 0 .$$

It is readily seen that with $\nabla g_0(x)$ replaced by $R(x) - \alpha$ and $\nabla \psi_0(x, Y)$ by $\psi(x, Y)$, we can apply Theorem 1 literally and thus prove convergence of the following stochastic approximation procedure:

$$x_{i+1} = \pi_x[x_i - a_i \nabla \psi_0(x_i, Y_i)] \tag{3.2.7}$$

under the conditions (3.2.2)–(3.2.3). Procedure (3.2.7) can be extended as

$$x_{i+1} = \pi_x[x_i - \Gamma_i \nabla \psi_0(x_i, Y_i)] , \tag{3.2.8}$$

where Γ_i is an $n \times n$ matrix. The convergence conditions of (3.2.8) are discussed in Tsypkin (1970).

Frequently, the exact value of the gradient $\nabla \psi_0(x, Y)$ is not available. To proceed we shall make several assumptions and definitions, assuming that $g(x)$ is a convex but not a differentiable function.

Definition 1 A *subgradient* (SG) of a function $g(x)$, $x \in R^n$, is a vector $\hat{\nabla} g(x)$ satisfying

$$g(y) - g(x) \geq \langle \hat{\nabla} g(x), y - x \rangle , \qquad \forall y \in R^n , \tag{3.2.9}$$

where $\langle \, , \rangle$ denotes the scalar product.

Definition 2 [Ermoliev (1969)] *A stochastic quasi-gradient* (*subgradient*) (SQG)‡ of a function $g(x)$ is a random vector $\tilde{\nabla} g(x, Y)$ satisfying

†For the p.d.f. $f_Y(y, x)$ see equations (3.2.20)–(3.2.22).
‡For an example of an SQG, see (3.2.20) and (3.2.21).

$$E[\tilde{\nabla} g(x_i, Y_i) \mid X_1 = x_1, \ldots, X_i = x_i] = \tilde{\nabla} g(x_i) + b_i\,, \qquad (3.2.10)$$

where b_i is a vector depending on $x_1, \ldots, x_i$. From now on we shall use for simplicity $\tilde{\nabla} g(x_i)$ instead of $\tilde{\nabla} g(x, Y)$, keeping in mind that $\tilde{\nabla} g(x_i)$ is a random vector.

Let $g(x)$ be a convex and continuous function in $x \in X \subseteq R^n$, where X is a convex, closed, and bounded region. Consider the iterative procedure

$$x_{i+1} = \pi_x[x_i - a_i \tilde{\nabla} g(x_i)]\,. \qquad (3.2.11)$$

Ermoliev (1969, 1983) found conditions on $g_0(x)$, $\tilde{\nabla} g_0(x)$, and a_i under which procedure (3.2.11) converges almost surely to the set of optimal solutions of the unconstrained problem (P1).

3.2.3 Constrained Stochastic Optimization Problems

Most methods for solving problem (P1) [see Ermoliev (1983)] are based on the penalty function method and on the Lagrange multiplier methods as described briefly below.

(a) Penalty Function Method Define the penalty function

$$\varphi(x, c) = g_0(x) + c \sum_{j=1}^{M} \max[0, g_j(x)]\,, \qquad c > 0\,.$$

The subgradient of $\varphi(x, c)$ with respect to x is given by

$$\hat{\nabla}\varphi(x, c) = \hat{\nabla} g_0(x) + \frac{c}{2} \sum_{j=1}^{M} \{\max\{0, g_j(x)\}\, \hat{\nabla} g_j(x)\}^2\,.$$

If the exact values of $g_j(x)$ and $\hat{\nabla} g_j(x)$, $j = 0, 1, \ldots, M$, are known, then *deterministic* iterative procedures [see Avriel (1976)] based on $\hat{\nabla}\varphi(x, c)$ can be used for minimizing $\varphi(x, c)$ and therefore for solving (P1). If, however, only statistical estimates of $g_j(x)$ and $\hat{\nabla} g_j(x)$, $j = 0, 1, \ldots, M$, are available, then $\hat{\nabla}\varphi(x, c)$ cannot be calculated exactly and *stochastic* algorithms (say, of the stochastic approximation type) can be used for minimizing $\varphi(x, c)$. Ermoliev (1983), for example, suggested the iterative procedure

$$x_{i+1} = \pi_x\left\{x_i - a_i\left[\tilde{\nabla} g_0(x_i) + c \sum_{j=1}^{M} \max(0, Z_i^j)\, \tilde{\nabla} g_j(x_i)\right]\right\},$$

where

$$Z_{i+1}^j = \rho_i \theta_j(x_i) + (1 - \rho_i) Z_i^j ,$$

and

$$E\{\theta_j(X_i \mid x_1, \ldots, x_i)\} = g_j(x_i) + \gamma_j(x_i) .$$

Assuming that

$$g_j(x_i) - g_j(x^*) \leq E\{\tilde{\nabla} g_j(X_i \mid x_1, \ldots, x_i)\} + \beta_j(x_i) ,$$

where $\tilde{\nabla} g$ is the SQG of g, and

$$a_i > 0 , \quad \sum_{i=0}^{\infty} a_i = \infty , \quad \rho_i > 0 , \quad \frac{a_i}{\rho_i} \to 0 ,$$

$$\sum_{i=0}^{\infty} \sum_{j=1}^{M} E\{a_i |\beta_j(X_i)| + \rho_i |\gamma_j(X_i)|\} < \infty ,$$

Ermoliev proved convergence of his algorithm almost surely.

Gupal (1977), (1979) adopted algorithm (3.2.11) for solving the constrained problem (P1), defining

$$\tilde{\nabla} g(x_i) = \begin{cases} \tilde{\nabla} g_0(x_i) & \text{if } \max_{1 \leq j \leq M} z_i^j \leq 0 , \\ \tilde{\nabla} g_k(x_i) & \text{if } z_i^k = \max_{1 \leq j \leq M} z_i^j > 0 . \end{cases} \tag{3.2.12}$$

Here $\tilde{\nabla} g_j(x_i)$ $(j = 0, 1, \ldots, M)$ is the SQG of $g_j(x)$ satisfying

$$E[\tilde{\nabla} g_j(X_i) \mid x_0, z_0, \ldots, x_i, z_i] = \tilde{\nabla} g_j(x_i) + b_j(x_i) , \tag{3.2.13}$$

$$z_{i+1}^j = z_i^j + \rho_i(\theta_i^j - z_i^j) , \qquad j = 1, \ldots, M , \tag{3.2.14}$$

$$E[\theta_i^j \mid x_0, z_0, \ldots, x_i, z_i] = g_j(x_i) , \qquad j = 1, \ldots, M , \tag{3.2.15}$$

and ρ_i, a_i are sequences defined below.

We present now a slightly modified version of Gupal's (1977) theorem.

Theorem 2 Let $x \in R^n$ be a convex, closed, and bounded set, $g_j(x)$, $j = 0, 1, \ldots, M$, continuous, convex functions, and all constraints of problem (P1) satisfy Slater's condition, that is

$$D = \{x: g_j(x) < 0, j = 1, \ldots, M\} \neq \emptyset .$$

Assume also that the following conditions hold:

$$\text{(i)} \quad a_i > 0 , \qquad \rho_i > 0 ,$$

$$\text{(ii)} \quad \sum_{i=0}^{\infty} a_i = \infty , \qquad \sum_{i=0}^{\infty} \rho_i = \infty ,$$

$$\text{(iii)} \quad \sum_{i=0}^{\infty} a_i^2 < \infty , \qquad \sum_{i=0}^{\infty} \rho_i^2 < \infty ,$$

$$\text{(iv)} \quad \frac{a_i}{\rho_i} \to 0 \quad \text{as } i \to \infty , \tag{3.2.16}$$

$$\text{(v)} \quad \|b_j(x_i)\| \leq \beta_i , \qquad \sum_{i=0}^{\infty} a_i \beta_i < \infty ,$$

$$\text{(vi)} \quad E[|\theta_i^j|^2] \leq C < \infty \qquad (j = 1, \ldots, M) ,$$

$$\text{(vii)} \quad E[\|\tilde{\nabla} g_j(X_i)\|^2 \leq C < \infty \qquad (j = 0, 1, \ldots, M).$$

Then, one of the limit points of $\{x_i\}$ belongs to the set of solutions of problem (P1) almost surely. If, in addition, $g_0(x)$ is strictly convex, then

$$\lim_{i \to \infty} x_i = x^* \quad \text{a.s.} ,$$

where x^* is a unique solution of the constrained problem (P1).

Note 1 It follows also from (3.2.12) that, for the unconstrained problem (P1),

$$\tilde{\nabla} g(x_i) = \tilde{\nabla} g_0(x_i) .$$

Note 2 If $\psi_j(x, y)$ $(j = 0, 1, \ldots, M)$ are convex with respect to x for any fixed y, then for the case $f_Y(y, x) = f_Y(y)$, we can take

$$\tilde{\nabla} g_j(x_i) = \hat{\nabla} \psi_j(x_i, Y_i) , \qquad j = 0, 1, \ldots, M , \tag{3.2.17}$$

and

$$\theta_i^j = \psi_j(x_i, Y_i) , \qquad j = 1, \ldots, M . \tag{3.2.18}$$

Note 3 If θ_i^j is unbounded, then the values z_i^j in (3.2.14) have to be normalized as

$$z^j_{i+1} = \begin{cases} z^j_i + \rho_i(\theta^j_i - z^j_i) = \bar{z}^j_{i+1} & \text{if } |\bar{z}^j_{i+1}| \leq c\,, \\ c \operatorname{sign}(\bar{z}^j_{i+1}) & \text{otherwise}\,, \end{cases} \tag{3.2.19}$$

where c is a sufficiently large constant.

Note 4 Assume that $g_j(x)$, $j = 0, 1, \ldots, M$, have uniformly bounded second derivatives not available explicitly. Consider the following SQGs [finite difference of the gradients of $\psi_j(x, Y)$]:

$$\tilde{\nabla} g_j(x_i) = \frac{1}{\beta_i} \sum_{s=1}^{n} [\psi_j(x_i + \beta_i e^s, Y_i^{(s)}) - \psi_j(x_i, Y_i^{(0)})] e^s\,, \tag{3.2.20}$$

and

$$\tilde{\nabla} g_j(x_i) = \frac{3r_i}{2\beta_i} \sum_{s=1}^{r_i} [\psi_j(x_i + \beta_i \Xi_i, Y_i^{(s)}) - \psi_j(x_i, Y_i^{(0)})] \Xi_i\,, \tag{3.2.21}$$

where e^s, $s = 1, \ldots, n$ is the orthonormal basis in R^n, $r_i \geq 1$ and $\Xi_i = (\Xi_i^{(1)}, \ldots, \Xi_i^{(n)})$, $i = 1, 2, \ldots$, are independent random vectors with independent components $\Xi_i^{(k)}$, $k = 1, \ldots, n$, each distributed uniformly over the interval $[-1, 1]$. Note that (3.2.20) and (3.2.21) require $n + 1$ and $r_i + 1$, $r_i = 1, 2, \ldots$, evaluations of $\psi_j(x, y)$, respectively. It is shown in Ermoliev (1983) that, for both (3.2.20) and (3.2.21), we have

$$E[\tilde{\nabla} g_j(X_i) \mid X_i = x_i] = \nabla g_j(x_i) + b_j(x_i)\,, \qquad j = 0, 1, \ldots, M\,, \tag{3.2.22}$$

where

$$\|b_j(x_i)\| < c\beta_i\,, \qquad c > 0\,. \tag{3.2.23}$$

Kiefer and Wolfowitz (1952) were the first to estimate the derivative of $g(x)$, $x \in R^1$, in a stochastic approximation procedure using the finite difference estimate

$$\tilde{\nabla} g(x_i) = \frac{1}{2\beta_i} [\psi(x_i + \beta_i, Y_i^{(1)}) - \psi(x_i - \beta_i, Y_i^{(0)})]\,. \tag{3.2.24}$$

The optimal choice of a_i and β_i [see Wasan (1969)] in (3.2.11) with

$\tilde{\nabla} g(x_i)$ defined as in (3.2.24) is $a_i = a/i$ and $\beta_i = \beta/i^{1/4}$, respectively, where a and β are positive constants.

(b) Lagrange Multiplier Method Define the Lagrange function

$$\varphi(x, \lambda) = g_0(x) + \sum_{j=1}^{M} \lambda^j g_j(x) . \tag{3.2.25}$$

Ermoliev (1983) considered the iterative procedure

$$\begin{cases} x_{i+1} = \pi_X\left\{x_i - a_i\left[\tilde{\nabla} g_0(x_i) + \sum_{j=1}^{M} \lambda_i^j \tilde{\nabla} g_j(x_i)\right]\right\}, \\ \lambda_{i+1}^j = \max\{0, \lambda_i^j + \rho_i \theta_j(x_i)\} , \end{cases} \tag{3.2.26}$$

where

$$E[\theta_j(x_i) \mid x_1, \ldots, x_i] = g_j(x_i) , \tag{3.2.27}$$

and $\tilde{\nabla} g_j(x_i)$ is as per (3.2.13). He gave conditions on $\tilde{\nabla} g_j$, ρ_i and a_i under which the above procedure converges almost surely to the optimal solution.

3.2.4 Asympotical Convergence and Stopping Rules

The asymptotical rate of convergence of SAA, for the unconstrained problem (P1) was considered by many authors [see Lai and Robbins (1979), Polyak and Tsypkin (1980a, b), Saks (1958), and Wasan (1969)].

Saks (1958) considered the following SAA:

$$x_{i+1} = x_i - \frac{a}{i} (\psi_i - \alpha) . \tag{3.2.28}$$

He found conditions on $R(x)$ and $\psi(x, Y)$ under which $i^{1/2}(x_i - \theta)$ is distributed asymptotically normal with mean zero and variance $V = a^2\sigma^2(2ab - 1)^{-1}$, where b is the derivative of $R(x)$ at θ, that is, $b = R'(\theta)$, and

$$\sigma^2 = \lim_{x \to \theta} \operatorname{Var}[\psi(x, Y)] .$$

Clearly, V achieves its minimum ($V = \sigma^2 b^{-2}$) when $a = b^{-1} = [R'(\theta)]^{-1}$.

Polyak and Tsypkin (1980a) considered the SAA

$$x_{i+1} = x_i - \frac{\Gamma}{i} \nabla\psi_0(x_i, Y_i) , \tag{3.2.29}$$

where Γ is an $n \times n$ matrix. They found conditions on $g_0(x)$ and $\psi_0(x, Y)$ under which the r.v. $i^{1/2}_{i\to\infty}(x_i - x^*) \sim N(0, V)$, where V is some covariance matrix. They showed that V is minimal† when $\Gamma = [\nabla^2 g_0(x^*)]^{-1}$, where $[\nabla^2 g_0(x^*)]^{-1}$ is the inverse of the Hessian of g_0 at x^*. In this case

$$V = [\nabla^2 g_0(x^*)]^{-1} D [\nabla^2 g_0(x^*)]^{-1} , \tag{3.2.30}$$

where

$$D = E[\nabla\psi_0(x^*, Y)\nabla^t\psi_0(x^*, Y) - \nabla g_0(x^*)\nabla^t g_0(x^*)] .$$

Generally $b = R'(\theta)$ (and $\Gamma = [\nabla^2 g_0(x^*)]^{-1}$) are unknown. They can, however, be estimated adaptively [see Hasminskii and Nevelson (1973), Lai and Robbins (1979), (1981), Tsypkin (1982a, b), and Venter (1967)].

Venter (1967) suggested estimating b by

$$\hat{b}_i = \frac{1}{i} \sum_{j=1}^{i} \tilde{\nabla} g_0(x_j) ,$$

where $\tilde{\nabla} g_0(x_j)$ is given in (3.2.22) and found conditions under which

$$\hat{b}_i \xrightarrow{\text{a.s.}} b \quad \text{as} \quad i \to \infty .$$

Lai and Robbins (1979) suggested estimating b by

$$\hat{b}_i = k \vee (K \wedge b_i) , \tag{3.2.31}$$

where $\vee$ and $\wedge$ denote maximum and minimum, respectively; k and K are lower and upper bounds of b, respectively; and b_i is some sequence (see opt. cit.). They showed that if

$$\hat{b}_i \xrightarrow{\text{a.s.}} b = R'(\theta) ,$$

then the asymptotical rate of convergence is still $V = \sigma^2 b^{-2}$. We shall use (3.2.31) as an estimate for b while simulating inventory models in Section 3.7.

†Since V is positive-definite, we say that $V_1 < V_2$ if for any vector $A \neq 0$, $A^t V_1 A < A^t V_2 A$, where t denotes the transpose operator.

Stopping rules for the Robbins-Monro procedure were considered by Sielken (1973) and later by Stroup and Braun (1982).

By Sielken's rule (1973), stochastic approximation may end as soon as the confidence interval [based on the asymptotical normality of $i^{1/2}(x_i - \theta)$] is as narrow as desired.

In particular, let K_δ be the upper $100(1-\delta)$ percentile of the standard normal distribution. For an approximate $100(1-2\delta)$ percent confidence interval of the form $(x_i - d, x_i + d)$, sampling would stop as soon as $K_\delta \cdot \sigma / i^{1/2}$ is less than d, where σ^2 is the asymptotical variance of $i^{1/2}(x_i - \theta)$, or equivalently as soon as i is greater than $K_\delta^2 \cdot \sigma^2/d^2$. Sielken also proposed estimators of σ^2 and $\beta = R'(\theta)$ and obtained sufficient conditions for them to converge with probability 1.

Stroup and Braun (1982) recommend a rule based on a prediction criterion of Akaike (1969). Let $\bar{y}_i$ be the average of two responses observed at the same input level x_i, and let s_n be the estimate of σ at step n. Define

$$u_n(k) = \sum_{i=n-k+1}^{n} \frac{\bar{y}_i^2}{ks_n^2} \qquad (n = k, k+1, \ldots; k = 1, 2, \ldots),$$

and as a stopping rule take $N_k = \inf(n\colon u_n(k) < c)$ for an appropriate positive constant c. They proved asymptotical normality of $N_k^{1/2}(x_{N_k} - \theta)$ and showed experimentally that their stopping procedure works well in a variety of situations.

3.2.5 Robust Adaptive Procedures

(a) Introduction The fact that SAAs are convergent procedures is insufficient to recommend them in practice. The most important feature of an algorithm is, of course, the rate of its convergence as characterized by different criteria. Examples of the latter are

(i) $E\|x_i - x^*\|^2$;

(ii) $E(g(x_i) - g(x^*))^2$;

(iii) $P(\|x_i - x^*\|^2 \le \varepsilon)$, $\varepsilon > 0$; and

(iv) $P(g(x_i) - g(x^*) \le \varepsilon)$.

Various algorithms of the stochastic approximation type have been proposed for minimizing (i)–(iv). Most of them deal [see Wasan (1969)] with choosing the step parameter a_i or the matrix in Γ_i in (3.2.7) or (3.2.8), respectively. Polyak and Tsypkin (1980a, b) and Tsypkin (1979, 1982a, b) showed that SAAs are *unstable* and highly sensitive to small deviations from the assumptions primarily from those on the underlying distribution F.

In this section we shall show how SAAs can be stabilized by reducing the level of uncertainty about F, that is, by assuming some prior information. This treatment is closely related to the notion of *robustness* (*robust estimates*) introduced by Huber (1964, 1972).

(b) Robust Statistical Estimates (Huber's Minimax Approach) Say we seek to estimate a location parameter θ from a sample $Y_1, \ldots, Y_N$, that is related to θ as

$$Y_i = \theta + \Xi_i , \qquad i = 1, \ldots, N ,$$

where Ξ_i are i.i.d. r.v.'s (noises) with c.d.f. F.

Let $T_N = T(Y_1, \ldots, Y_N)$ be an estimate of θ. It is well known that the sample mean

$$\bar{\theta}_N = \frac{1}{N} \sum_{i=1}^{N} Y_i \tag{3.2.32}$$

minimizes the sum of squares

$$\sum_{i=1}^{N} (Y_i - T_N)^2 .$$

However, it was also observed for a long time that the sample mean is highly sensitive to small deviations from the assumption on F. Let H be the *true* distribution of Ξ, expressed as

$$H(z) = (1 - \varepsilon)F(z) + \varepsilon P(z) , \qquad \varepsilon > 0 , \tag{3.2.33}$$

where $P(z)$ is some unknown distribution. Denoting by σ_0^2 and σ_1^2 the variances associated with F and P respectively, the variance of the sample mean is

$$\operatorname{Var} \bar{\theta}_N = E(\bar{\theta}_N - \theta)^2 = \frac{\sigma^2}{N} ,$$

where $\sigma^2 = (1 - \varepsilon)\sigma_0^2 + \varepsilon\sigma_1^2$.

Let $\varepsilon = 0.1$ and $\sigma_0^2 = 1$. If

(a) $\sigma_1^2 = 1$, then $\sigma^2 = 1$;
(b) $\sigma_1^2 = 10$, then $\sigma^2 = 1,9$;
(c) $\sigma_1^2 = 100$, then $\sigma^2 = 10,9$.

Thus, for fixed ε and N, the variance of $\bar{\theta}_N$ increases rapidly with σ_1^2. As an extreme case, if $P(z)$ is a Cauchy distribution, then

$$\sigma^2 = \infty\,, \quad \text{since } \sigma_1^2 = \infty\,,$$

and $\bar{\theta}_N$ is an inconsistent estimate. The above example illustrates the instability or nonrobustness of the mean square estimator. In the search for remedies to this drawback, an important role was played by Tukey (1960). Huber (1964, 1972, 1977, 1981), was the founder of the new approach known as *robust estimation*; on that occasion Huber introduced several such estimates of a location parameter θ and proved their consistency and optimality in the minimax since.

We shall consider only one of these, the so-called M-estimate.

Definition [Huber (1964)] Any estimate T_N defined by a minimization problem of the form

$$\sum_{i=1}^{N} \varphi(Y_i, T_N) = \min!\,, \tag{3.2.34}$$

or by an implicit equation

$$\sum_{i=1}^{N} \psi(Y_i, T_N) = 0\,, \tag{3.2.35}$$

where φ is an arbitrary convex function and $\psi(Y, \theta) = (\partial/\partial\theta)\varphi(Y, \theta)$, is called an M-estimate of θ. [Note that for $\varphi(z) = -\log h(z)$, where $h(z) = H'(z)$, we obtain the ordinary maximum likelihood estimate.]

For the location parameter, formulas (3.2.34) and (3.2.35) can be written, respectively, as

$$\sum_i \varphi(Y_i - T_N) = \min! \tag{3.2.36}$$

and

$$\sum \psi(Y_i - T_N) = 0\,. \tag{3.2.37}$$

Presenting (3.2.37) as

$$\sum_i w_i(Y_i - T_N) = 0 ,$$

where

$$w_i = \frac{\psi(Y_i - T_N)}{Y_i - T_N} ,$$

we obtain

$$T_N = \frac{\sum_i w_i Y_i}{\sum w_i} . \tag{3.2.38}$$

It was hoped that replacement of the square $\Sigma_i(Y_i - T_N)^2$ by a less rapidly increasing function φ would yield a less sensitive estimate. Indeed, assuming that the true distribution $H(z)$ of the noise Ξ belongs to the class of ε-contaminated distributions,

$$R = \{H(z)\colon H(z) = (1 - \varepsilon)F(z) + \varepsilon P(z)\} , \tag{3.2.39}$$

where $F(z)$ is some known c.d.f., $P(z)$ is an arbitrary unknown c.d.f., and $0 \leq \varepsilon \leq 1$ is a known number, Huber (1964) formulated conditions for consistency, asymptotical normality, and asymptotical optimality in the minimax sense of his M-estimates (3.2.36) and (3.2.37) on the class R. Asymptotical optimality in the minimax sense signifies that there exists a c.d.f. $H^* = (1 - \varepsilon)F + \varepsilon P^*$ ($H^* \in R$) and a function $\varphi = \varphi^*$ such that the asymptotical variance of the M-estimate satisfies

$$\mathrm{Var}(\varphi^*, H) \leq \mathrm{Var}(\varphi^*, H^*) \leq \mathrm{Var}(\varphi, H^*) \tag{3.2.40}$$

for any $H \in R$. Thus, choosing $\varphi(z) = \varphi^*(z)$ in (3.2.36), he obtained an M-estimate whose asymptotical variance $\mathrm{Var}(\varphi^*, H)$ for any $H \in R$ does not exceed the minimal possible guaranteed quantity $\mathrm{Var}(\varphi^*, H^*) = \min_\varphi \max_H \mathrm{Var}(\varphi, H)$. Moreover, the p.d.f. h^* has Fisher's information

$$\Phi(z) = \int \frac{[h'(z)]^2}{h(z)}\, dz \tag{3.2.41}$$

that is minimal among all distributions on the class R.

Methods for finding φ^* that determine minimax M-estimates [minimize Fisher's information $\Phi(z)$] for a given class of distributions H where considered by many authors [e.g., Huber (1981) and Polyak and Tsypkin (1979)]. Table 1 lists several classes of distributions R and the associated optimal function ψ^*, where $\psi^* = (\varphi^*)'$.

Huber (1981) considered robust covariance matrices and other topics related to robustness in statistics. Consider for example the problem of estimating the vector θ of a linear regression function from observations $Y_1, \ldots, Y_N$ to which it is related by

$$Y_i = \sum_{j=1}^{n} x_{ij}\theta_j + \Xi_i\,, \qquad i = 1, \ldots, N\,,$$

where $x = (x_1, \ldots, x_n)$ is a known vector. Huber defined different classes of distributions R [see (3.2.39)] and found conditions on φ such that his M-estimate (3.2.34) and (3.2.35) is consistent, asymptotically normal, and optimal in the minimax sense [see (3.2.40)].

Table 1 Some Classes of Distributions and the Associated Optimal Function ψ^*

R	Class of distribution R	$\psi^*(z)$
R_1	$\int z^2 p(z)\, dz \le \sigma^2 < \infty$	$\psi^*(z) = \dfrac{z}{\sigma^2}$
R_2	$h(z)$ continuous at zero and $h(0) \ge \beta \ge 0$	$\psi^*(z) = 2\beta \operatorname{sign} z$
R_3	$\int_{-a}^{a} h(z)\, dz = 1$	$\psi^*(z) = \tan \dfrac{\pi z}{2a}$
R_4	$h(z) = (1-\varepsilon)f(z) + \varepsilon P(z)$, where $f(z) = \begin{cases} \dfrac{1}{2d} & \lvert z\rvert \le d\,, \\ 0 & \lvert z\rvert > 0\,, \end{cases}$ and $P(z)$ is an arbitrary p.d.f.	$\psi(z) = \begin{cases} 0\,, & \lvert z\rvert \le d \\ 1 - \dfrac{\varepsilon}{\varepsilon d} \operatorname{sign} z\,, & \lvert z\rvert > d \end{cases}$
R_5	$h(z) = (1-\varepsilon)f(z) + \varepsilon P(z)$, where $f(z)$ is a normal density with $\mu = 0$ and Var Z $= \sigma^2$ and $P(z)$ is an arbitrary p.d.f.	$\psi(z) = \max\left\{\dfrac{z}{\sigma^2}, \dfrac{\varepsilon}{\sigma^2} \operatorname{sign}(z)\right\}$

(c) Robust Adaptive Algorithms Consider the unconstrained problem (P1) with $F_Y(y, x) = F_Y(y)$ assuming different levels of uncertainty about the c.d.f. F_Y. Consider also the following generalized version of SAA of (3.2.8), namely,

$$x_{i+1} = x_i - \Gamma_i Q[\nabla\psi(x_i, Y_i)] , \tag{3.2.42}$$

where Γ_i is an $n \times n$ matrix and $Q: R^n \rightarrow R^n$ some vector function. Clearly, if $Q(y) = y$, we obtain the SAA (3.2.8). In the particular case where $\Gamma_i = a_i \Gamma$, we shall write

$$x_{i+1} = x_i - a_i \varphi[\nabla\psi(x_i, Y_i)] . \tag{3.2.43}$$

where $\varphi(z) = \Gamma Q(z)$. It is also clear that choosing different transformations $\varphi: R^n \rightarrow R^n$, we obtain different algorithms.

The natural question at this juncture is how to choose Γ_i and Q for the algorithm (3.2.42) to be optimal in some sense [say, in the minimization of criteria (i)–(ii) or maximization of (iii)–(iv)]. There are two approaches with this respect: the local (nonasymptotical) one, whereby the choice is such that the algorithm (3.2.42) is optimal at *each iteration*, and the asymptotical one, whereby the asymptotical variance (covariance matrix) is minimized. Polyak (1979) and Al'ber and Shilman (1981) considered some theoretical aspects related to the first approach; Polyak and Tsypkin (1980a, b) and Tsypkin (1982a, b) presented a rigorous treatment for the second approach. We shall deal only with the second approach. The results presented here† are taken from Polyak and Tsypkin (1980a, b) and Tsypkin (1982a, b).

It turns out that if $\Gamma_i = (1/i)\Gamma$, then the optimal matrix Γ (the one minimizing the asymptotical variance (covariance matrix) is associated with the Fisher information matrix [see (3.2.50)], and the optimal transformation Q is $Q(z) = \nabla \log h(z)$, where h is a p.d.f. defined below.

Before establishing these results, consider the iterative procedure

$$x_{i+1} = x_i - a_i \varphi(Z_i) , \tag{3.2.44}$$

where‡ $Z_i = \nabla g(x_i) + \Xi_i$, and Ξ_i, $i = 1, 2, \ldots$ are i.i.d. random vectors (noises) with $E(\Xi_i) = 0$. We assume first that the distribution of the noise is

†Note that the results are valid only asymptotically. To the best of our knowledge, the important problem of choosing a_i and φ such that a prescribed accuracy for the criteria (i)–(iv) is ensured in a minimum number of iterations is still an open one.

‡Z is usually called the perturbed value of the gradient and Ξ the random noise.

completely known. For this case Polyak and Tsypkin (1980a) considered algorithm (3.2.44) and proved that if

(i) The function $g(x)$ is differentiable and its gradient satisfies the Lipschitz condition, namely,

$$\|\nabla g(x_1) - \nabla g(x_2)\| \le L\|x_2 - x_1\| , \qquad L > 0 .$$

There is a unique stationary point x^* such that $\nabla g(x^*) = 0$ and $g(x) \to \infty$ as $\|x\| \to \infty$.

(ii) The random vectors (noises) $\Xi_1, \Xi_2, \ldots$ are independent and symmetrically distributed with respect to zero.

(iii) The function φ: $R^n \to R^n$ is measurable odd [i.e., $\varphi(z) = -\varphi(-z)$], and uniformly monotonic, namely, for all z and z_1

$$\langle \varphi(z) - \varphi(z_1), z - z_1 \rangle \ge \mu(\|z - z_1\|) , \quad \inf_{\varepsilon \le \tau \le \varepsilon^{-1}} \mu(\tau) > 0, \epsilon > 0 ,$$

where $\langle\ ,\ \rangle$ denotes the inner product.

(iv) Either $\|\varphi(z)\| \le k$ for all $z \in R^n$, that is, φ is bounded, or $\|\varphi(z)\| \le k\ (1 + \|z\|)$, that is, φ is not steeper than a linear function. The variance of the noise satisfies

$$\mathrm{Var}\|\Xi\| = E\|\Xi\|^2 \le k[g(x) - g(x^*) + 1] ,$$

(v) $a_i \ge 0$, $\Sigma_{i=1}^{\infty} a_i = \infty$, and $\Sigma_{i=1}^{\infty} a_i^2 < \infty$.

Then, $x \to x^*$ almost surely.

If, in addition,

(vi) $g(x)$ is twice differentiable in x^*, and the matrix $\nabla g^2(x^*)$ is positive-definite;

(vii) the noises $\Xi_1, \Xi_2, \ldots$ are distributed identically with p.d.f. h;

(viii) the function $T(a) = E\varphi(z + \Xi) = \int \varphi(a + z)h(z)\, dz$ is differentiable at 0;†

(ix) the matrix $B = \frac{1}{2}I - T'(0)\nabla g(x^*)$ is stable (I denoting the unit matrix) and the matrix $A = \int \varphi(z)\varphi'(z)h(z)\, dz > 0$;

(x) $a_i = 1/i$, that is, (3.2.44) can be written as

$$x_{i+1} = x_i - \frac{1}{i}\varphi(Z_i) , \qquad Z_i = \nabla g(x_i) + \Xi_i , \tag{3.2.45}$$

†This condition is satisfied if, for example, φ is differentiable and its derivatives are bounded; in this case $T'(0) = E(\nabla\varphi) = \int \nabla\varphi(z)h(z)\, dz$.

then asymptotically $i^{1/2}(x_i - x^*) \sim N(0, \Sigma_\varphi)$, where Σ_φ is the solution of the matrix equation

$$B\Sigma_\varphi + \Sigma_\varphi B^t = -A ,$$

and t denotes the transpose operator.

Thus, the asymptotic rate of convergence of algorithm (3.2.45) defined by the covariance matrix Σ depends on properties of $g(x)$, the c.d.f. H of the noise Ξ, and the transformation φ. To emphasize the dependence of Σ of H and φ, we shall subsequently write $\Sigma(\varphi, H)$ instead of Σ_φ.

A natural question that arises is whether in the class of algorithms (3.2.45) there exists a transformation φ: $R^n \rightarrow R^n$ minimizing the asymptotic covariance matrix† $\Sigma(\varphi, H)$ of the random vector $i^{1/2}(x_i - x^*) \sim N(0, \Sigma(\varphi, H))$. The answer is affirmative and is given by Polyak and Tsypkin (1980a), who proved that if the c.d.f. H is known in advance, then under some conditions‡ the minimum $\Sigma(\varphi, H)$ is achieved for $\varphi = \varphi^0$ defined as

$$\varphi^0(z) = \nabla^2 g(x^*)^{-1} \Phi(h)^{-1} \nabla \log h(z) , \tag{3.2.46}$$

where

$$\Phi(h) = \int \frac{\nabla \log h(z) \, \nabla^t \log h(z)}{h(z)} \, dz \tag{3.2.47}$$

is the *Fisher information matrix* (assumed to be finite and positive-definite) and A^{-1} is the inverse of A.

The algorithm (3.2.45) with the optimal transformation φ_0 is

$$x_{i+1} = x_i - \frac{1}{i} \nabla^2 g(x^*)^{-1} \Phi(h)^{-1} \nabla \log h(Z_i) , \tag{3.2.48}$$

where $Z_i = \nabla g(x_i) + \Xi_i$, and the asymptotic covariance matrix for (3.2.48) is

$$\Sigma(\varphi^0, H) = \nabla^2 g(x^*)^{-1} \Phi(h)^{-1} \nabla^2 g(x^*)^{-1} . \tag{3.2.49}$$

†Since Σ is positive- (nonnegative) definite, we say that $\Sigma_1 \leq \Sigma_2$ if, for any vector $x \in R^n$, $x^t \Sigma_1 x \leq x^t \Sigma_2 x$.

‡Essentially Polyak's and Tsypkin's conditions for (3.2.45) to be asymptotically optimal in the sense of the minimum covariance matrix (3.2.49) are: $h(z)$ is symmetric and differentiable for almost all z; $\log h(z)$ is a convex function that increases no faster than a quadratic one; and $\varphi(a) = \int \log \varphi(z - a) \, dH(z)$ is differentiable at 0, $\varphi'(0) = \Phi(h)$ and $0 < \Phi(h) < \infty$.

Thus, for given H,

$$\Sigma(\varphi^0, H) \leq \Sigma(\varphi, H),$$

and therefore the algorithm (3.2.48) is optimal in the class of algorithms (3.2.45) in the sense of the asymptotic rate of convergence.

Representing $\varphi(z)$ as $\varphi(z) = \Gamma Q(z)$ [see (3.2.43)], we see from (3.2.46) that the optimal Γ and Q are, respectively,

$$\Gamma^0 = \nabla^2 g(x^*)^{-1}\Phi(h), \tag{3.2.50}$$

and

$$Q^0(z) = \nabla \log h(z). \tag{3.2.51}$$

For the one-dimensional case we obtain

$$x_{i+1} = x_i - \frac{a}{i}\frac{h'(z_i)}{h(z_i)}, \qquad a = [\Phi(h)g''(x^*)]^{-1},$$

$$\Phi(h) = \int \frac{[h'(z)]^2}{h(z)}\, dz, \tag{3.2.52}$$

and the minimal variance is

$$\Sigma(\varphi^0, H) = \sigma^2(\varphi^0, H) = \frac{1}{[g''(x^*)]^2\Phi(h)}. \tag{3.2.53}$$

Formulas (3.2.52) and (3.2.53) were obtained earlier by Anbar (1973) and Abdelhamid (1973).

Consider now several examples.

(a) Univariate Case

(i) *Normal distribution*

$$h(z) = \frac{1}{(2\pi)^{1/2}\sigma}\exp\left(-\frac{z^2}{2\sigma^2}\right), \qquad -\infty < z < \infty. \tag{3.2.54}$$

We have $-(d/dz)\log h(z) = z\sigma^{-2}$, $\Phi(h) = \sigma^{-2}$. The optimal algorithm is

$$x_{i+1} = x_i - \frac{1}{ig''(x^*)} Z_i, \qquad \text{where } Z_i = g'(x_i) + \Xi_i \tag{3.2.55}$$

and is called the *linear algorithm*. The minimal variance is

$$\sigma^2(\varphi^0, H) = [g''(x^*)]^{-2}\sigma^2 .$$

(ii) *Laplace distribution*

$$h(z) = \beta \exp(-2\beta|z|) , \qquad -\infty < z < \infty . \tag{3.2.56}$$

We have $-(d/dz) \log h(z) = 4\beta \text{ sign } z$, where

$$\text{sign } z = \begin{cases} -1 , & z > 0 , \\ 1 , & z < 0 , \end{cases}$$

and $\Phi(h) = 2\beta$. *The optimal algorithm is*

$$x_{i+1} = x_i - \frac{1}{2\beta i g(x^*)} \text{ sign } Z_i , \qquad Z_i = g'(x_i) + \Xi_i \tag{3.2.57}$$

and is called the *sign algorithm*. Finally

$$\sigma^2(\varphi^0, H) = [2\beta g''(x^*)]^{-2} .$$

(b) Multivariate Case

(i) Let the components $\Xi = (\Xi^{(1)}, \ldots, \Xi^{(n)})$ of the noise be independent; then $h(z) = \Pi_{j=1}^n h^{(j)}(z)$, where $h_j(z)$ are the marginal p.d.f.'s. In this case the optimal algorithm (3.2.48) can be written component-wise and therefore reduced to the univariate case.

(ii) *Normal distribution*

$$\Xi \sim N(0, S), \text{ where } S \text{ is the covariance matrix.} \tag{3.2.58}$$

We have

$$-\log h(z) = S^{-1}z , \qquad \Phi(h) = S^{-1} .$$

The optimal algorithm is the multidimensional version of the linear one [(3.2.55)], that is,

$$x_{i+1} = x_i - \frac{1}{i} \nabla^2 g(x^*)^{-1} Z_i , \tag{3.2.59}$$

where $Z_i = \nabla g(x_i + \Xi_i)$. The minimal covariance matrix is

$$\Sigma(\varphi^0, H) = \nabla^2 g(x^*)^{-1} S \, \nabla^2 g(x^*)^{-1} .$$

Therefore, if the noise is normal and conditions (i)–(x) are satisfied, the SAA [see also (3.2.42)]

$$x_{i+1} = x_i - \Gamma_i Z_i , \qquad Z_i = \nabla g(x_i) + \Xi_i ,$$

where $\Gamma_i = (1/i) \, \nabla^2 g(x^*)^{-1}$, is optimal.

The optimal algorithm (3.2.48) requires complete knowledge of the distribution of the noise Ξ, which is rarely the case in practice. Moreover, it can be readily shown [see Tsypkin (1979, 1982a)] that (3.2.48) is highly sensitive to small deviations from the assumptions of the c.d.f. H.

Let now a class R of distributions of Ξ, rather than an exact distribution of Ξ be given.

In analogy with (3.2.40) we say that (3.2.45) is asymptotically optimal in the minimax sense if there exist a distribution H^* $(H^* \in R)$ and a transformation φ^* such that

$$\Sigma(\varphi^*, H) \leq \Sigma(\varphi^*, H^*) \leq \Sigma(\varphi, H^*) \tag{3.2.60}$$

holds for any $H \in R$. Clearly, if (3.2.60) holds, then the transformation $\varphi = \varphi^*$ guarantees that the asymptotic covariance matrix $\Sigma(\varphi^*, H)$ does not exceed the minimal possible matrix

$$\Sigma(\varphi^*, H^*) = \min_{\varphi} \max_{H} \Sigma(\varphi, H)$$

on the class $H \in R$.

The following theorem is due to Polyak and Tsypkin (1980b).

Theorem 3 Let R be a class of distributions, and let h^* be such that

$$0 < \Phi(h^*) \leq \Phi(h) < \infty , \qquad \text{for all } h \in R , \tag{3.2.61}$$

where $\Phi(h)$ is Fisher's information matrix. Let also conditions (i)–(x) hold. Then, the algorithm (3.2.45) with $\varphi = \varphi^*$, where

$$\varphi^*(z) = \nabla^2 g(x^*)^{-1} \Phi(h^*)^{-1} \, \nabla \log h^*(z) \tag{3.2.62}$$

is optimal on the class R, and

$$\Sigma(\varphi^*, H^*) = \nabla^2 g(x^*)^{-1} \Phi(h^*)^{-1} \, \nabla^2 g(x^*)^{-1} , \tag{3.2.63}$$

where $\Sigma(\varphi^*, H^*)$ satisfies (3.2.60). It follows from Theorem 3 that in order to guarantee the minimax optimality of (3.2.45), we have first to find *the last favorable c.d.f.* H^* $(H^* \in R)$ satisfying (3.2.60) and then the optimal transformation $\varphi^*(z)$ according to (3.2.62). Choosing the transformation φ according to (3.2.62) makes the algorithm (3.2.45) *stable* or *robust*. [Recall that (3.2.60) is actually the minimax principle of robustness of Huber (1964)]. Thus, for a given class R of noises Ξ the optimal transformation φ^* leads to stabilization of the algorithm (3.2.45) and therefore ensures robustness of the estimate x_i of the optimal solution x^*. Clearly, in the particular case where the exact distribution of the noise Ξ is known (i.e., the class R consists of a single element), formulas (3.2.62) and (3.2.63) reduce to (3.2.46) and (3.2.49), respectively.

Reverting now to the algorithms (3.2.42) and (3.2.43), let $\nabla\psi(x, Y)$ be an unbiased estimate of $\nabla g(x)$ representable as

$$\nabla\psi(x, Y) = \nabla g(x) + \Xi(x, Y)\,, \tag{3.2.64}$$

where $\Xi(x, Y) \in R^n$ is the random noise depending generally on both x and Y.

Denote $\Xi^* = \nabla\psi(x^*, Y)$. Polyak and Tsypkin (1980b) showed that formulas (3.2.46)–(3.2.51) are valid for the algorithms (3.2.42) and (3.2.43) with $Z = \nabla\psi(x, Y)$ if the p.d.f. h of Ξ^* is *known in advance* and the conditions (i)–(x) are satisfied.

Therefore, where complete information on the noise $\Xi = \nabla\psi(x^*, Y)$ is available, we can use the optimal adaptive algorithm (3.2.48) instead of (3.2.42) or (3.2.43) and obtain a higher rate of asymptotic convergence.

Let now the distribution of the noise $\Xi = \nabla\psi(x, Y)$ be unknown as such but known to belong to a certain class R. Let also the conditions of Theorem 3 be satisfied. It follows then directly from Polyak and Tsypkin (1980b) that formulas (3.2.62) and (3.2.63) are valid for the algorithms (3.2.42) and (3.2.43). The same authors (1979, 1982a) considered different classes of the distribution of Ξ and found for each of them the least favorable distribution H^* and the optimal transformation φ^*. Several such classes† and the associated robust (stable) algorithms optimal in the minimax since (3.2.60) are considered below.

(i) *Class of finite variance distributions*

$$R_1 = \left\{h\colon \int_{-\infty}^{\infty} z^2 h(z)\, dz < \sigma^2\right\}.$$

†The classes of distribution R_1, R_2, and R_3 are taken from Table 1.

The least favorable density is the normal (3.2.54) and the optimal algorithm is the linear one (3.2.55).

(ii) *Class of nondegenerative distributions*

$$R_2 = \{h\colon h(z) \text{ is continuous at } 0,\ h(0) \geq \beta > 0\}\ .$$

The least favorable density is the Laplace (3.2.56) and the optimal algorithm is the sign one (3.2.57).

(iii) *Class of approximately uniform distributions*

$$R_3 = \{h\colon h(z) = (1-\varepsilon)f(z) + \varepsilon p(z)\}\ ,$$

where

$$f(z) = \begin{cases} \dfrac{1}{2d}\ , & |z| \leq d\ , \\ 0\ , & |z| > d\ , \end{cases}$$

and $p(z)$ is an arbitrary density. The optimal transformation is

$$\varphi^*(z) = \begin{cases} \dfrac{1-\varepsilon}{\varepsilon d} \operatorname{sign} z & \text{if } |z| > d\ , \\ 0 & \text{if } |z| \leq d\ , \end{cases}$$

and the optimal algorithm

$$x_{i+1} = \begin{cases} x_i - \dfrac{1-\varepsilon}{\varepsilon d} \operatorname{sign} Z_i\ , & |z_i| > d\ , \\ x_i\ , & |z_i| \leq d\ , \end{cases}$$

where $Z_i = g'(x_i) + \Xi_i$.

(iv) *Class of finite variance-covariance distributions*

$$R_4 = \left\{h\colon \int zz'h(z)\,dz \leq S\right\}\ ,$$

where $S > 0$ is a given matrix. The least favorable density is the normal and the optimal algorithm is the linear one (3.2.59). Clearly, R_3 is the multivariate analog of R_1.

Consider again the problem of minimization of

$$g(x) = E\{\psi(x, Y)\} = \int \psi(x, y)\, dF_Y(y)\,. \qquad (3.2.65)$$

The sample mean,

$$\hat{g}(x) = \frac{1}{N} \sum_{i=1}^{N} \psi(x, Y_i)\,, \qquad (3.2.66)$$

is an unbiased estimate of $g(x)$ at any x. Clearly, the unknown solution x^* can be approximated by minimizing $\hat{g}(x)$ analytically or numerically (e.g., by the Newton-Raphson method). We shall call the methods for minimizing $\hat{g}(x)$ nonrecursive in order to distinguish them from the recursive SAAs, say, (3.2.42) or (3.2.43). At first glance, it might seem that nonrecursive methods converge to x^* faster than the recursive ones, but such is not the case: Polyak and Tsypkin (1980a) showed that their asymptotic covariance matrices and the asymptotic rate of convergence are the same.

Consider now the problem of finding the root θ of $g(x) = E\{\psi(x, Y)\} = 0$, where $g(x)$ is a vector function. In this case algorithm (3.2.45) can be written as

$$x_{i+1} = x_i - \frac{1}{i}\, \varphi(Z_i)\,,$$

where $Z_i = g(x_i) + \Xi_i$.

Polyak and Tsypkin (1979, 1980a, b) showed that under conditions similar to (i)–(x), formulas analogous to (3.2.46)–(3.2.51), (3.2.62) and (3.2.63) hold. In particular, the minimal asymptotic covariance matrix [see also (3.2.63)]

$$\Sigma(\varphi^*, H^*) = \nabla g(\theta)^{-1} \Phi(h^*)^{-1}\, \nabla g(\theta)^{-1}\,.$$

It is worth noting that optimal transformations (3.2.46) and (3.2.62) require knowledge of the inverse of the matrix $\nabla^2 g(x^*)$, which is rarely known. Polyak and Tsypkin (1979, 1980a, b) proposed several techniques for approximation of $\nabla^2 g(x^*)$. For example, while applying (3.2.45) to minimize $g(x)$ in (3.2.65), they suggested approximating $\nabla^2 g(x^*)$ by $(1/i)[\Sigma_{j=1}^{i} \nabla^2 \psi(x_i, Y_j)]$.

3.3 APPLICATIONS (INVENTORY MODELS)†

3.3.1 Unconstrained Single-Commodity Models

(a) Single-Period One-Commodity Inventory Model Without Setup Cost

This model can be described as follows: Items are produced (purchased) for a single period at a cost of c dollars per item. There are a *holding* cost function $H(\cdot)$ associated with inventory left over at the end of a period and a *depletion* (*shortage*) cost function $P(\cdot)$ associated with unmet demand. The demand Y is a continuous random variable with c.d.f. $F_Y(y)$. Given an initial inventory r, the problem is to determine the additional inventory, say, $x - r$, to be ordered so as to minimize the total expected cost, that is,

$$\begin{aligned}\min_{x \geq r} g(x) &= \min_{x \geq r} c(x-r) + \int_0^x H(x-y) f_Y(y)\, dy \\ &\quad + \int_x^\infty P(y-x) f_Y(y)\, dy \\ &= \min_{x \geq r} [c(x-r) + G(x)] , \end{aligned} \tag{3.3.1}$$

where

$$G(x) = \int_0^x H(x-y) f_Y(y)\, dy + \int_x^\infty P(y-x) f_Y(y)\, dy . \tag{3.3.2}$$

The solution for the optimal inventory policy with $g(x)$ to be strictly convex‡ is [see Hillier and Lieberman (1974)]

$$\begin{cases} \text{if } r \leq x^* , & \text{order } x^* - r , \\ \text{if } r > x^* , & \text{do not order} , \end{cases} \tag{3.3.3}$$

where x^* is the value of x that satisfies

$$g'(x) = G'(x) + c = 0 . \tag{3.3.4}$$

In the particular case where

$$H(x-y) = h(x-y) , \qquad h > 0 , \tag{3.3.5}$$

†This section is based on Rubinstein and Kreimer (1985).

‡A sufficient condition for $g(x)$ to be strictly convex is that both the holding and shortage costs be strictly increasing convex functions and $f_Y(y) > 0$.

and

$$P(y-x)=p(y-x), \qquad p>0, \tag{3.3.6}$$

that is, the holding and shortage cost functions are linear, the optimal inventory policy x^* satisfies [see Hillier and Lieberman (1974)]

$$F_Y(x^*)=\frac{p-c}{p+h}, \qquad p>c \tag{3.3.7}$$

and therefore equals

$$x^*=F_Y^{-1}\left(\frac{p-c}{p+h}\right), \tag{3.3.8}$$

where F_Y^{-1} denotes the inverse of F_Y.

The problem of finding the root x^* of (3.3.4) generally cannot be solved analytically, and we have to resort to numerical procedures such as the Newton-Raphson method, represented by the formula

$$x_{i+1}=x_i-\frac{g'(x_i)}{g''(x_i)}=x_i-\frac{G'(x_i)+c}{G''(x_i)}, \tag{3.3.9}$$

which converges to x^* under certain condition on g, g', and g'' [see Avriel (1976)].

If $g(x)$ is convex but not differentiable, then the following subgradient (see (3.2.9)) algorithm can be used:

$$x_{i+1}=x_i-\alpha_i\hat{\nabla}g(x_i), \qquad \alpha_i>0,\ \alpha_i\to 0, \tag{3.3.10}$$

where $\hat{\nabla}g(x_i)=\hat{\nabla}G(x_i)+c$.

Consider now the same model assuming the c.d.f. $F_Y(y)$ of the demand is unknown but a sample $Y_1,\ldots,Y_N$ from it is available. In this case we shall distinguish the following two inventory models: the *off-line* and *on-line* models. For the off-line model we assume that the sample $Y_1,\ldots,Y_N$ is *known in advance* and the optimal ordering policy x^* has to be found (estimated) on the basis of this sample. By contrast, for the on-line model we assume that the sample $Y_1,\ldots,Y_N$ is *unknown in advance*, but each realization Y_i, $i=1,\ldots,N$, appears (follows) after the corresponding order x_i has been placed. In order words, at each instant of time i we observe the pair $\{x_i, Y_i\}$, $i=1,\ldots,N$. If, additionally, storage of the pairs $\{x_i, Y_i\}$ is available, then x_{i+1} can be updated on the basis of all previous data $\{x_j, Y_j\}$, $j=1,\ldots,i$.

As an example of the on-line model consider the well-known newsboy problem in which a newsboy orders a certain amount of newspapers each day. Assume that the demand is unknown. Clearly, at day $i+1$ the newsboy can make his ordering decision x_{i+1} on the basis of all the previous data $\{x_j, Y_j\}$, $j = 1, \ldots, i$.

Before proceeding with the off-line and on-line models, let us note that $G(x)$ [see (3.3.2)] can be written as

$$G(x) = E[L(x, Y)], \tag{3.3.11}$$

where

$$L(x, y) = \begin{cases} H(x-y) & \text{if } y < x, \\ P(y-x) & \text{if } y \geq x, \end{cases} \tag{3.3.12}$$

and equation (3.3.4) can be written as

$$g'(x) = c + \int_0^x H'(x-y) f_Y(y)\, dy - \int_x^\infty P'(y-x) f_Y(y)\, dy$$
$$+ H(0) f_Y(x) - P(0) f_Y(x) = 0, \tag{3.3.13}$$

provided the operators integration and differentiation are interchangeable. We shall assume further that $H(0) = P(0) = 0$, that is, there is no holding or depletion cost if there is no inventory in the warehouse. Under this assumption, (3.3.13) can be rewritten as

$$g'(x) = E[\tilde{\nabla} L(x, Y) + c] = 0, \tag{3.3.14}$$

where

$$\tilde{\nabla} L(x, y) = \begin{cases} H'(x-y) & \text{if } y < x, \\ -P'(y-x) & \text{if } y \geq x. \end{cases} \tag{3.3.15}$$

The second derivative of $g(x)$ is

$$g''(x) = \int_0^x H''(x-y) f_Y(y)\, dy + \int_x^\infty P''(y-x) f_Y(y)\, dy$$
$$+ [H'(0) + P'(0)] f_Y(x). \tag{3.3.16}$$

Formula (3.3.16) can be also rewritten as

$$g''(x) = E[\tilde{\nabla}^2 L(x, Y)] ,$$

where

$$\tilde{\nabla}^2 L(x, y) = [H'(0) + P'(0)] f_Y(x) + \begin{cases} H''(x-y) & \text{if } y < x , \\ P''(y-x) & \text{if } y \geq x . \end{cases} \tag{3.3.17}$$

In the particular case where $H'(0) = P'(0)$ [e.g., $H(x) = hx^2$, $P(x) = px^2$], we obtain

$$g''(x) = \int_0^x H''(x-y) f_Y(y)\, dy + \int_x^\infty P''(y-x) f_Y(y)\, dy = E[\tilde{\nabla}^2 L(x, Y)] ,$$

where

$$\tilde{\nabla}^2 L(x, Y) = \begin{cases} H''(x-y) & \text{if } y < x , \\ P''(y-x) & \text{if } y \geq x . \end{cases}$$

1. Off-Line Model With a sample $Y_1, \ldots, Y_N$ at our disposal, the unknown function $g(x)$ can be estimated as

$$\hat{g}_N(x) = \frac{1}{N} \sum_{i=1}^{N} \psi(x, Y_i) = c(x-r) + \frac{1}{N} \sum_{i=1}^{N} L(x, Y_i) .$$

We shall call $\hat{g}_N(x)$ the *empirical loss function.*

Clearly, we would like to find an approximate solution, say, $\hat{x}_N$, of x^* by solving

$$\min_{x \geq r} \left[\hat{g}_N(x) = \frac{1}{N} \sum_{i=1}^{N} \psi(x, Y_i) \right]$$

analytically or numerically.

Unfortunately, the conventional optimization methods [see Avriel (1976)] may fail to find the minimum of $\hat{g}_N(x)$, since $L(x, Y_i)$ is not continuously differentiable at points $x = Y_i$, $i = 1, \ldots, N$. To overcome this difficulty, we shall adopt Zang's (1980) technique, according to which the original function is smoothed out in some neighborhoods of the kink (discontinuity) points. It is shown there that

- **(a)** these neighborhoods may be of arbitrarily small size, thereby leaving the original objective function unchanged at the rest of R^n;
- **(b)** the maximal possible difference between the optimal values of the approximate problem and the original one is determined a priori by fixing the value of a single parameter; and

(c) the approximations introduced preserve properties such as convexity and continuous differentiability, provided that each function composing the original problem has the same properties.

To demonstrate this technique, consider smoothing out the absolute value operator in $g(r) = |r|$ by

$$g_1(\beta, r) = \begin{cases} -r & \text{if } r \le -\beta , \\ \dfrac{r^2 + \beta^2}{2\beta} & \text{if } -\beta \le r \le \beta , \\ r & \text{if } r \ge \beta , \end{cases}$$

or by

$$g_2(\beta, r) = \begin{cases} -r & \text{if } r \le -\beta , \\ \dfrac{-r^4}{8\beta^3} + \dfrac{3r^2}{4\beta} + \dfrac{3\beta}{8} & \text{if } -\beta \le r \le \beta , \\ r & \text{if r} \ge \beta . \end{cases}$$

It is readily seen that $g_1(\beta, r)$ and $g_2(\beta, r)$ are continuously differentiable once and twice, respectively; they replace the original absolute value only for $-\beta \le r \le \beta$; and as $\beta \to 0$, both $g_1(\beta, r)$ and $g_2(\beta, r)$ tend to $g(r)$ as a limit.

Applying $g_1(\beta, r)$ and $g_2(\beta, r)$ to $L(x, y)$, we obtain the following approximations:

$$A_1(\beta, L(x, Y)) = \begin{cases} L(x, y) & \text{if } L(x, y) \ge \beta , \\ \dfrac{L^2(x, y) + \beta^2}{2\beta} & \text{if } L(x, y) \le \beta , \end{cases}$$

and

$$A_2(\beta, L(x, y)) = \begin{cases} L(x, y) & \text{if } L(x, y) \ge \beta , \\ -\dfrac{L^4(x, y)}{8\beta^3} + \dfrac{3L^2(x, y)}{4\beta} + \dfrac{3\beta}{8} & \text{if } L(x, y) \le \beta . \end{cases}$$

It is readily seen that $A_1(\cdot)$ and $A_2(\cdot)$ are continuously differentiable once and twice, respectively, if $H(\cdot)$ and $P(\cdot)$ are so and $H(0) = 0$ and $P(0) = 0$. We also have

$$\lim_{\beta \to 0} A_j(\beta, L) = L , \qquad j = 1, 2 .$$

We consider further applications of $A_2(\cdot)$ only. The approximation of $\hat{g}_N(x)$ with the aid of $A_2(\cdot)$ can now be written as

$$\hat{G}_N(\beta, x) = c(x - r) + \frac{1}{N}\sum_{i=1}^{N} A_2(\beta, L(x, Y_i)) .$$

Clearly, $\hat{G}_N(\beta, x)$ is convex and twice continuously differentiable since $H(\cdot)$ and $P(\cdot)$ are so.

Unlike the original function $\hat{g}_N(x)$, the smooth approximation $\hat{G}_N(\beta, x)$ can be minimized using efficient second-order techniques such as the Newton-Raphson method, which can be written as

$$x_{i+1} = x_i - \frac{\hat{G}_N'(\beta, x_i)}{\hat{G}_N''(\beta, x_i)} . \tag{3.3.18}$$

In the particular case where $H(x - y) = h(x - y)$ and $P(y - x) = p(y - x)$, we have [see (3.3.8)]

$$\hat{x}_N = \min \hat{F}_N^{-1}\left(\frac{p - c}{p + h}\right), \qquad p > c ,$$

where $\hat{F}_N$ is the empirical c.d.f. of the r.v. Y. If $g(x)$ is convex but not differentiable, then subgradient algorithm (3.3.10) with SG $\hat{\nabla} g(x)$ replaced by SQG $\tilde{\nabla} g_N(x) = c + \Sigma_{j=1}^{N} \tilde{\nabla} L(x, Y_j)$, where

$$\tilde{\nabla} L(x, y) = \begin{cases} \hat{\nabla} H(x - y) & \text{if } y < x , \\ -\hat{\nabla} P(y - x) & \text{if } y \geq x , \end{cases} \tag{3.3.19}$$

can be used. The convergence of this algorithm follows directly from Gupal's Theorem 2.

2. *On-Line Model* We consider two modifications of this model.

(A) *A model without memory.* For this model storage of the data $\{x_j, Y_j\}$, $j = 1, 2, \ldots$, is *not available* and therefore x_{i+1} can be updated *only* by making use of the current pair $\{x_i, Y_i\}$.

(B) *A model with memory.* For this model storage of the data is *available*, that is, x_{i+1} can be updated on the basis of all the previous information $\{x_j, Y_j\}$, $j = 1, 2, \ldots, i$.

(*A*) First let $g(x)$ be strictly convex and differentiable. In this case we can apply the following Robbins–Monro procedure:

$$x_{i+1} = x_i - a_i[\tilde{\nabla} L(x_i, Y_i) + c] . \tag{3.3.20}$$

Proposition 1 Let

(i) H and P be convex, strictly increasing, and continuously differentiable functions such that $H(0) = P(0) = 0$ and $f_Y(y) > 0$;

(ii) there exists a positive number d such that, for all x,

$$E\{[\tilde{\nabla}L(x, Y)]^2\} \leq d(1 + x^2),$$

where $\tilde{\nabla}L$ is as per (3.3.15)—then procedure (3.3.20) converges almost surely and in the mean square to x^* if

$$\sum_{i=1}^{\infty} a_i = \infty, \qquad \sum_{i=1}^{\infty} a_i^2 < \infty.$$

Proof It suffices to show that conditions (i) and (ii) are equivalent, respectively, to the Gladyshev conditions (3.2.4) and (3.2.5), which in our case can be written as

$$\inf g'(x)(x - x^*) > 0 \tag{3.2.4)'$$

for each $1 > \varepsilon > 0$, $\varepsilon < |x - x^*| < \varepsilon^{-1}$. There exists a positive number d such that, for all x,

$$E\{[\tilde{\nabla}L(x, Y)]^2\} \leq d(1 + x^2). \tag{3.2.5)'$$

Now, condition (3.2.4)′ holds if $g(x)$ is strictly convex (i.e., unimodal), which follows directly from (i); condition (3.2.5)′ is literally identical with condition (ii). Q.E.D.

Note 5 It can be readily shown that condition (ii) will be satisfied for $\tilde{\nabla}L$ if both H' and P' satisfy (ii).

Assume now that H and P are strictly convex but not differentiable. Assume also that $Y \leq M < \infty$, that is, the demand is bounded. In this case the SQG algorithm (3.2.11) of Section 2.2.

$$x_{i+1} = \pi_X\{x_i - a_i[\tilde{\nabla}L(x_i, Y_i) + c]\}, \tag{3.3.21}$$

where $X = [r, M]$ and $\tilde{\nabla}L(x, Y)$ is as per (3.3.19) can be applied.

Convergence of x_i to x^* follows from Gupal's Theorem 2 if the following condition is satisfied:

$$\int_0^x [\hat{\nabla}H(x - y)]^2 f_Y(y)\, dy + \int_x^M [\hat{\nabla}P(y - x)]^2 f_Y(y)\, dy \leq d. \tag{3.3.22}$$

Note 6 Equation (3.3.22) obviously holds for convex H and P.

Let both H and P be strictly increasing convex with bounded second derivatives with respect to x, which cannot be found explicitly, and let the demand be bounded again, that is, $Y \le M < \infty$. In this case $g(x)$ also has a bounded second derivative,

$$|g''(x)| = \left| \int_0^x H''(x-y) f_Y(y)\, dy + \int_x^M P''(y-x) f_Y(y)\, dy + [H'(0) + P'(0)] f_Y(x) \right| < \infty , \tag{3.3.23}$$

Here, again the procedure (3.2.11) with SQG in the form of finite differences [see (3.2.20)] can be used and is

$$x_{i+1} = \begin{cases} \pi_X\left[x_i - a_i\left(c + \dfrac{H(x_i + \beta_i - Y_i') - H(x_i - Y_i'')}{\beta_i}\right)\right], & \text{if } Y_i' < x_i + \beta_i,\ Y_i'' < x_i , \\ \pi_X\left[x_i - a_i\left(c + \dfrac{H(x_i + \beta_i - Y_i') - P(Y_i'' - x_i)}{\beta_i}\right)\right], & \text{if } Y_i' < x_i + \beta_i,\ Y_i'' \ge x_i , \\ \pi_X\left[x_i - a_i\left(c + \dfrac{P(Y_i' - x_i - \beta_i) - H(x_i - Y_i'')}{\beta_i}\right)\right], & \text{if } Y_i' \ge x_i + \beta_i,\ Y_i'' < x_i , \\ \pi X\left[x_i - a_i\left(c + \dfrac{P(Y_i' - x_i - \beta_i) - P(Y_i'' - x_i)}{\beta_i}\right)\right], & \text{if } Y_i' \ge x_i + \beta_i,\ Y_i'' \ge x_i . \end{cases} \tag{3.3.24}$$

Here $X = [0, M]$ and Y_i', Y_i'' are samples from the p.d.f. $f_Y(y)$. It readily follows now that in order for x_i to converge to x^*, the following conditions have to be satisfied:

(a) condition (ii) of Proposition 1;

(b) $a_i > 0 , \quad \beta_i > 0 , \quad \sum_{i=0}^{\infty} a_i = \infty , \quad \sum_{i=0}^{\infty} a_i^2 < \infty , \quad \sum_{i=0}^{\infty} a_i \beta_i < \infty ;$

(3.3.25)

(B) for this modification of the on-line model (example newsboy problem) we can use algorithm (3.3.18) with N replaced by i, namely,

$$x_{i+1} = x_i - \frac{\hat{G}_i'(\beta, x_i)}{\hat{G}_i''(\beta, x_i)} . \tag{3.3.26}$$

Obviously the convergence conditions of (3.3.18) and (3.3.26) are the same. Simulation results verifying convergence of (3.3.18) and (3.3.26) are given in Section 3.7.

So far we assumed that the distribution of the demand $F_Y(y)$ is independent of the ordering policy. There are many situations, however, where such dependence exists, for example, situations where sales may be affected by the size of the stock, as in the case of consumer goods (say shoes) offered in a variety of sizes, colors, and so on, or food products displayed in quantities on the shelves at supermarkets or food stores.

To be more specific, assume that the p.d.f. f_Y of the demand depends on the ordering policy x. We denote this dependence by $f_Y(y, x)$. Note also that for different x the p.d.f.'s f can have different forms. In order to find the optimal policy x^* in this case we have to solve again problem (3.3.1) with $G(x)$ replaced by

$$\tilde{G}(x) = \int_0^x H(x-y) f_Y(y, x)\, dx + \int_x^\infty P(y-x) f_Y(y, x)\, dx = E[L(x, Y)] ,$$

where $L(x, Y)$ is as per (3.3.12).

If $g(x)$ is strictly convex† and $H(0) = P(0) = 0$, the optimal policy x^* is obtainable from the solution of [see also (3.3.14)]

$$g'(x) = \tilde{G}'(x) + c = 0 , \tag{3.3.27}$$

where

$$\tilde{G}'(x) = \int_0^x [H'(x-y) f_Y(y, x) + H(x-y) f_Y'(y, x)]\, dy$$
$$- \int_x^\infty [P'(y-x) f_Y(y, x) - P(y-x) f_Y'(y, x)]\, dy = E[\tilde{\nabla}\tilde{L}(x, Y)] ,$$

†Note that when $f_Y(y, x) \neq f_Y(y)$, the function $g(x)$ is generally not convex even when both H and P are convex. In this case methods of multiextremal optimization [see chapter 5] must be used.

and

$$\tilde{\nabla}\tilde{L}(x, y) = \begin{cases} H'(x - y) + H(x - y)\dfrac{f'_Y(y, x)}{f_Y(y, x)} & \text{if } y < x , \\ -P'(y - x) + P(y - x)\dfrac{f'_Y(y, x)}{f_Y(y, x)} & \text{if } y \geq x . \end{cases}$$

provided the operators integration and differentiation are interchangeable.

If the p.d.f. $f_Y(y, x)$ is *known*, then x^* can be found by solving (3.3.27) analytically or numerically [e.g., by the Newton-Raphson procedure (3.3.18)]. If the p.d.f. $f_Y(y, x)$ is *unknown*, then neither the Newton-Raphson method nor the Robbins-Monro SAA (3.3.20) with $\tilde{\nabla}L$ replaced by $\tilde{\nabla}\tilde{L}$ is applicable, since $\tilde{\nabla}\tilde{L}$ requires knowledge of both $f_Y(y, x)$ and $f'_Y(y, x)$. In this case, however, the SQG algorithm (3.3.24) may be used for both off-line and on-line models.†

(b) Model with Fixed Ordering (Setup) Cost This model differs from the model (a) in that a fixed (setup) cost K is charged for placing an order. In this case the expected cost for stocking a quantity $x \geq r$ is

$$g(x) = \begin{cases} K + c(x - r) + G(x) & \text{if } x > r , \\ G(r) & \text{if } x = r , \end{cases}$$

where

$$G(x) = \int_0^x H(x - y) f_Y(y)\, dy + \int_x^\infty P(y - x) f_Y(y)\, dy .$$

It is shown in Hillier and Lieberman (1974) that the minimum of $g(x)$ is achieved if we rule according to the so-called (S, s) policy, which can be written as

$$\begin{cases} \text{order } S - r & \text{if } s > r , \\ \text{do not order} & \text{if } s \leq r , \end{cases} \tag{3.3.28}$$

where S is the solution of the problem

$$\min_{x \geq r} [K + c(x - r) + G(x)] \tag{3.3.29}$$

†Note that now in (3.3.24) Y'_i and Y''_i are samples from $f_Y(y, x_i + \beta_i)$ and $f_Y(y, x_i)$, respectively.

and s is the smallest value that satisfies

$$cs + G(s) = K + cS + G(S) . \tag{3.3.30}$$

Clearly, the solution of (3.3.29) is identical to that of (3.3.4), that is, $S = x^*$.

Let us show how to estimate S and s from (3.3.29) and (3.3.30), respectively, when the p.d.f. $f_Y(y)$ is *unknown* but a sample $Y_1, \ldots, Y_N$ from it is available. We again consider the off-line and on-line models separately.

1. Off-Line Model Since S coincides with x^*, we can estimate S by algorithms proposed for estimating x^*. To find s, we need to solve (3.3.30) with respect to s. Denote

$$K + cS + G(S) = \alpha . \tag{3.3.31}$$

Actually, α is unknown since the p.d.f. $f_Y(y)$ and $G(S)$ are. Both $G(S)$ and α can, however, be estimated via their sample means

$$\bar{G}(S) = \frac{1}{N} \sum_{i=1}^{N} L(S, Y_i) \tag{3.3.32}$$

and

$$\bar{\alpha} = K + cS + \bar{G}(S) , \tag{3.3.33}$$

respectively.

Now s can be estimated by solving $cs + \bar{G}(s) = \bar{\alpha}$ [where $\bar{G}(s)$ is as in (3.3.32) with S replaced by s] analytically or numerically. For example, the following Newton-Raphson type algorithm can be used†:

$$s_{i+1} = \min\left(\bar{S}_N, s_i - \frac{Q(s_i)}{Q'(s_i)} \right) , \tag{3.3.34}$$

where

$$Q(s_i) = (cS_i - \bar{\alpha}) + \frac{1}{N} \sum_{j=1}^{N} A_2(\beta, L(s_i, Y_j)) ,$$

†In (3.3.34) the operator min is essentially a projection operator onto $X = [0, \bar{S}_N]$.

$\bar{S}_N$ is the estimate of S, and $A_2(\beta, L(s, Y))$ is Zang's (1980) (see also Zang and Tishler (1982)) smooth approximation of $L(s, Y)$.

2. On-Line Model We consider for simplicity the model without memory: In this case both quantities S_i, s_i must be updated simultaneously just after the demand Y_i is observed. That is, for the on-line model we have to estimate S and s according to the scheme

$$Y_1 \to (S_1, s_1)\,, \qquad Y_2 \to (S_2, s_2), \ldots, Y_N \to (S_N, s_N)\,. \tag{3.3.35}$$

Before presenting the algorithm for this model, let us rewrite (3.3.30) as

$$K + cS + E[L(S, Y)] = \alpha\,, \tag{3.3.36}$$

where

$$\alpha = cs + E[L(s, Y)]\,. \tag{3.3.37}$$

If S were known, we might estimate α by applying the following Robbins-Monro procedure to (3.3.36):

$$\alpha_{i+1} = \alpha_i - a_i'[K + cS + L(S, Y_i) - \alpha_i]\,. \tag{3.3.38}$$

In analogy, if both S and α were known, we might estimate s by applying the following Robbins-Monro procedure† to (3.3.37):

$$s_{i+1} = \min\{S, s_i - a_i''[cs_i + L(s_i, Y_i) - \alpha]\}\,. \tag{3.3.39}$$

Clearly, both α_i and s_i converge to α and s, respectively, if the conditions of Proposition 1 are satisfied.

Since neither S nor α are known, the algorithms (3.3.38) and (3.3.39) are unrealizable. There is, however, another approach for updating S_i, α_i, and s_i, namely, the triplet S_i, α_i, s_i, $i = 1, 2, \ldots, N$, can be updated simultaneously at each iteration in the order

$$S_1 \to \alpha_1 \to s_1\,, \qquad S_2 \to \alpha_2 \to s_2, \ldots, \qquad S_N \to \alpha_N \to s_N\,.$$

Bearing in mind (3.3.20), (3.3.38), and (3.3.39), we can present the

†If $\int_0^x H^2(x-y)f_Y(y)\,dy + \int_x P^2(y-x)f_Y(y)\,dy < \infty$ and $\int_0^x H(x-y)f_Y(y)\,dy + \int_x^\infty P(y-x)f_Y(y)\,dy < \infty$ for $x \in [0, S]$, then $E[cs + L(s, Y)]^2 < \infty$ since $X = [0, S]$ is bounded, and we have $s_i \to s$ a.s.

simultaneous Robbins-Monro procedure as

$$S_{i+1} = S_i - a_i[\tilde{\nabla}L(S_i, Y_i) + c] , \qquad (3.3.40a)$$

$$\alpha_{i+1} = \alpha_i - a_i'[K + cS_{i+1} + L(S_{i+1}, Y_i) - \alpha_i] , \qquad (3.3.40b)$$

$$s_{i+1} = \min\{S_{i+1}, s_i - a_i''(cs_i + L(s_i, Y_i) - \alpha_{i+1}\} . \qquad (3.3.40c)$$

Note that (3.3.40a) is identical with (3.3.20) since $S = x^*$; (3.3.40b) and (3.3.40c) differ from (3.3.38) and (3.3.39) in that α and S are replaced, respectively, by α_{i+1} and S_{i+1}. It is readily seen that the convergence conditions of (3.3.40b) and (3.3.40c) follows from those of (3.3.38) and (3.3.39), respectively.

Consider now the case where the demand depends on the ordering policy, that is, we have $f_Y(y, x)$ instead of $f_Y(y)$. As was pointed out earlier, the parameter x^* (and therefore S) cannot be estimated with the aid of (3.3.20) [or (3.3.40a)] since $\tilde{\nabla}\tilde{L}(x, y)$ [see (3.3.27)] requires knowledge of $f_Y(y, x)$ and $f_Y'(y, x)$. In this case, however, we can estimate S by using, say, the SQG algorithm (3.3.24). The parameters α and s can be still estimated with the aid of (3.3.40b) and (3.3.40c), respectively, where L is as per (3.3.12).

Note 7 For the above models we have assumed that the p.d.f. of the demand $f_Y(y)$ [or $f_Y(y, x)$] is unknown, besides the sample $Y_1, \ldots, Y_N$. It would be interesting to find robust (in Huber's minimax sense; see Section 3.2.5) adaptive algorithms assuming that a sample from $f_Y(\cdot)$ is available and $f_Y(\cdot)$ belongs to a certain class of distributions.

From here on we consider on-line models without memory only.

3.3.2 Constrained Single-Commodity Models

(a) Single-Period One-Commodity Inventory Models with No Setup Cost

We consider here two modified versions of the standard single-period one-commodity model, including some constraints.

In Proposition A.1 of the Appendix 2 we show that if $G(x)$ [see (3.3.2)] is strictly convex and the constraints define a closed set, then the optimal inventory policy is given by the rule defined in (A2.9), which extends the standard inventory policy (3.3.3) and is similar to it.

Assume that the warehouse capacity equals V and that one unit of an item needs v square feet of floor space. This constraint can be written as

$$vx \le V .$$

Assume, additionally, that the inventory at hand x has to be greater than or equal to the mean value of the demand Y, that is,

$$x \ge E[Y] . \tag{3.3.41}$$

Mathematically this problem can be written as†

$$\min_{x \ge r} \{g_0(x) = E[\psi_0(x, Y)]\} = \min_{x \ge r} \{c(x - r) + E[L(x, Y)]\} , \tag{3.3.42}$$

subject to

$$g_1(x) = E[\psi_1(x, Y)] = E[Y - x] \le 0 ,$$

$$x \in X = \left[0, \frac{V}{v}\right] , \tag{3.3.43}$$

where $L(x, Y)$ is given in (3.3.12)

Consider several particular cases. Assume first that $H(t)$ and $P(t)$ are differentiable functions. The stochastic quasi-gradient of g_0 and g_1 are then, respectively,

$$\tilde{\nabla} g_0(x) = c + \hat{\nabla} L(x, y) \tag{3.3.44}$$

$$\tilde{\nabla} g_1(x) = -1 , \tag{3.3.45}$$

where

$$\tilde{\nabla} L(x, y) = \begin{cases} H'(x - y) & \text{if } y < x , \\ -P'(y - x) & \text{if } y \ge x . \end{cases} \tag{3.3.46}$$

In this case Gupal's recursive procedure (3.2.11)–(3.2.15) can be rewritten as

$$x_{i+1} = \min\left[\frac{V}{v}, x_i - a_i \tilde{\nabla} g(x_i)\right] , \tag{3.3.47}$$

where

$$\tilde{\nabla} g(x_i) = \begin{cases} \tilde{\nabla} g_0(x_i) & \text{if } z_i \le 0 , \\ \tilde{\nabla} g_1(x_i) & \text{if } z_i > 0 . \end{cases} \tag{3.3.48}$$

†Existence of (s, S) policy follows directly from Appendix 2.

and

$$z_{i+1} = z_i + \rho_i(y_i - x_i - z_i) . \tag{3.3.49}$$

Assume, additionally, that $E[|\tilde{\nabla} g_0(x)|^2] < \infty$ or, equivalently, that $E[|\tilde{\nabla} L(x, Y)|^2] < \infty$, $E(Y) < \infty$ and $E(Y^2) < \infty$ [see conditions (vi) and (vii) of Theorem 2]. The convergence of (3.3.47)–(3.3.49) to the optimal solution follows directly from Theorem 2.

Assume now that $H(t)$ and $P(t)$ are not differentiable. Then we can replace the derivatives H' and P' in (3.3.46) by their subgradients $\hat{\nabla} H$ and $\hat{\nabla} P$, respectively, and convergence of (3.3.47)–(3.3.49) follows again from Theorem 2.

Finally, assume that H' and P' cannot be found explicitly [for example, $H(t)$ and $P(t)$ are unknown but their values can be observed at each point t] and their second derivatives are bounded. Then the finite difference estimator for H' and P' [see (3.2.20) or (3.2.21)] can be used. Convergence again follows directly from Theorem 2.

(b) Single-Period One-Commodity Inventory Models with Fixed Ordering (Setup) Cost Consider again model (a), assuming additionally that a fixed (setup) cost K is charged for placing an order. In Proposition A2 we prove that under assumption of Proposition 1 the optimal policy for both models (a) and (b) is the (S^*, s) policy [see (A2.20)], where S^* is the optimal solution of the corresponding model without setup cost ($K = 0$) and s is the root of the equation (A2.16). The SAAs for finding S^* and s are (3.3.47) and (3.3.40c), respectively.

3.3.3 Multicommodity Models

Assume that the warehouse capacity equals V and that one unit of the item of kth kind, $k = 1, \ldots, n$, needs v^k square feet of floor space. This constraint can be written as

$$\sum_{k=1}^{n} v^k x^k \leq V .$$

Assume also that $P(y \leq x) \geq a$, which is the same as†

$$E[\alpha - \chi_{[y \leq x]}(Y)] \leq 0 ,$$

where

†The constraint reads "the probability that the inventory level x exceeds the demand Y is more than or equal to α" (for example, $\alpha = 0.95$).

$$\chi_{[Y \le x]}(y) = \begin{cases} 1 & \text{if } y \le x, \\ 0 & \text{otherwise}. \end{cases}$$

Mathematically this version reads as

$$\min_{x \ge r} g_0(x) = \min_{x \ge r} E[\psi_0(x, Y)] = \min_{x \ge r} \left(\sum_{k=1}^{n} c^k(x^k - r^k) + E[L(x, Y)] \right) \tag{3.3.50}$$

subject to

$$g_1(x) = E[\psi_1(x, Y)] = E[\alpha - \chi_{[y \le x]}(Y)] \le 0, \tag{3.3.51}$$

$$x \in X = \left\{ x: \sum_{k=1}^{n} v^k x^k \le V \right\}. \tag{3.3.52}$$

In order to solve (3.3.50)–(3.3.52), we can again apply Gupal's procedure (3.2.11)–(3.2.15).

Consider two particular cases:

1. Assume that $g_1(x)$ is convex and has bounded second derivatives [that is the case when $F_Y(y)$ is concave and has bounded second derivatives]. The stochastic quasi-gradient $\tilde{\nabla} g_1(x)$ can then be found by using finite differences, that is,

$$\tilde{\nabla} g_1(x_i) = \sum_{s=1}^{n} \frac{\chi(x_i^s, Y_i^{(0)}) - \chi(x_i^s + \beta_i e^s Y^{(s)})}{\beta_i} e^s. \tag{3.3.53}$$

In the case Gupal's recursive procedure is

$$x_{i+1} = \pi_X[x_i - a_i \tilde{\nabla} g(x_i)], \tag{3.3.54}$$

where†

† To find a projection of any point $w \in R^n$ onto the hyperplane (3.3.52), it suffices to find a point $x \in X$ satisfying

$$\frac{w^k - x^k}{v^k} = \frac{w^m - x^m}{v^m}, \qquad m, k = 1, \ldots, n.$$

Substituting the last equalities in (3.3.52), we obtain

$$x^m = w^m + v^m \frac{V - \sum_{k=1}^{n} v^k w^k}{\sum_{k=1}^{n} (v^k)^2}, \qquad k, m = 1, \ldots, n.$$

$$\tilde{\nabla}g(x_i) = \begin{cases} c + \nabla_x L(x_i, Y_i) & \text{if } z_i \le 0 , \\ \tilde{\nabla}g_1(x_i) & \text{if } z_i > 0 , \end{cases} \tag{3.3.55}$$

and

$$z_{i+1} = z_i + \rho_i[\alpha - \chi_{[y \le x_i]}(y_i^{(0)}) - z_i] . \tag{3.3.56}$$

Reasoning as in Section 3.3.2, we conclude that procedure (3.3.54)–(3.3.56) converges to the optimal solution if

$$E[\|\nabla L(x, Y)\|^2] < \infty .$$

If $L(x, y)$ is not differentiable, we replace the gradient $\nabla L(\cdot)$ by the subgradient $\hat{\nabla} L(\cdot)$ with all other data in (3.3.54)–(3.3.56) unchanged. Convergence then follows again from Theorem 2.

2. Consider the case where $g_1(x)$ is separable, that is $g_1(x) = \Sigma_{k=1}^n g_1^k$, where

$$g_1^k(x) = E[\alpha_k - \chi_{[y^k \le x^k]}(Y^k)] \le 0 , \qquad k = 1, \ldots, n , \tag{3.3.57}$$

and the demands $Y^1, \ldots, Y^n$ are independent. We have then

$$\tilde{\nabla}g(x_i) = \begin{cases} c + \hat{\nabla}_x L(x_i, Y_i) & \text{if } \max\limits_{1 \le j \le n} z_i^j \le 0 , \\ \tilde{\nabla}g_k(x_i) & \text{if } z_i^k = \max\limits_{1 \le j \le n} z_i^j > 0 , \end{cases} \tag{3.3.58}$$

where

$$z_{i+1}^j = z_i^j + \rho_i[\alpha - \chi_{[y^j \le x_i^j]}(y_i^j) - z_i^j] , \qquad j = 1, \ldots, n , \tag{3.3.59}$$

$$\tilde{\nabla}g_k(x_i) = e^k \begin{cases} 0, & \text{if } Y_i^{(0)} \le x_i^k, \quad Y_i^{(k)} \le x_i^k + \beta_i , \\ \dfrac{1}{\beta_i}, & \text{if } Y_i^{(0)} \le x_i^k, \quad Y_i^{(k)} > x_i^k + \beta_i , \\ -\dfrac{1}{\beta_i}, & \text{if } Y_i^{(0)} > x_i^k, \quad Y_i^{(k)} \le x_i^k + \beta_i , \\ 0, & \text{if } Y_i^{(0)} > x_i^k, \quad Y_i^{(k)} > x_i^k + \beta_i . \end{cases}$$

If $\max_{1 \le j \le n} z_i^j > 0$, then $\tilde{\nabla}g(x_i)$ in (3.3.58) can be written as

$$\tilde{\nabla}g(x_i) = \sum_{j \in J} \tilde{\nabla}g_j(x_i), \quad \text{where } J = \{j: z_i^j > 0\} . \tag{3.3.60}$$

Finally, if $L(x, y)$ is separable, that is, $L(x, y) = \Sigma_{k=1}^{n} L_k(x^k, y^k)$, then each item can be treated separately, since in this case our n-dimensional model splits into n separate one-dimensional models.

3.4 OTHER APPLICATIONS

In this section we apply stochastic approximation to product mix and to allocation models.

3.4.1 Product Mix Model

Description of this model is taken almost verbatim from Johnson and Montgomery (1974). Assume that a company produces and sells a number of products over a certain period of time. Let

$x = (x^1, \ldots, x^n)$ be the vector of products produced in the period;
b^q be the amount of resource q, $q = 1, \ldots, Q$, available during the period;
a^{kq} be the number of units of resource q required to produce one unit of product k;
r^k be the unit selling price of item k if sold during the sale ($r^k > c^k$);
$\tilde{r}^k$ be the unit disposal value of item k if not sold during the sale ($\tilde{r}^k < r^k$);
c^k be the unit variable cost of producing a unit of product k;
y^k be the demand for item k;
$y = (y^1, \ldots, y^n)$ be the random vector of demands with unknown p.d.f.

$$f_Y(y) = \prod_{k=1}^{n} f_{Y^k}(y^k)\,;$$

B be the budget of the company.

The objective is

$$\max_x E[\psi_0(x, Y)] = \sum_{k=1}^{n} \left\{(r^k - c^k)x^k + (r^k - \tilde{r}^k) \cdot \left[\int_0^{x^k} (y^k - x^k) f_{Y^k}(y^k)\, dy^k\right]\right\} \tag{3.4.1}$$

subject to

$$\sum_{k=1}^{n} a^{kq}x^k \le b^q, \qquad q = 1, \ldots, Q\,, \tag{3.4.2}$$

$$\sum_{k=1}^{n} c^k x^k \le B\,, \tag{3.4.3}$$

$$x^k \ge 0, \qquad k = 1, \ldots, n\,. \tag{3.4.4}$$

It is readily seen that $\psi_0(x, y)$ is a concave function when $f_Y(y) > 0$ for all y. If $f_Y(y)$ is continuous, then $g_0(x) = E[\psi_0(x, Y)]$ is strictly concave.

Denote $\tilde{\nabla} g_0(x) = \{\tilde{\partial} g_0(x)/\partial x^k, k = 1, \ldots, n\}$. Then the kth component of the SQG vector $\tilde{\nabla} g_0(x)$ is given by

$$\frac{\tilde{\partial} g_0(x)}{\partial x^k} = \begin{cases} \hat{r}^k - c^k & \text{if } Y^k \le x^k\,, \\ r^k - c^k & \text{if } Y^k > x^k\,, \end{cases} \tag{3.4.5}$$

Gupal's recursive procedure (3.2.11)–(3.2.15) for this problem can be written as

$$x_{i+1} = \pi_X[x_i + a_i\, \tilde{\nabla} g(x_i)]\,, \tag{3.4.6}$$

where

$$\tilde{\nabla} g(x_i) = \begin{cases} \tilde{\nabla} g_0(x_i) & \text{if } \max\limits_{1\le s\le M} z_i^s \le 0\,, \\ \tilde{\nabla} g_j(x_i) & \text{if } z_i^j = \max\limits_{1\le s\le M} z_i^s > 0\,. \end{cases} \tag{3.4.7}$$

Convergence of (3.4.6)–(3.4.7) follows now for the same reason as for models of section 3.3.3.

3.4.2 Allocation Model

This model is associated with an urban service system and is taken from Sivazlian and Stanfel (1975).

Assume that the urban population is sufficiently dense so that the demand requirements for a service can be viewed as continuously distributed over the area of the city. We refer the system to the set of Cartesian coordinates y_1, y_2 and denote

$$f_Y(y) = f_{Y_1, Y_2}(y^1, y^2)\,,$$

the p.d.f. of the demand at point $y = (y^1, y^2)$. Let $x = (x^1, x^2)$ be the location of the source point. Assume that the cost to meet the requirement at a given point y is proportional to the demand density $f_Y(y)$ at this point and its distance from the source x. Our aim is to find the optimal location x^* that minimizes the average total cost,

$$g(x) = E[\psi(x, Y)] = \int_\Omega [(y^1 - x^1)^2 + (y^2 - x^2)^2]^{1/2} f_{Y^1, Y^2}(y^1, y^2)\, dy^1\, dy^2 . \tag{3.4.9}$$

over a given domain Ω. Since both $\psi(x, y) = [(y^1 - x^1)^2 + (y^2 - x^2)^2]^{1/2}$ and $g(x)$ are convex, the minimum of $g(x)$ can be obtained by solving

$$\nabla g(x) = E[\nabla \psi(x, Y)] = 0 , \tag{3.4.10}$$

where

$$\nabla \psi(x, y) = \left\{ \frac{x^1 - y^1}{[(x^1 - y^1)^2 + (x^2 - y^2)^2]^{1/2}}, \frac{x^2 - y^2}{[(x^1 - y^1)^2 + (x^2 - y^2)^2]^{1/2}} \right\} . \tag{3.4.11}$$

The Robbins-Monro procedure (3.2.1) for this model is

$$x_{i+1} = x_i - a_i \nabla \psi(x_i, Y_i) , \tag{3.4.12}$$

where $Y_1, Y_2, \ldots$ is a sample from $f_Y(y)$, and $\nabla\psi(x, y)$ is given in (3.4.11). Convergence of (3.4.12) to x^* follows directly from Theorem 1, since $g(x)$ is strictly convex and $E[\|\nabla\psi\|]^2 = 1$.

3.5 VARIANCE REDUCTION

In the preceding sections we applied the SQG (3.2.20) to several adaptive algorithms [see, e.g., (3.3.21) and (3.3.24)], assuming that $Y_i^{(s)}$ and $Y_i^{(0)}$, $s = 1, 2, \ldots, i = 1, 2, \ldots,$ are independent samples each from $F_Y(y)$. Clearly (3.2.20) is an unbiased estimate of the finite difference gradient

$$\hat{\nabla} g(x_i) = \frac{1}{\beta_i} \sum_{s=1}^{n} [g(x_i + \beta_i e^s) - g(x_i)] e^s . \tag{3.5.1}$$

In this section we shall consider several modified versions of the SQG (3.2.20) and show that under certain conditions on ψ they are more accurate than (3.2.20).

CASE 1 $\psi(x, y)$ is monotonic in y for any x. Consider the following modified version of (3.2.20):

$$\tilde{\nabla}^{(1)} g(x_i) = \frac{1}{\beta_i} \sum_{s=1}^{n} [\psi(x_i + \beta_i e^s, Y_i^{(s)}) - \psi(x_i, Y_i^{(s)})] e^s \qquad (3.5.2)$$

with a single r.v. $Y_i^{(s)}$ instead of the two $[Y_i^{(s)}$ and $Y_i^{(0)}]$ originally used. We call $\tilde{\nabla}^{(1)} g$ the common random variate (CRV) estimate of $\hat{\nabla} g$.

We shall now show that the estimate (3.5.2) is more accurate than (3.2.20) in the sense that

$$\operatorname{Var} \sum_{s=1}^{n} \tilde{\nabla}_s^{(1)} g(x_i) \leq \operatorname{Var} \sum_{s=1}^{n} \tilde{\nabla}_s g(x_i), \qquad i = 1, 2, \ldots. \qquad (3.5.3)$$

Here

$$\tilde{\nabla}_s g(x_i) = \frac{1}{\beta_i} [\psi(x_i + \beta_i e^s, Y_i^{(s)}) - \psi(x_i, Y_i^{(0)})] \qquad (3.5.4)$$

and

$$\tilde{\nabla}_s^{(1)} g(x_i) = \frac{1}{\beta_i} [\psi(x_i + \beta_i e^s, Y_i^{(s)}) - \psi(x_i, Y_i^{(s)})] \qquad (3.5.5)$$

represent the partial derivatives of the SQGs (3.2.20) and (3.5.2), respectively.

Clearly, a more accurate estimate at each iteration leads to acceleration of convergence of the entire SQG procedure. Another advantage of $\tilde{\nabla}^{(1)} g$ is that it requires at each iteration only n independent r.v.'s $Y_i^{(1)}, \ldots, Y_i^{(n)}$ instead of the $n+1$ r.v.'s $Y_i^{(0)}, \ldots, Y_i^{(n)}$ required by $\tilde{\nabla} g$.

Let us prove (3.5.3) first for the one-dimensional case, that is, when $n = 1$. Denote

$$h_1(Y^{(1)}) = \psi(x + \beta, Y^{(1)}) \quad \text{and} \quad h_0(Y^{(0)}) = \psi(x, Y^{(0)}).$$

It follows then directly from Proposition 1 (Chapter 1) that

$$\min_{F_{Y(1),Y(0)}} \operatorname{Var}[h_1(Y^{(1)}) - h_0(Y^{(0)})] = \operatorname{Var}[h_1(Y) - h_0(Y)], \qquad (3.5.6)$$

where $Y = Y^{(1)} = Y^{(0)}$. Thus, within the set of all joint two-dimensional c.d.f.'s $F_{Y(1),Y(0)}$, the variance of $h_1(Y^{(1)}) - h_0(Y^{(0)})$ is minimized when the common random variate (CRV) $Y^{(1)} = Y^{(0)} = Y$ is used in both h_1 and h_0.

Obviously

$$\operatorname{Var} \tilde{\nabla}^{(1)} g \leq \operatorname{Var} \tilde{\nabla} g ,$$

and therefore for the one-dimensional case the estimate $\tilde{\nabla} g^{(1)}$ is more accurate than $\tilde{\nabla} g$.

Consider now the n-dimensional case. Define

$$\tilde{\nabla}_s^{(2)} g(x_i) = \frac{1}{\beta_i} [\psi(x_i + \beta_i e^s, Y_{i,1}^{(s)}) - \psi(x_i, Y_{i,2}^{(s)})] , \tag{3.5.7}$$

where $Y_{i,1}^{(s)}$ and $Y_{i,2}^{(s)}$ are independent samples each from $F_Y(y)$. Clearly, if we take $Y_{i,1}^{(s)} = Y_{i,2}^{(s)} = Y_i^{(s)}$, then (3.5.7) reduces to (3.5.5). But by Proposition 1 (Chapter 1), $\tilde{\nabla}_s^{(1)} g(x_i)$ is more accurate than $\tilde{\nabla}_s^{(2)} g(x_i)$, that is,

$$\operatorname{Var} \tilde{\nabla}_s^{(1)} g(x_i) \leq \operatorname{Var} \tilde{\nabla}_s^{(2)} g(x_i) , \qquad s = 1, \ldots , n . \tag{3.5.8}$$

Clearly, (3.5.3) will be proved if we establish

$$\operatorname{Var} \sum_{s=1}^{n} \tilde{\nabla}_s^{(2)} g(x_i) \leq \operatorname{Var} \sum_{s=1}^{n} \tilde{\nabla}_s g(x_i) . \tag{3.5.9}$$

To prove (3.5.9), we need the following lemma.

Lemma Let $V_1, \ldots , V_n$ be independent r.v.'s and $Z_0, Z_1, \ldots , Z_n$ independent identically distributed r.v.'s. Assume that V_j and Z_k, $j = 1, \ldots , n$, $k = 0, 1, \ldots , n$, are also independent, each having a finite variance. Then

$$\operatorname{Var} \sum_{s=1}^{n} (V_s - Z_s) < \operatorname{Var} \sum_{s=1}^{n} (V_s - Z_0) . \tag{3.5.10}$$

Proof By independence of V_s and Z_s, we have

$$\operatorname{Var} \sum_{s=1}^{n} (V_s - Z_s) = \sum_{s=1}^{n} \operatorname{Var} V_s + n \operatorname{Var}(Z) .$$

Now

$$\operatorname{Var} \sum_{s=1}^{n} (V_s - Z_0) = \sum_{s=1}^{n} \operatorname{Var} V_s + n \operatorname{Var} Z + 2 \sum_{i<j} \operatorname{Cov}(V_i - Z_0)(V_j - Z_0) .$$

Taking into account that $\operatorname{Cov}(V_i - Z_0)(V_j - Z_0) = \operatorname{Var} Z_0 > 0$, $i, j = 1, \ldots , n$, we obtain immediately (3.5.10). Q.E.D.

Denote $V_s = \psi(x_i + e^s,\ Y_{i,1}^{(s)})$, $s = 1, \ldots, n$; $Z_s = \psi(x_i, Y_{i,2}^{(s)})$, $s = 1, \ldots, n$, and $Z_0 = \psi(x_i, Y_i^{(0)})$.

The proof of (3.5.9) and therefore of (3.5.3) follows now directly from (3.5.10).

Thus, if the function $\psi(x, y)$ is monotonic in y for each x, the SQG (3.5.2) is more accurate than (3.2.20) and therefore can be recommended in practice.

Note The inequality (3.5.3) actually holds for any SQG represented in the form of finite differences provided that $\psi(x, y)$ is monotonic in y for any x.

Consider, for example, the SQG

$$\tilde{\nabla}^{(1)} g(x_i) = \frac{3r_i}{2\beta_i} \sum_{s=1}^{r_i} [\psi(x_i + \beta_i \Xi_i, Y_i^{(s)}) - \psi(x_i, Y_i^{(s)})]\Xi_i \,. \tag{3.5.11}$$

Here $r_i \geq 1$ and $\Xi_i = (\Xi_i^{(1)}, \ldots, \Xi_i^{(n)})$, $i = 1, 2, \ldots$, are independent random vectors with i.i.d. components $\Xi_i^{(k)}$, $k = 1, \ldots, n$, each distributed uniformly over the interval $[-1, 1]$. Clearly, the CRV estimate (3.5.11) is more accurate than its counterpart (3.2.21) of Section 2.2.

CASE 2 $\psi(x, y)$ is *not monotonic* in y, but subintervals of monotonicity of ψ in y are known. In this case use of Procedure 2 (Chapter 1) allows for efficient SQGs (see inventory model below).

We now present two applications of the modified versions of stochastic quasi-gradients. The first is associated with Case 1 and deals with the production planning model; the second is associated with Case 2 and deals with the single-period one-commodity inventory model without setup cost.

(a) Production Planning Assume that at the beginning of each period a manufacturer has to order a quantity $x = (x^{(1)}, \ldots, x^{(n)})$ of certain products he or she would like to produce for sale during that period. The cost of production is denoted by $c(x)$. Consider the case in which the demand for these products during each period is a random vector $Y = (Y^{(1)}, \ldots, Y^{(n)})$ whose probability distribution F_Y is unknown in advance. For each period it is assumed that if a quantity x is produced and the demand is Y, then the benefit is $\varphi(x, Y)$. The manufacturer would like to produce a quantity x that maximizes

$$g(x) = -c(x) + \int_0^\infty \varphi(x, y)\, dF_Y(y)\,.$$

In order to find the optimal policy x^*, we can use the SQG algorithm (3.2.11) with, say, $\tilde{\nabla}g(x_i)$, as per (3.2.20); namely,

$$\tilde{\nabla}g(x_i) = \frac{1}{\beta_i}\left[\sum_{s=1}^{n} \psi(x_i + \beta_i e^s, Y_i^{(s)}) - \psi(x_i, Y_i^{(0)})\right] e^s .$$

Here $\psi(x, Y) = -c(x) + \varphi(x, Y)$.

Let $\varphi(x, y)$ be *monotonic* in each component y_s, $s = 1, \ldots, n$, for any x. In this case the CRV estimate $\tilde{\nabla}^{(1)}g(x_i)$ [(3.5.2)] is more accurate than $\tilde{\nabla}g(x_i)$, and therefore the algorithm (3.2.11) with $\tilde{\nabla}g$ replaced by $\tilde{\nabla}g^{(1)}$ will converge faster to x^*.

(b) Inventory Model We shall consider only the single-period one-commodity inventory model, assuming the c.d.f. $F_Y(y)$ of the demand is unknown, but a sample $Y_1, \ldots, Y_N$ from it is available. Our results to the multicommodity model can be extended by analogous arguments.

Consider the SQG algorithms (3.3.21) and (3.3.24). Assume that for each x functions $H(x - y)$ and $P(y - x)$ *are monotonic nonincreasing* and *monotonic nondecreasing* in y, respectively, and that $H(0) = P(0) = 0$ (see Fig. 1).

Clearly,

$$L(x, y) = \begin{cases} H(x - y) & \text{if } x < y , \\ P(y - x) & \text{if } y \geq x , \end{cases}$$

is unimodal with respect to y for any x. The functions $L(x, y)$ and $L(x + \beta, y)$ have their minima at points x and $x + \beta$, respectively (see Fig. 1). Three subintervals of monotonicity (not necessarily in the same direction) of $L(x, y)$ and $L(x + \beta, y)$ can be distinguished:

1. $(0, x)$: both $L(x, y) = H(x - y)$ and $L(x + \beta, y) = H(x + \beta - y)$ monotonic nonincreasing;

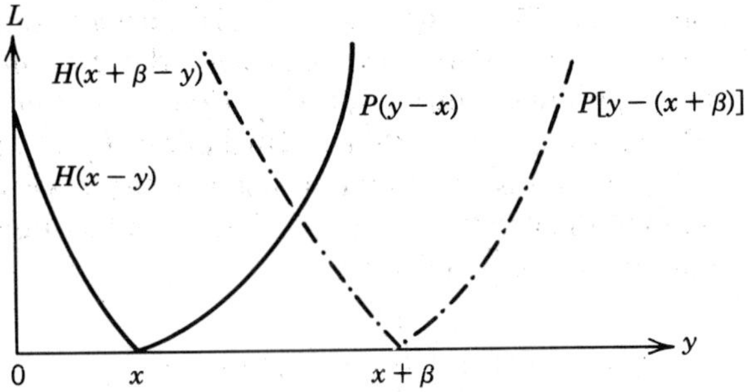

Fig. 1 The cost functions H and P.

2. $(x, x+\beta)$: $L(x+\beta, y) = H(x+\beta-y)$ monotonic nonincreasing and $L(x, y) = P(y-x)$ monotonic nondecreasing; and
3. $(x+\beta, \infty)$: both $L(x, y) = P(y-x)$ and $L(x+\beta, y) = P(y-(x+\beta))$ monotonic nondecreasing.

Figure 2 presents the $m^2 = 9$ rectangles of monotonicity of $L(x, y)$ and $L(x+\beta, v)$, with the corresponding diagonals indicating the character of monotonicity (in the same or in the opposite direction) of g^* and h^* defined as

$$g^*(U_1) = \sum_{i=1}^{3} g^*(U_i)I_i = L(x+\beta, Y^{(1)})$$

and

$$h^*(U_2) = \sum_{j=1}^{3} h^*(U_2)I_j = L(x, Y^{(0)}),$$

respectively. Here

$$I_k(u) = \begin{cases} 1 & \text{if } u \in t_k, \\ 0 & \text{otherwise}, \end{cases}$$

where t_k is the k-th subinterval of monotonicity.

If $F_Y(y)$ is known in advance, Procedure 2 of Chapter 1 yields an SQG estimate of $\tilde{\nabla}g$ more accurate than (3.3.24) in the sense of (3.5.3). And even if $F_Y(y)$ is unknown, the situation is not "hopeless." Indeed, it follows from Fig. 2 that if both $Y^{(1)}$ and $Y^{(0)}$ belong either to the subinterval $(0, x)$ or to $(x+\beta, \infty)$, then CRV ($Y^{(1)} = Y^{(0)} = Y$) can be used and variance reduction achieved (see Proposition 3, Chapter 1). This is so because in the extreme left-hand rectangle $t_1 \times t_1$ and $t_3 \times t_3$, both g^*

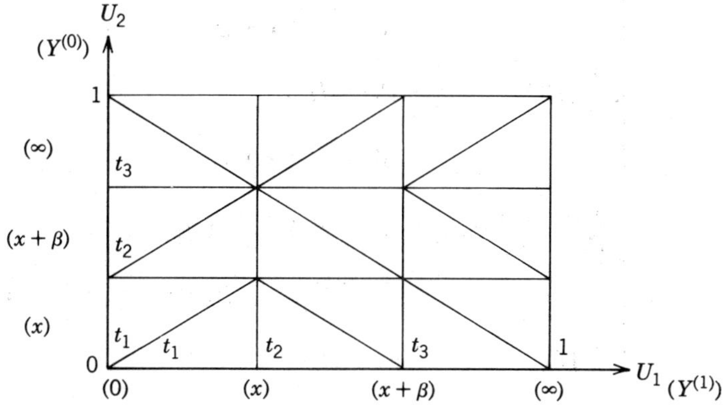

Fig. 2 Rectangles of monotonicity of $L(x, Y)$ and $L(x+\beta, Y)$ [$g^*(U_1)$ and $h^*(U_2)$].

and h^* are monotonic in the same direction. Note that application of Procedure 2 of Chapter 1 to the other rectangles necessitates knowledge of $F_Y(y)$.

Denote by A the event that both $Y^{(1)}$ and $Y^{(0)}$ belong either to $(0, x)$ or to $(x + \beta, \infty)$, we have

$$\begin{aligned} P(A) &= P\{[Y^{(1)} \in (0, x),\ Y^{(0)} \in (0, x)] \\ &\quad \cup [Y^{(1)} \in (x + \beta, \infty),\ Y^{(0)} \in (x + \beta, \infty)]\} \\ &= P\{(U_1 \in t_1, U_2 \in t_1) \cup (U_1 \in t_3, U_2 \in t_3)\} = t_1^2 + t_3^2 . \end{aligned}$$

Clearly, $P(A) \geq \frac{1}{2}$ as $\beta \to 0$.

Thus when $F_Y(y)$ is unknown we can use the common random variable $(Y^{(1)} = Y^{(0)} = Y)$ if both demands $Y^{(1)}$ and $Y^{(0)}$ belong either to the subinterval $(0, x)$ or to $(x + \beta, \infty)$ and two independent random variables $Y^{(1)}$ and $Y^{(0)}$ otherwise. The modified version of the SQG (3.3.24) can be written as

$$\tilde{\nabla} g^{(3)}(x_i) = c + \begin{cases} \dfrac{1}{\beta_i}[H(x_i + \beta_i - Y_i^{(1)}) - H(x - Y_i^{(1)})] \\ \qquad \text{if } Y_i^{(1)}, Y_i^{(0)} < x , \\ \dfrac{1}{\beta_i}[H(x_i + \beta_i - Y_i^{(1)}) - H(x - Y_i^{(0)})] \\ \qquad \text{if } x < Y_i^{(1)} < x + \beta,\ Y_i^{(0)} < x , \\ \dfrac{1}{\beta}[H(x_i + \beta_i - Y_i^{(1)}) - P(Y_i^{(0)} - x)] \\ \qquad \text{if } Y_i^{(1)} < x + \beta,\ Y_i^{(0)} \geq x , \\ \dfrac{1}{\beta}[P(Y_i^{(1)} - x_i - \beta_i) - H(x - Y_i^{(0)})] \\ \qquad \text{if } Y_i^{(1)} \geq x + \beta,\ Y_i^{(0)} < x , \\ \dfrac{1}{\beta}[P(Y_i^{(1)} - x_i - \beta_i) - P(Y_i^{(0)} - x_i)] \\ \qquad \text{if } Y^{(1)} \geq x + \beta,\ x - \beta < Y_i^{(0)} < x + \beta , \\ \dfrac{1}{\beta}[P(Y_i^{(1)} - x - \beta) - P(Y_i^{(1)} - x_i)] \\ \qquad \text{if } Y_i^{(1)}, Y_i^{(0)} \geq x + \beta . \end{cases}$$

The effect of variance reduction by using $\tilde{\nabla}^{(3)} g$ instead of $\tilde{\nabla} g$ [see (3.3.24)] can be substantial, especially when β is small.

3.6 SENSITIVITY AND OPTIMIZATION OF COMPUTER SIMULATION MODELS

This section is based on Rubinstein (1976).

(i) Preliminary Results Consider the following optimization problem:

$$\min_{x \in X} \{g_0(x) = \int \varphi(y) f_Y(y, x)\, dy\} \tag{3.6.1}$$

subject to

$$g_j(x) \leq 0\,, \qquad j = 1, \ldots, M\,, \tag{3.6.2}$$

where x and y are n- and m-dimensional vectors, respectively, and X is a given region. We assume that the p.d.f. $f_Y(y, x)$ depends on a vector of parameters x. We also assume that p.d.f. $f_Y(y, x)$ is completely known, $f_Y(y, x) \neq 0$, if $x \in X$ and $f_Y(y, x)$ is twice differentiable with respect to x. Let us first find the sensitivities that is gradient $\nabla g_0(x)$ and the Hessian $\nabla^2 g_0(x)$ of $g_0(x)$.

The partial derivative of $g_0(x)$ with respect to x_j, $j = 1, \ldots, n$, is

$$\begin{aligned}
\frac{\partial g_0(x)}{\partial x_j} &= \frac{\partial}{\partial x_j} \int \varphi(y) f_Y(y, x)\, dy = \int \varphi(y) \frac{\partial f_Y(y, x)}{\partial x_j}\, dy \\
&= \int \varphi(y) \frac{\partial f_Y(y, x)}{\partial x_j} \frac{f_Y(y, x)}{f_Y(y, x)}\, dy \\
&= \int \varphi(y) \frac{\partial \ln f_Y(y, x)}{\partial x_j} f_Y(y, x)\, dy \\
&= E\left(\varphi(Y) \frac{\partial \ln f_Y(Y, x)}{\partial x_j}\right),
\end{aligned}$$

and therefore the gradient of $g(x)$ can be presented as†

$$\nabla g_0(x) = E[\varphi(Y) \nabla \ln f_Y(Y, x)]\,. \tag{3.6.3}$$

†It is assumed in (3.6.3) that the operators expectation and differentiation are interchangeable.

Examples

(a) Let $Y \sim N(\mu, \Sigma)$, that is,

$$f_Y(y, \mu) = \frac{1}{(2\pi)^{m/2}|\Sigma|^{1/2}} \exp[-\tfrac{1}{2}(y-\mu)^t\Sigma^{-1}(y-\mu)], \quad (3.6.4)$$

where t denotes the transpose operator. We have therefore

$$\nabla g_0(\mu) = E\{\varphi(Y)\Sigma^{-1}(Y-\mu)\}. \quad (3.6.5)$$

In the particular case where $\Sigma = \sigma^2 I$ we obtain

$$\nabla g_0(\mu) = \sigma^2 E\{\varphi(Y)(Y-\mu)\}. \quad (3.6.6)$$

(b) Let Y_k, $k = 1, \ldots, m$, be independent r.v.'s each distributed $\exp(\lambda_k)$, that is,

$$f_{Y_k}(y_k, \lambda_k) = \lambda_k \exp(-\lambda_k Y_k), \qquad k = 1, \ldots, m,\ \lambda_k > 0,\ y_k > 0. \quad (3.6.7)$$

We have

$$\ln f_{Y_k}(y_k, \lambda_k) = \ln \lambda_k - \lambda_k y_k, \qquad \nabla \ln f_Y(y, \lambda) = \lambda^{-1} - y,$$

where $\lambda = (\lambda_1, \ldots, \lambda_m)$, $\lambda^{-1} = (\lambda_1^{-1}, \ldots, \lambda_m^{-1})$ and $y = (y_1, \ldots, y_m)$, and therefore

$$\nabla g_0(\lambda) = E[\varphi(Y)(\lambda^{-1} - Y)]. \quad (3.6.8)$$

(c) Let Y_k, $k = 1, \ldots, m$, be independent r.v.'s each distributed Laplace (double exponential), that is

$$f_{Y_k}(y_k, \lambda_k) = \lambda_k \exp(-2\lambda_k|y_k|), \qquad k = 1, \ldots, m,$$

$$\lambda_k > 0,\ -\infty < y_k < \infty. \quad (3.6.9)$$

We have

$$\ln f_{Y_k}(y_k, \lambda_k) = \ln \lambda_k - 2\lambda_k|y_k|,$$

$$\nabla \ln f_Y(y, \lambda) = \lambda^{-1} - 2|y|, \quad \text{where } \lambda^{-1} = (\lambda_1^{-1}, \ldots, \lambda_m^{-1})$$

and $|y| = (|y_1|, \ldots, |y_n|)$ and therefore

$$\nabla g_0(\lambda) = E[\varphi(Y)(\lambda^{-1} - 2|Y|)]. \quad (3.6.10)$$

(d) Let Y_k, $k = 1, \ldots, m$, be independent r.v.'s each distributed Rayleigh, that is,

$$f_{Y_k}(y_k, \sigma_k^2) = \frac{y_k}{\sigma_k^2} \exp\left(\frac{-y_k^2}{2\sigma_k^2}\right).$$

We have

$$\ln f_{Y_k}(y_k, \sigma_k^2) = \ln y_k - \ln \sigma_k^2 - \frac{y_k^2}{2\sigma_k^2},$$

$$\frac{\partial \ln f_{Y_k}(y_k, \sigma_k^2)}{\partial(\sigma_k^2)} = -\frac{1}{\sigma_k^2} + \frac{y_k^2}{2\sigma_k^4},$$

and therefore

$$\nabla g_0(\sigma^2) = \sum_{k=1}^{m} E\left[\varphi(Y)\left(\frac{Y_k^2}{2\sigma_k^4} - \frac{1}{\sigma_k^2}\right)\right] e^k,$$

where e^k, $k = 1, \ldots, m$, is the orthonormal basis in R^m.

(e) Let Y_k, $k = 1, \ldots, m$, be independent r.v.'s each distributed Weibull, that is,

$$f_{Y_k}(y_k, \alpha_k, \beta_k) = \frac{\alpha_k}{\beta_k^{\alpha_k}} y_k^{\alpha_k - 1} \exp\left(\frac{-y_k}{\beta_k}\right)^{\alpha_k}, \qquad y_k \geq 0,\ \alpha_k > 0, \quad \beta_k > 0,\ k = 1, \ldots, m.$$

We have

$$\ln f_{Y_k}(y_k, \alpha_k, \beta_k) = \ln \frac{\alpha_k}{\beta_k^{\alpha_k}} + (\alpha_k - 1) \ln y_k - (y_k/\beta_k)^{\alpha_k}$$

$$= \ln \alpha_k - \alpha_k \ln \beta_k + (\alpha_k - 1) \ln y_k - \left(\frac{y_k}{\beta_k}\right)^{\alpha_k}, \qquad k = 1, \ldots, m,$$

$$\frac{\partial \ln f_{Y_k}(y_k, \alpha_k, \beta_k)}{\partial \alpha_k} = \frac{1}{\alpha_k} - \ln \beta_k + \ln y_k - \left(\frac{y_k}{\beta_k}\right)^{\alpha_k} \ln \frac{y_k}{\beta_k}$$

$$= \frac{1}{\alpha^k} + \ln \frac{y_k}{\beta_k}\left[1 - \left(\frac{y_k}{\beta_k}\right)^{\alpha_k}\right],$$

$$\frac{\partial \ln f_{Y_k}(y_k, \alpha_k, \beta_k)}{\beta_k} = -\frac{\alpha_k}{\beta_k} + \alpha_k y_k^{\alpha_k} \beta_k^{-\alpha_k - 1} = -\frac{\alpha_k}{\beta_k}\left[1 - \left(\frac{y_k}{\beta_k}\right)^{\alpha_k}\right]$$

and therefore

$$\nabla g_0(\alpha, \beta) = \sum_{k=1}^{m} E\left(\varphi(Y)\left\{\frac{1}{\alpha_k} + \ln\frac{Y_k}{\beta_k}\left[1 - \left(\frac{Y_k}{\beta_k}\right)^{\alpha_k}\right]\right\} e_1^k\right.$$
$$\left. + \varphi(Y)\left\{-\frac{\alpha_k}{\beta_k}\left[1 - \left(\frac{Y_k}{\beta_k}\right)^{\alpha_k}\right]\right\} e_2^k\right),$$

where e_1^k, e_2^k, $k = 1, \ldots, m$, is orthonormal basis in R^{2m}.

In many practical situations (see below) calculation of $\nabla g_0(x)$ according to (3.6.3) [even with known p.d.f. $f_Y(y, x)$] is unfeasible, and we have to resort to the Monte Carlo methods. An unbiased estimate of $\nabla g_0(x)$ is its arithmetic mean, that is,

$$\bar{\nabla} g_0(x) = \frac{1}{N}\sum_{j=1}^{N} \varphi(Y_j)\nabla \ln f_Y(Y_j, x) . \qquad (3.6.11)$$

For examples (a)–(e) we have, respectively,

(a) Normal:

$$\bar{\nabla} g_0(x) = \frac{1}{N}\sum_{j=1}^{N} \varphi(Y_j)\Sigma^{-1}(Y_j - x) .$$

(b) Exponential:

$$\bar{\nabla} g_0(\lambda) = \frac{1}{N}\sum_{j=1}^{N}\sum_{k=1}^{m} \varphi(Y_j)\left(\frac{1}{\lambda_k} - Y_{jk}\right) e^k .$$

(c) Laplace:

$$\bar{\nabla} g_0(\lambda) = \frac{1}{N}\sum_{j=1}^{N}\sum_{k=1}^{m} \varphi(Y_j)\left(\frac{1}{\lambda_k} - 2|Y_{jk}|\right) e^k .$$

(d) Rayleigh:

$$\bar{\nabla} g_0(\sigma^2) = \frac{1}{N}\sum_{j=1}^{N}\sum_{k=1}^{m} \varphi(Y_j)\left(\frac{Y_{jk}^2}{2\sigma_k^4} - \frac{1}{\sigma_k^2}\right) e^k .$$

(e) Weibull:

$$\bar{\nabla} g_0(\alpha, \beta) = \frac{1}{N}\sum_{j=1}^{N}\sum_{k=1}^{m} \left(\varphi(Y_j)\left\{\frac{1}{\alpha_k} + \ln\frac{Y_{jk}}{\beta_k}\left[1 - \left(\frac{Y_{jk}}{\beta_k}\right)^{\alpha_k}\right]\right\} e_1^k\right.$$
$$\left. + \varphi(Y_j)\left\{-\frac{\alpha_k}{\beta_k}\left[-\left(\frac{Y_{jk}}{\beta_k}\right)^{\alpha_k}\right]\right\} e_2^k\right).$$

The estimate (3.6.11) is generally more accurate than the following finite difference SQG estimates:

$$\tilde{\nabla} g_0(x) = \frac{1}{\beta N} \sum_{j=1}^{N} \sum_{s=1}^{n} \{\varphi[Y_j^{(s)}(x + \beta e^s)] - \varphi[Y_j^{(s)}(x)]\} e^s , \quad (3.6.12)$$

where e^s, $s = 1, \ldots, n$, is the orthonormal basis in R^n, or

$$\tilde{\nabla} g_0(x) = \frac{3r}{2\beta N} \sum_{j=1}^{N} \sum_{s=1}^{r} \{\varphi[Y_j^{(s)}(x + \beta h^s)] - \varphi[Y_j^{(s)}(x)]\} h^s , \quad (3.6.13)$$

where $r \geq 1$ and $h^1, \ldots, h^r$ are realizations of the random vectors whose components are independent and uniformly distributed over the interval $[-1, 1]$. Note that in (3.6.12) [and (3.6.13)] $y_j^{(s)}(x + \beta e^s)$ and $y_j^{(s)}(x)$ are generated by using the common random numbers.

Let us find the Hessian $\nabla^2 g_0(x) = \{\partial^2 g_0(x)/\partial x_j\, \partial x_k\}, j, k = 1, \ldots, n$, of $g_0(x)$, assuming that $f_Y(y, x)$ is twice differentiable and

$$\nabla^2 \ln f_Y(y, x) = \left\{ \frac{\partial^2 \ln f_Y(y, x)}{\partial x_j\, \partial x_k} \right\} \quad \text{exist for all } j, k = 1, \ldots, n .$$

We have

$$\begin{aligned} \frac{\partial^2 g_0(x)}{\partial x_j\, \partial x_k} &= \frac{\partial}{\partial x_k} \int \varphi(y) \frac{\partial \ln f_Y(y, x)}{\partial x_j} f_Y(y, x)\, dy \\ &= \int \varphi(y) \left[\frac{\partial^2 \ln f_Y(y, x)}{\partial x_j\, \partial x_k} + \frac{\partial \ln f_Y(y, x)}{\partial x_j} \frac{\partial \ln f_Y(y, x)}{\partial x_k} \right] f_Y(y, x)\, dy \\ &= E\left\{ \varphi(Y) \left[\frac{\partial^2 \ln f_Y(Y, x)}{\partial x_j\, \partial x_k} + \frac{\partial \ln f_Y(Y, x)}{\partial x_j} \frac{\partial \ln f_Y(y, x)}{\partial x_k} \right] \right\} \end{aligned}$$

and therefore the Hessian

$$\nabla^2 g_0(x) = E\{\varphi(Y)[\nabla^2 \ln f_Y(Y, x) + \nabla \ln f_Y(Y, x) \nabla \ln f_Y(Y, x)^t]\} . \quad (3.6.14)$$

Let us now compute $\nabla^2 g_0(x)$ for the examples

(a) $Y \sim N(\mu, \Sigma)$. We have

$$\nabla^2 g_0(\mu) = E\{\varphi(Y)[\Sigma^{-1}(Y - \mu)(Y - \mu)^t \Sigma^{-1} - \Sigma^{-1}]\} .$$

For the particular case $\Sigma = \sigma^2 I$ we obtain

$$\nabla^2 g_0(\mu) = \frac{1}{\sigma^4} E\{\varphi(Y)[(Y-\mu)(Y-\mu)' - \sigma^2 I]\} .$$

(b) $Y_k \sim \lambda_k \exp(-\lambda_k y_k)$. We have

$$\nabla^2 g_0(\lambda) = E\Big\{\varphi(Y)\Big[-\delta_{jk}\frac{1}{\lambda_k^2} + \Big(\frac{1}{\lambda_j} - Y_j\Big)\Big(\frac{1}{\lambda_k} - Y_k\Big)\Big]\Big\}_{jk} ,$$

$$j, k = 1, \ldots, m ,$$

where δ_{jk} is the Kronecker delta, that is,

$$\delta_{jk} = \begin{cases} 1, & j = k , \\ 0, & j \neq k . \end{cases}$$

(c) $Y_k \sim \lambda_k \exp(-2\lambda_k |y_k|)$. We have

$$\nabla^2 g_0(\lambda) = E\Big\{\varphi(Y)\Big[-\delta_{jk}\frac{1}{\lambda_k^2} + \Big(\frac{1}{\lambda_j} - 2|Y_j|\Big)\Big(\frac{1}{\lambda_k} - 2|Y_k|\Big)\Big]\Big\}_{jk} ,$$

$$j, k = 1, \ldots, m .$$

(d) $Y_k \sim (y_k/\sigma_k^2)\exp(-y_k^2/2\sigma_k^2)$. We have

$$\nabla^2 g_0(\sigma^2) = E\Big\{\varphi(Y)\Big[\delta_{jk}\Big(\frac{1}{\sigma_k^4} - \frac{y_k}{\sigma_k^6}\Big) + \Big(\frac{y_j}{2\sigma_j^4} - \frac{1}{\sigma_j^2}\Big)\Big(\frac{y_k}{2\sigma_k^4} - \frac{1}{\sigma_k^2}\Big)\Big]\Big\}_{jk} ,$$

$$j, k = 1, \ldots, m .$$

(e) $f_{Y_k}(y_k) = (\alpha_k/\beta_k^{\alpha_k}) y^{\alpha_k - 1} \exp(-y_k/\beta_k)^{\alpha_k}$. We have

$$\frac{\partial^2 g_0(\alpha, \beta)}{\partial \alpha_j\, \partial \alpha_k} = E\Big(\varphi(Y)\Big\{-\delta_{jk}\Big[\frac{1}{\alpha_k^2} + \Big(\frac{Y_k}{\beta_k}\Big)^{\alpha_k} \ln^2 \frac{Y_k}{\beta_k}\Big]$$

$$+ \Big[\frac{1}{\alpha_j} + \ln\frac{Y_j}{\beta_j}\Big[1 - \Big(\frac{Y_j}{\beta_j}\Big)^{\alpha_j}\Big]\Big]\Big[\frac{1}{\alpha_k} + \ln\Big(\frac{Y_k}{\beta_k}\Big)^{\alpha_k}\Big[1 - \Big(\frac{Y_k}{\beta_k}\Big)^{\alpha_k}\Big]\Big]\Big\}\Big) ,$$

$$j, k = 1, \ldots, m ,$$

$$\frac{\partial^2 g_0(\alpha, \beta)}{\partial \beta_j \, \partial \beta_k} = E\left(\varphi(Y)\left\{\delta_{jk}\frac{\alpha_k}{\beta_k^2}\left[1-(1+\alpha_k)\left(\frac{Y_k}{\beta_k}\right)^{\alpha_k}\right]\right.\right.$$

$$\left.\left. + \frac{\alpha_j}{\beta_j}\left[1-\left(\frac{Y_j}{\beta_j}\right)^{\alpha_j}\right]\frac{\alpha_k}{\beta_k}\left[1-\left(\frac{Y_k}{\beta_k}\right)^{\alpha_k}\right]\right\}\right), \qquad j, k = 1, \ldots, m,$$

and

$$\frac{\partial^2 g_0(\alpha, \beta)}{\partial \alpha_j \, \partial \beta_k} = E\left(\varphi(Y)\left[\delta_{jk}\frac{1}{\beta_k}\left[1-\left(\frac{Y_k}{\beta_k}\right)^{\alpha_k}\left(1+\alpha_k \ln\frac{Y_k}{\beta_k}\right)\right]\right.\right.$$

$$\left.\left. + \left\{\frac{1}{\alpha_j} + \ln\frac{Y_j}{\beta_j}\left[1-\left(\frac{Y_j}{\beta_j}\right)^{\alpha_j}\right]\right\}\frac{\alpha_k}{\beta_k}\left[1-\left(\frac{Y_k}{\beta_k}\right)^{\alpha_k}\right]\right]\right),$$

$$j, k = 1, \ldots, m.$$

An unbiased estimate of $\nabla^2 g_0(x)$ in (3.6.14) is

$$\bar{\nabla}^2 g_0(x) = \frac{1}{N}\sum_{j=1}^{N} \{\varphi(Y_j)[\nabla^2 \ln f_Y(Y_j, x)$$

$$+ \nabla \ln f_Y(Y_j, x) \nabla \ln f_Y(Y_j, x)^t]\}. \tag{3.6.15}$$

If, for example, $Y \sim N(\mu, \Sigma)$, we obtain

$$\bar{\nabla}^2 g_0(x) = \frac{1}{N}\sum_{j=1}^{N} \{\varphi(Y_j)[\Sigma^{-1}(Y_j - x)(Y_j - x)^t \Sigma^{-1} - \Sigma^{-1}]\}.$$

It is important to note that both sensitivity estimates $\nabla g_0(x)$ and $\nabla^2 g_0(x)$ were estimated *simultaneously* from a *single* sample $Y_1, \ldots, Y_n$. We next present several motivating examples involving optimization in simulation.

(ii) Motivating Examples

1. Design of a Stochastic Network Consider a problem of design, a stochastic network with a minimal shortest path. The shortest path of a network can be expressed [see (1.1.22)] as

$$\varphi(Y) = \min_{j=1,\ldots,p} \sum_{i \in L_j} Y_i, \tag{3.6.16}$$

where L_j is the jth complete path from a source to a sink in the network,

Y_i, $i = 1, \ldots, m$, are the durations of activities (links), and p is the number of complete paths in the network.

Let $F_{Y_i}(Y_i, \theta_i)$, $i = 1, \ldots, m$, be the c.d.f. of the duration of activity i, the vector of parameters θ_i, $i = 1, \ldots, m$, being assumed to be controllable (e.g., can be chosen within the interval $[\theta_i^{(1)}, \theta_i^{(2)}]$).

Consider now the following problem. Find a vector $\theta = (\theta_1, \ldots, \theta_m)$ that *minimizes* the expected shortest path

$$g_0(\theta) = \int \varphi(y)\, dF_Y(y, \theta) \tag{3.6.17}$$

subject to

$$g_j(\theta) \le 0\,, \qquad j = 1, \ldots, M\,,$$

$$\theta^{(1)} \le \theta \le \theta^{(2)}\,. \tag{3.6.18}$$

Denoting by c_i the cost per unit increase in θ_i, where $\theta_i = E(Y_i)$, one of the natural constraints would be

$$\sum_{i=1}^{m} c_i \theta_i \le b\,. \tag{3.6.19}$$

2. ***Design of a Reliability System*** Given a coherent reliability systems, its lifetime can be written [see (1.1.23)] as

$$\varphi(Y) = \max_{j=1,\ldots,p} \min_{i \in L_j} Y_i\,, \tag{3.6.20}$$

where L_j and Y_i are as per (3.6.16). For a fixed θ the mean lifetime of the system is as per (3.6.17). Let the parameters θ_i, $i = 1, \ldots, m$, be controllable, and we seek to design a most reliable system satisfying some (say, cost) constraints. Mathematically, our problem would read: Find a vector $\theta = (\theta_1, \ldots, \theta_m)$ that *maximizes* the mean lifetime (3.6.17) of the reliability system subject to constraints (3.6.19).

3. ***Queueing Decision-Making Model*** In queueing models, optimization is applied to such design parameters as service rates, number of parallel services, and maximum allowable queue length. We consider here the simplest GI/G/1 model with arrival rate λ and service rate μ.

Assume that the service rate μ is controllable. We seek the value of μ that minimizes the expected cost of waiting and service per unit time. Let c_1 be the cost per unit increase of μ, c_2 the cost of waiting customer, and

$W(\mu)$ the wainting time of a customer in the queue with fixed μ. Consider the optimization problem

$$\min_{\mu} \{g_0(\mu) = c_1\mu + c_2 E[W(\mu)]\} \tag{3.6.21}$$

subject to $\rho = \lambda/\mu < 1$. Here ρ is the traffic density.

For the simplest M/M/1 queue [see Hillier and Lieberman (1974)] we have

$$E(W) = \frac{\lambda}{\mu - \lambda},$$

and thus the optimal service rate μ^* is

$$\mu^* = \lambda + \left(\frac{c_2\lambda}{c_1}\right)^{1/2}.$$

For the GI/G/1 queue the mean waiting time $E(W)$ is not available analytically. But resorting to Monte Carlo simulation, we can estimate it through its sample mean

$$\hat{W} = \frac{1}{N}\sum_{i=1}^{N} W_i,$$

where $W_i = \max\{0, W_{i-1} - A_i + S_{i-1}\}$ and A_i and S_i are, respectively, the interarrival and service times of the ith customer. An unbiased estimate of $g_0(\mu)$ is therefore

$$\hat{g}_0(\mu) = c_1\mu + c_2\hat{W}(\mu).$$

In the preceding model the number of customers R was unlimited. Now let R be restricted. Clearly, the larger it is, the smaller will be the number of lost customers. Denoting by c_3 the cost of additional accommodation per unite time and c_4 the cost per lost customer, the extended version of (3.6.21) reads [see Hillier and Lieberman (1974)]

$$\min_{\mu,R} \{g_0(\mu, R) = c_1\mu + c_2E[W(\mu, R)] + c_3R + c_4\lambda P_R\}, \tag{3.6.22}$$

where λP_R is the number of lost customers per unit time (P_R being the steady-state probability of the system being in state R). Note that for this model P_R is likewise unknown and has to be estimated by simulation. For fixed μ and R an unbiased estimate of $g(\mu, R)$ is

$$\hat{g}_0(\mu, R) = c_1\mu + c_2\hat{W}(\mu, R) + c_3R + c_4\lambda\hat{P}_R .$$

Clearly, none of the problems (3.6.17)–(3.6.21) lend themselves to analytic solution, since the objective functions, gradients, and Hessians are obtained through Monte Carlo simulation. They can, however, be solved by means of stochastic approximation algorithms.

(iii) Adaptive Algorithms Consider the unconstrained problem (3.6.1), which we shall solve by applying the SAA

$$x_{i+1} = x_i - a_i[\nabla^2 g_0(x^*)]^{-1}\tilde{\nabla} g_0(x_i) , \tag{3.6.23}$$

where $\nabla^2 g_0(x^*)$ is the Hessian of g_0 at the optimal point x^*. The SQG $\tilde{\nabla} g_0(x)$ is defined as in (3.6.11) or as in (3.6.12) and (3.6.13) according as $f_Y(y, x)$ is differentiable or not with respect to x. The Hessian $\nabla^2 g_0(x^*)$ is generally unknown but can be estimated adaptively [ex. Lai and Robbins (1979)]. If $f_Y(y, x)$ is twice differentiable, than at the i-th iteration $\nabla^2 g_0(x^*)$ can be estimated by $1/i\ \Sigma_{j=1}^{i}\ \tilde{\nabla}^2 g_0(x_j)$, where $\tilde{\nabla} g_0^2(x)$ is given in (3.6.15).

Consider the GI/G/1 queueing model (3.6.21). For it, the algorithm (3.6.23) reads

$$\mu_{i+1} = \mu_i - \frac{a_i}{g''(\mu^*)}\,[c_1 + c_2\,\nabla\hat{W}(\mu_i)] , \tag{3.6.24}$$

where $\nabla\hat{W}$ is computed according to (3.6.11).†

For the constrained problem (3.6.1) and (3.6.2) we can apply Gupal's algorithm (3.2.11)–(3.2.15). Again, according as $f_Y(y, x)$ is differentiable with respect to x or not, we use for $\tilde{\nabla} g_0(x)$ the SQG (3.6.11) or (3.6.12) and (3.6.13), respectively.

Returning to the stochastic network and the reliability models let Y_k, $k = 1, \ldots, m$, be independent r.v.'s each distributed $\exp(\lambda_k)$. Substituting (3.6.16) and (3.6.20) in (3.6.8), we obtain

$$\nabla g_0(\lambda) = E\Big[\min_{j=1,\ldots,p}\ \sum_{i\in L_j} Y_i(\lambda^{-1} - Y)\Big] ,$$

and

$$\nabla g_0(\lambda) = \hat{E}[\max_{j=1,\ldots,p}\ \min_{i\in L_j} Y_i(\lambda^{-1} - Y)]$$

†In Chapter 4 we shall present the so called perturbation analysis approach for estimating $\nabla\hat{W}(\mu)$. The proposed approach associated with estimates (3.6.11) and (3.6.15) is called the *score function* approach (see Rubenstein, 1986).

for the shortest path and reliability model, respectively. Here $\lambda = (\lambda_1, \ldots, \lambda_m)$ and $Y = (Y_1, \ldots, Y_m)$. Unbiased estimates of $\nabla g_0(x)$ for the two models are

$$\tilde{\nabla} g_0(\lambda) = \frac{1}{N} \sum_{s=1}^{N} \left(\min_{j=1,\ldots,p} \sum_{i \in L_j} Y_i^{(s)} (\lambda^{-1} - Y^{(s)}) \right),$$

and

$$\tilde{\nabla} g_0(\lambda) = \frac{1}{N} \sum_{s=1}^{N} \left(\max_{j=1,\ldots,p} \min_{i \in L_j} Y_i^{(s)} - Y^{(s)}) \right).$$

respectively. For more details on this issue see Rubinstein (1986).

3.7 SIMULATION RESULTS

In this section we present simulation results for several unconstrained and constrained inventory models. In all examples we assume that the holding cost function $H(x) = 9x$, the shortage cost function $P(x) = 45x^2$, the cost per commodity unit $c = 20$, and the fixed ordering (setup) cost $K = 40$. The unconstrained problem is: Minimize

$$g_0(x) = E[\psi_0(x, Y)] = \begin{cases} 40 + 20x + E[L(x, Y)] & \text{if } x > 0, \\ E[L(x, Y)] & \text{if } x = 0, \end{cases} \tag{3.7.1}$$

where

$$L(x, y) = \begin{cases} 9(x - y) & \text{if } y < x, \\ 45(x - y)^2 & \text{if } y \geq x. \end{cases}$$

We assumed that the demand Y is distributed uniformly in the interval $(0, r)$, that is, $Y \sim U(0, r)$, where $1 \leq r \leq 20$, and that a sample $Y_1, \ldots, Y_N$ from this distribution is available. We simulated both the off-line and the on-line models.

The step parameter a_i was chosen adaptively. Algorithm 1 presents an adaptive procedure for choosing a_i, $i = 1, \ldots, N$, while estimating S according to (3.3.20).

Algorithm 1

1. $X \leftarrow X_0$, $a \leftarrow a_0 = b^{-1} = [g_0''(x^*)]^{-1}$, $i \leftarrow 1$, $r \leftarrow 1$.
2. Generate Y' and Y''.

3. $X' = X - a[\tilde{\nabla}L(X, Y') + c]$ and $X'' = X' - a[\tilde{\nabla}L(X', Y'') + c]$.
4. If $X'' \geq X' \geq X$ or $X'' \leq X' \leq X$, go to step 6.
5. $r \leftarrow r + 1$, $a \leftarrow a_0/r$, $X \leftarrow X''$.
6. $r \leftarrow r$, $a \leftarrow a$, $X \leftarrow X''$.
7. $i \leftarrow i + 1$.
8. If the stopping criterion (3.7.5) is not satisfied, go to step 2.
9. Otherwise, stop.

To apply Algorithm 1 for other stochastic approximation procedures, we need only to rewrite steps 1 and 3. For example, for estimating s according to (3.3.40c), steps 1 and 3 must be rewritten as

1. $a_0 = b^{-1} = [g_0''(s)]^{-1}$ and
3. $s' = \min\{S, x - a'''(cx + L(x, Y) - \alpha)\}$ and $s'' = \min\{S, x' - a'''(cx' + L(x', Y) - \alpha)\}$,

respectively.

According to our procedure, at each iteration i we generate simultaneously two independent demands Y', Y'' and perform sequentially, according to (3.3.20), two iterations (one double iteration, see step 3) with the same step parameter a. Further, we check the monotonicity condition (see step 4)

$$X'' \geq X' \geq X \quad \text{or} \quad X'' \leq X' \leq X . \tag{3.7.2}$$

If the monotonicity condition holds, then the value of the step parameter a remains unchanged (see step 6), and we perform the following double iteration (see step 3) with the same a. If, however, the monotonicity condition (3.7.2) is violated, then the step parameter is reduced according to the harmonic sequence

$$a_0, \tfrac{1}{2}a_0, \tfrac{1}{3}a_0, \ldots . \tag{3.7.3}$$

Assume that we have obtained, by our adaptive procedure, the sequence

$$\underbrace{a_0, a_0,}_{a_1} \ \underbrace{\tfrac{1}{2}a_0, \tfrac{1}{2}a_0, \tfrac{1}{2}a_0,}_{a_2} \ \underbrace{\tfrac{1}{3}a_0,}_{a_3} \ \underbrace{\tfrac{1}{4}a_0}_{a_4} . \tag{3.7.4}$$

Here the monotonicity condition (3.7.2) is violated at the second iteration of the first two double iterations of the stochastic approximation, holds for the third and for the fourth, and is again violated for the fifth, sixth, and seventh. We shall call a sequence of the type (3.7.4) an *adaptive harmonic sequence*, distinguishable from the *nonadaptive harmonic sequence* in (3.7.3).

Thus, according to our adaptive procedure, the step parameter remains the same for a subsequent pair of iterations if at the preceding pair S was approached monotonically from the left ($X'' \geq X' \geq X$) or from the right ($X'' \leq X' \leq X$) and is reduced according to the harmonic sequence (3.7.3) whenever the monotonicity condition (3.7.2) is violated.

Algorithm 1 assumes $a_0 = b^{-1} = [g_0''(x^*)]^{-1}$. Such a choice of a_0 yields minimal asymptotic variance [see section 3.2.4]

$$V = \sigma^2(g_0''(x^*))^{-2}, \quad \text{where } \sigma^2 = \text{Var}[\psi'(x^*, Y)] .$$

In most practical situations $g_0''(x^*)$ is unknown and has to be estimated from the data. We estimated $g_0''(x^*)$ by using (3.2.31) assuming $k = 0.01$, $K = 100$, and [see (3.3.17)]

$$b_i = [H'(0) + P'(0)]\hat{f}_Y(x_i) + \frac{1}{i} \sum_{j=1}^{i} \tilde{\nabla}^2 L(x_i, Y) ,$$

where†

$$\tilde{\nabla}^2 L(x, y) = \begin{cases} H''(x - y) , & y \leq x , \\ P''(y - x) , & y > x , \end{cases}$$

and $\hat{f}_Y(x_i)$ is the empirical density function based on the observations $Y_1, \ldots, Y_i$. Clearly, $a_0 = [g_0''(x^*)]^{-1}$ in Algorithm 1 has to be replaced by its estimate (3.2.31). With this modification, the sequence (3.7.4) can be written as

$$a_0^{(1)}, a_0^{(2)}, \tfrac{1}{2}a_0^{(3)}, \tfrac{1}{2}a_0^{(4)}, \tfrac{1}{2}a_0^{(5)}, \tfrac{1}{3}a_0^{(6)}, \tfrac{1}{4}a_0^{(7)} ,$$

where $a_0^{(i)} = \hat{b}_1^{-1}$.

Stopping Criterion

Denote $\Delta x^*(\delta) = [(1 - \delta)x^*, (1 + \delta)x^*]$, where $0 < \delta < 1$ and x^* is the optimal solution. Let I_{x^*} be the minimal number of iterations such that, for $i \geq I_{x^*}$, the point x_i falls in the neighborhood $\Delta x^*(\delta)$ and remains there for the next T (large) iterations of stochastic approximation. Mathematically, this reads

$$x_i \in \Delta x^*(\delta) \quad \text{for } I_{x^*} \leq i < I_{x^*} + T . \tag{3.7.5}$$

†Note that $b_i \to b = [g_0''(x^*)]^{-1}$ a.s., where $g_0''(x) = [H'(0) + P'(0)]f_Y(x) + E[\tilde{\nabla}^2 L(x, Y)]$.

Our stopping criterion is condition (3.7.5). Note that I_{x^*} is an r.v. and it depends on x_0, on the specific problem (P1), and on the specific SAA.

We run model (3.7.1) for different δ (0.01, 0.03, 0.05, and 0.1) assuming $T = 100$. The number of replications was equal 20.

We would like to emphasize that stopping rule (3.7.5) is chosen artificial. The reason for using it is only to define some δ neighborhoods of x^* and then to run stochastic approximation to see how fast it reaches them.

(a) Unconstrained One-Commodity Inventory Model with Fixed Setup Cost Let $Y \sim U(0, 10)$. The optimal solution of (3.7.1) is readily found to be $S = 7.56$, $\alpha = 238.71$, and $s = 5.87$.

Table 2 presents the minimal number of iterations needed for both off-line and on-line procedures to hit certain δ neighborhoods of s and S, according to criterion (3.7.5). We set $S_0 = \alpha_0 = s_0 = 0$ and $\beta = 0.01$ in (3.3.18), (3.3.26), and (3.3.34).

It follows from Table 2 that for relatively large neighborhoods ($\delta = 0.05$, 0.1) all off-line and on-line algorithms perform rather well (I is small), but for relatively small ones ($\delta = 0.03$, 0.01) their rate of convergence decreases rapidly (I is rather large). It also follows from Table 2 that for large δ off-line algorithms are faster than their on-line counterparts. This advantage disappears, however, as δ decreases. This is readily understandable if we recall [see Polyak and Tsypkin (1980a)] that the asymptotic rate of convergence [see (3.2.30)] is the same for off-line and on-line algorithms.

We also simulated the model (3.7.1) assuming dependence of the demand Y on the ordering policy x, that is, $Y \sim f_Y(y, x)$. Particularly, we considered $Y \sim U(0, r + x)$, $r \geq 1$, $x > 0$. We observed that the speed of convergence are of the same order for the on-line models without memory with $Y \sim U(0, r + x)$, $1 \leq r \leq 5$, and $Y \sim U(0, 10)$ (see Table 2), respectively.

Note As we mentioned earlier, generally, $g_0(x)$ is not convex. However, in this particular example

$$g_0'' = \frac{3r^2(10r + 3)}{(x + r)^3} > 0$$

for any $r > 0$, and therefore $g_0(x)$ is convex.

(b) Constrained One-Commodity Model with No Setup Cost Consider now the problem (3.7.1) with $K = 0$, $Y \sim U(0, 10)$, and constraints of type

Table 2 Minimal Number I of Iterations Needed for (3.7.5) To Hold as Function of δ for the Unconstrained Model

Model	Number of Iterations I	$\delta = 0.01$	$\delta = 0.03$	$\delta = 0.05$	$\delta = 0.1$	Algorithm
Off-line	I_S	794	127	12	6	(3.3.18)
	I_s	420	117	11	7	(3.3.34)
On-line	I_S	721	164	42	15	(3.3.26)
with memory	I_s	409	131	27	11	(3.3.34)
Off-line without	I_S	759	163	108	24	(3.3.40a)
memory (stochastic approximation)	I_s	411	119	73	27	(3.3.40c)

Table 3 Minimal number I_{S*} of iterations needed for (3.7.5) to hold as function of δ for different constrained models

Model	Constraints	Optimal S^*	I_{S*}		
			$\delta = 0.01$	$\delta = 0.05$	$\delta = 0.1$
1	$g_1(x) = E[Y - x] \leq 0$, $x \leq V/v = \infty$, $5 \leq x \leq \infty$	$S^* = S = 7.56$	1380	214	29
2	$g_1(x) = E[Y - x] \leq 0$, $x \leq V/v \leq 6$, $5 \leq x \leq 6$	$S^* = S_1^* = 6$	217	62	14
3	$g_1(x) = E[1.6Y - x] \leq 0$, $x \leq V/v = 9$, $8 \leq x \leq 9$	$S^* = S_2^* = 8$	1212	165	21
4	$P[Y \leq x] \geq 0.4$, $x \leq V/v = 8$, $4 \leq x \leq 8$	$S^* = S = 7.56$	1415	231	35
5	$P[Y \leq x] \geq 0.4$, $x \leq V/v = 6$, $4 \leq x \leq 6$,	$S^* = S_1 = 6$	912	183	31
6	$P[Y \leq x] \geq 0.8$, $x \leq V/v = 9$, $8 \leq x \leq 9$	$S^* = S_2 = 8$	4671	582	63

(3.3.43) and (3.3.51). Table 3 presents the minimal number of iterations needed for (3.7.5) to hold as function of δ. For models 1–3 we used the algorithm (3.3.47)–(3.3.49), and for models 4–6 we used the finite differences algorithm (3.3.53)–(3.3.56). We set $X_0 = 0$, $\rho_i = a_i^{3/4}$, and $\beta_i = a_i^{1/4}$ (e.g., models 4–6). The optimal values S^*, S_1^*, and S_2^* were calculated according to (A2.9). Here, as for the unconstrained model, the SQG algorithms (3.3.47)–(3.3.49) and (3.3.53)–(3.3.56) hit moderate neighborhoods δ rather fast, but their rate of convergence decreases substantially as δ decreases. We observed also that when common random numbers are used in the finite differences estimate $\tilde{\nabla} g_1(x_i)$ in (3.3.53), the rate of convergence increases up to 20%.

APPENDIX 1: PROOF OF THEOREM 1

The proof is taken almost verbatim from Gladyshev (1965). We assume without loss of generality that $\alpha = \theta = 0$ and omit the sign of projection π_X. Rewriting then (3.2.1) as $x_{i+1} = x_i - a_i \psi_i$, squaring x_{i+1} and taking the conditional expectation for given $x_1, \ldots, x_i$, we obtain, together with condition (3.2.5),

$$E(x_{i+1}^2 \mid x_1, \ldots, x_i) = x_i^2 - 2a_i x_i R(x_i) + a_i^2 E(\psi_i^2 \mid x_1, \ldots, x_i)$$
$$\le x_i^2 - 2a_i x_i R(x_i) + a_i^2 d(1 + x_i^2) . \qquad \text{(A1.1)}$$

Substituting (3.2.4) in (A1.1), we have

$$E(x_{i+1}^2 \mid x_1, \ldots, x_i) \le x_i^2(1 + da_i^2) + da_i^2 ,$$

or, equivalently,

$$E(z_{i+1} \mid x_1, \ldots, x_i) \le z_i , \qquad \text{(A1.2)}$$

where

$$z_i = x_i^2 \prod_{k=i}^{\infty} (1 + da_k^2) + \sum_{k=i}^{\infty} da_k^2 \prod_{m=k+1}^{\infty} (1 + da_m^2) . \qquad \text{(A1.3)}$$

Taking now the conditional expectation on both sides of (A1.2) for given $z_1, \ldots, z_i$, we get

$$E(z_{i+1} \mid z_1, \ldots, z_i) \le z_i . \qquad \text{(A1.4)}$$

Inequality (A1.4) shows that z_i is a supermartingale, where

$$Ez_{i+1} \le Ez_i \le \cdots \le Ez_1 < \infty . \qquad \text{(A1.5)}$$

According to the theory of supermartingales, the sequence z_i converges a.s. Hence, by (A1.3) and (3.2.3), the sequence x_i^2 converges a.s. to some r.v. ξ. It remains to show that $P(\xi = 0) = 1$.

From (A1.5), (A1.3), and (3.2.3) we have the boundedness of the sequence Ex_i^2. Further, taking the expectation on both sides of (A1.1) and summing the first i inequalities, we obtain

$$Ex_{i+1}^2 \le Ex_1^2 + \sum_{k=1}^{i} (da_k^2 + da_k^2 Ex_k^2) - \sum_{k=1}^{i} 2a_k Ex_k R(x_k) .$$

From the last inequality we have

$$\sum_{k=1}^{\infty} a_k Ex_k R(x_k) < \infty , \qquad \text{(A1.6)}$$

since the sequence Ex_i^2 is bounded and condition (3.2.3) is fulfilled. Using

(3.2.2) and the fact that $xR(x) \geq 0$ [see (3.2.4)], we obtain from (A1.6) that, for some subsequence i_k,

$$x_{i_k} R(x_{i_k}) \to 0 \quad \text{a.s.} \tag{A1.7}$$

Now since $x_{i_k} \to \xi$ a.s., it follows, directly by comparing (A1.7) with (3.2.4) that $\xi = 0$ with probability 1.

Note It can be shown that the sequence x_i converges to θ not only almost surely but also in the mean square.

APPENDIX 2: CONSTRAINED ONE-COMMODITY INVENTORY MODELS

(a) With No Setup Cost Consider the problem

$$\min g(x) = \min[c(x-r) + G(x)] \tag{A2.1}$$

subject to

$$r \leq x \in D\,, \tag{A2.2}$$

where D is a closed set in R^1 and $G(x)$ is strictly convex and given by (3.3.2). For example, D can be defined as in (3.3.43) or (3.3.51) or more generally as $D = \{x\colon\ g_j(x) = E[\psi_j(x, Y)] \leq 0,\ j = 1, \ldots, M\}$. It is clear that if $D = R^1$, the problem (A2.1) and (A2.2) coincides with the conventional problem (3.3.1) for which the optimal policy is given by (3.3.3).

Let D^* be the set of solutions of the problem

$$\min_{x \in D} \{cx + G(x)\}\,. \tag{A2.3}$$

Denote by S the solution of

$$\min_{x \in R^1} \{cx + G(x)\}\,, \tag{A2.4}$$

which is unique since $G(x)$ is strictly convex (see Fig. 3). Denote

$$\begin{aligned} S_1^* &= \max\{x\colon\ x \in D,\ x \leq S\}\,,\\ S_2^* &= \min\{x\colon\ x \in D,\ x \geq S\}\,. \end{aligned} \tag{A2.5}$$

These values exist since D is closed.

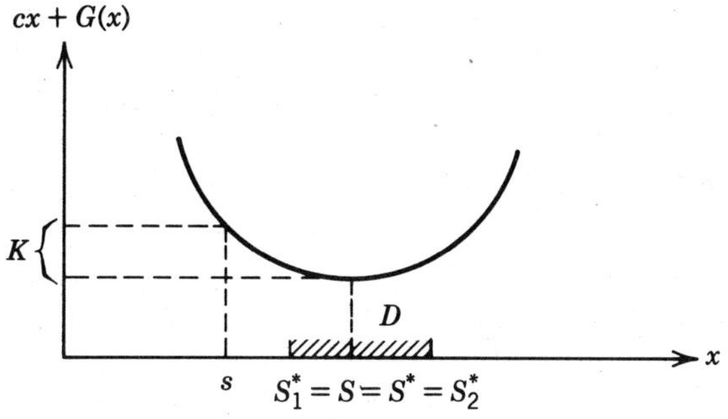

Fig. 3.

Consider the following two cases:

1. $S \in D$. It is obvious then (see Fig. 3) that D^* consists of only one point S^* and that $S^* = S = S_1^* = S_2^*$.
2. $S \overline{\in} D$. In this case it is clear that if

$$cS_1^* + G(S_1^*) < cS_2^* + G(S_2^*)\,, \tag{A2.6}$$

then D^* consists of only one point S^* such that $S^* = S_1^*$ (see Fig. 4). If the inequality (A2.6) is reversed, that is,

$$cS_1^* + G(S_1^*) > cS_2^* + G(S_2^*)\,, \tag{A2.7}$$

then again D^* consists of only one point $S^* = S_2^*$. If

$$cS_1^* + G(S_1^*) = cS_2^* + G(S_2^*)\,, \tag{A2.8}$$

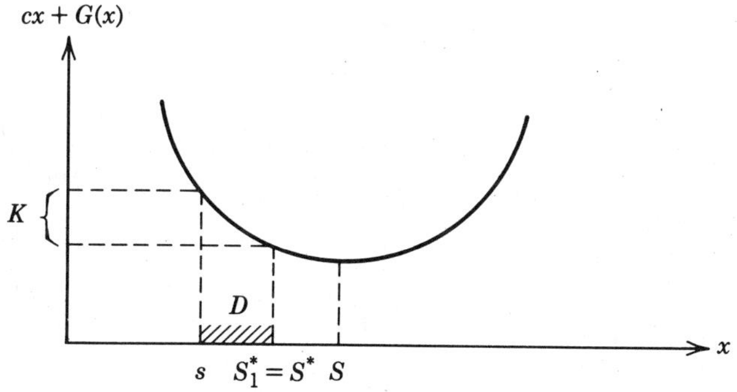

Fig. 4.

then D^* consists of two points S_1^* and S_2^*, and therefore either S_1^* or S_2^* is an optimal solution for (A2.1) and (A2.2).

Summarizing the above, we have:

Proposition A1 The optimal solution of the problem (A2.1) and (A2.2) under the assumption that D is closed and $G(x)$ is strictly convex is

$$\begin{array}{ll} \text{order } (S^* - r) & \text{if } S^* > r\,, \\ \text{do not order} & \text{if } S^* \leq r\,, \end{array} \tag{A2.9}$$

where S^* coincides with either S_1^* or S_2^* as defined in (A2.5), namely, $S^* = S_1^*$, $S^* = S_2^*$, and S^* is equal to any of S_1^*, S_2^*, depending on whether (A2.6), (A2.7), or (A2.8), respectively, holds.

Corollary If $D = [a, b]$ $(b \leq \infty)$, then the set D^* consists of only one point, and

$$S^* = \begin{cases} b & \text{if } b < S\,, \\ S & \text{if } a \leq S \leq b\,, \\ a & \text{if } a > S\,, \end{cases} \tag{A2.10}$$

where S is the unique minimum of function (A2.4).

(b) With Fixed Ordering (Setup) Cost This problem differs from the preceding one in that a fixed (setup) cost K is charged for placing an order and can be written as

$$\min g(x) = \min[G(r),\ K + c(x - r) + G(x)] \tag{A2.11}$$

subject to

$$r \leq x \in D\,, \tag{A2.12}$$

where D is a closed set in R^1 and $G(x)$ is strictly convex and is given by (3.3.2). It is clear that in the particular case where $D = R^1$, the optimal policy of (A2.11) and (A2.12) is the conventional (S, s) policy given in (3.3.28).

To find the optimal inventory policy for (A2.11) and (A2.12) for arbitrary closed D, consider the problem

$$\min_{x \in D} \{K + c(x - r) + G(x)\}\,. \tag{A2.13}$$

Obviously the set D^* of the solutions of (A2.13) is the same as in (A2.3), that is, it consists of at most two elements S_1^* and S_2^* given by (A2.5). Consider separately three possible cases, that is, $D^* = \{S_1^*\}$, $D^* = \{S_2^*\}$, and $D^* = \{S_1^*, S_2^*\}$, respectively.

(a) $D^* = \{S_1^*\}$. In this case $cx + \bar{G}(x)$ monotonically decreases on the interval $[0, S_1^*] \subseteq [0, S]$, and it is readily seen (see Fig. 4) that the optimal solution x^* can be written as

$$x^* = \begin{cases} r & \text{if } r \in D \quad \text{and} \quad G(r) \le K + c(S_1^* - r) + G(S_1^*)\,, \\ S_1^* & \text{if } r \overline{\in} D \quad \text{or} \quad G(r) > K + c(S_1^* - r) + G(S_1^*)\,. \end{cases} \tag{A2.14}$$

Let us rewrite

$$G(r) \le K + c(S_1^* - r) + G(S_1^*)$$

as

$$cr + G(r) \le K + cS_1^* + G(S_1^*)\,. \tag{A2.15}$$

Let s be the root of the equation

$$cx + G(x) = K + cS_1^* + G(S_1^*) \tag{A2.16}$$

on the interval $[0, S_1^*]$. Noting that s is unique (since $G(x)$ is strictly convex) and since (A2.15) holds for $r \ge s$ (see Fig. 4), we readily conclude that (A2.14) can be rewritten as

$$x^* = \begin{cases} r & \text{if } (r \in D \quad \text{and} \quad S_1^* \ge s) \text{ or } r \ge S_2^*\,, \\ S_1^* & \text{if } r \overline{\in} D \quad \text{or} \quad \text{if } r < s\,. \end{cases} \tag{A2.17}$$

Therefore the optimal policy is

$$\begin{aligned} &\text{order } S_1^* - r && \text{if } r \overline{\in} D \quad \text{or} \quad r < s\,, \\ &\text{do not order if} && (r \in D \quad \text{and} \quad S_1^* \ge r \ge s) \text{ or } r \ge S_2\,. \end{aligned} \tag{A2.18}$$

(b) $D^* = \{S_2^*\}$. In this case (see Fig. 5) $cx + G(x)$ monotonically decreases on the interval $[0, S]$ and monotonically increases on the interval $[S, S_2^*]$. In analogy to the previous case the optimal policy is

$$\begin{aligned} &\text{order } S_2^* - r && \text{if } r \overline{\in} D \quad \text{or} \quad r < s\,, \\ &\text{do not order} && \text{if } r \in D \quad \text{and} \quad r \ge s\,, \end{aligned} \tag{A2.19}$$

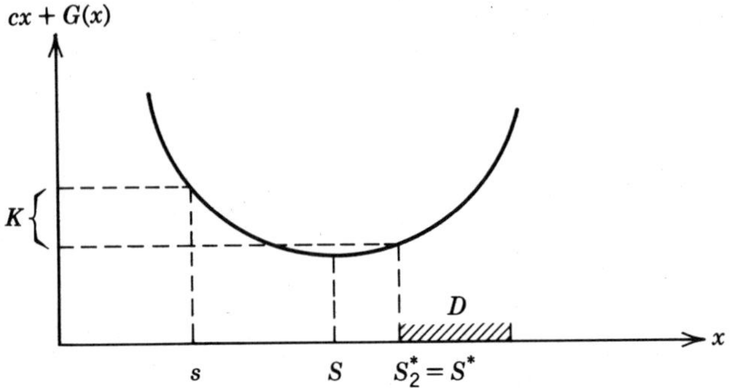

Fig. 5.

where s is the unique root of equation (A2.16) with S_1^* replaced by S_2^* on the interval $[0, S]$.

(c) $D^* = \{S_1^*, S_2^*\}$. In this case as in the previous ones (see Fig. 6), $cx + G(x)$ monotonically decreases on the interval $[0, S]$ and increases on the interval $[S, S_2^*]$. Clearly, either of the two policies (A2.18) or (A2.19) is optimal.

Summarizing the three cases, we obtain the following:

Proposition A2 The optimal solution of the problem (A2.11) and (A2.12) under the assumption that D is closed and $G(x)$ is strictly convex is

$$\begin{cases} \text{order } S^* - r & \text{if } r \overline{\in} D \text{ or } r < s\,, \\ \text{do not order} & \text{if } r \in D \text{ and } r \geq s\,, \end{cases} \tag{A2.20}$$

where S^* is chosen as in Proposition A1, that is, S^* is one of at most two possible numbers S_1^* and S_2^* defined in (A2.5) and s is the smallest root of

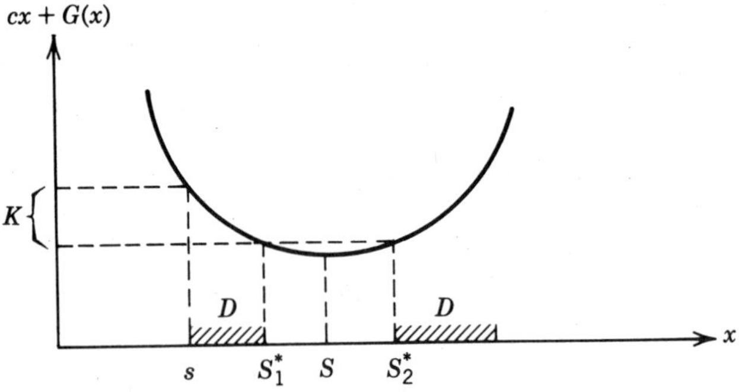

Fig. 6.

the equation

$$cx + G(x) = K + cS^* + G(S^*) \tag{A2.21}$$

on the interval $[0, \min\{S, S^*\}]$. The algorithm for finding the (S^*, s) policy follows.

Algorithm

1. Find S, the unique solution of (A2.4).
2. If $S \in D$, then $S^* = S$; go to step 8.
3. Find $S_1^* = \max\{x\colon x \in D, x \leq S\}$ [see (A2.5)].
4. Find $S_2^* = \min\{x\colon x \in D, x \geq S\}$ [see (A2.5)].
5. If $cS_1^* + G(S_1^*) < cS_2^* + G(S_2^*)$, then $S^* = S_1^*$; go to step 8.
6. If $cS_2^* + G(S_2^*) < cS_1^* + G(S_1^*)$, then $S^* = S_2^*$; go to step 8.
7. If $cS_1^* + G(S_1^*) = cS_2^* + G(S_2^*)$, then $S^* = S_1^*$ or $S^* = S_2^*$.
8. Find s, the unique solution of (A2.16) on the interval $[0, S]$.

REFERENCES

Abdelhamid, S. N. (1973). Transformation of Observations in Stochastic Approximation, *Michigan State University*, Technical Report.

Akaike, M. (1969). Fitting Autoregressive Models for Prediction, *Ann. Instit. Stat. Math.*, **21**, 243–247.

Al'ber Ya. I. and S. V. Shilman (1981). Nonasymptotic Estimation of the Convergence Speed of Stochastic Iterative Algorithms, *Automat. Remote Control*, No. 1, 41–52.

Anbar D. (1973). On Optimal Estimation Methods Using Stochastic Approximation Procedures, *Ann. Stat.*, **1**, No. 6, 1175–1184.

Arrow, K. J., S. Karlin, and H. Scarf (1958). *Studies in the Mathematical Theory of Inventory and Production*, Stanford University Press, Stanford, CA.

Avriel, M. (1976). *Nonlinear Programming Analysis and Methods*, Prentice-Hall, Englewood Cliffs, NJ.

Dvoretsky, A. (1956). On Stochastic Approximation, in *Proceedings of the Third Berkeley Symposium on Mathematical Statistics and Probability*, Vol. 1, pp. 39–55.

Ermoliev, Yu. M. (1969). On the Method of Generalized Stochastic Gradients and Quasi-Fejer Sequences, *Cybernetics*, **5**, No. 2, 208–220.

Ermoliev, Yu. M. (1983). Stochastic Quasigradient Methods and Their Applications to System Optimization, *Stochastic*, **9**, 1–36.

Ermoliev, Yu. M. and Mirzoakhmedov (1976). Direct Methods of Stochastic Programming in Inventory Control Problems, *Cybernetics*, No. 6.

Gal, Sh., R. Rubinstein, and A. Ziv (1984). On the Optimality and Efficiency of Common Random Numbers, *Math. Comp. Simul.*, **XXVI**, 502–512.

Gladyshev, E.G. (1965). On Stochastic Approximation, *Theory of Probability Applic.*, **10**, No. 2, 275–278.

Goodwin, G. C., P. J. Ramadge and P. E. Caines (1981). Discrete Time Stochastic Adaptive Control, *SIAM J. Cont. Opt.*, **19**, 829–853.

Gupal, A. M. (1977). Direct Method of Solution of Stochastic Programming Problems, *Automat. Remote Control, No.* 4, 502–506.

Gupal, A. M. (1979). *Stochastic Methods of Nonsmooth Optimization*, Naukova Dumka, Kiev (in Russian).

Handbook of Operations Research (1978), Vol. 2, Van Nostrand Reinhold, New York.

Hillier, F. S. and G. J. Lieberman (1974). *Introduction to Operations Research*, Holden Day, San Francisco, CA.

Huber, P. J. (1964). Robust Estimation of a Location Parameter, *Ann. Math. Stat.*, **35**, No. 1, 23–101.

Huber, P. J. (1972). Robust Statistics: A Review, *Ann Math. Stat.*, **43**, 1041–1067.

Huber, P. J. (1977). *Robust Statistical Procedures*, SIAM, Philadelphia.

Huber, P. J. (1981). *Robust Statistics*, Wiley, New York.

Johnson, L. A. and D. C. Montgomery (1979). Operations Research in Production Planning, *Scheduling and Inventory Control*, Wiley, New York.

Kiefer, J. and J. Wolfowitz (1952). Stochastic Estimation of the Maximum of Regression Function, *Ann. Math. Stat.*, **23**, 462–466.

Kushner, H. I. and D. S. Clark (1978). *Stochastic Approximation. Methods for Constrained and Unconstrained Systems*, Springer-Verlag, New York.

Lai, T. L. and H. Robbins (1979). Adaptive Design or Stochastic Approximation, *Ann. Stat.*, **7**, No. 6, 1196–1221.

Lai, T. L. and H. Robbins (1981). Consistency and Asymptotic Efficiency of Slope Estimates in Stochastic Approximation Schemes, *Zeitschrift Wahrscheinlichkeitstheorie*, **56**, 329–360.

Nevelson, M. B. and R. Z. Hasminskii (1973). *Stochastic Approximation and Recursive Estimation*, American Mathematical Society.

Polyak B. T. (1979). Subgradients Methods: A Survey of Soviet Research, in *Nonsmooth Optimization*, edited by Lemarechal and Mifflin, Pergamon Press, New York, pp. 5–30.

Polyak B. T. and Ya. Z. Tsypkin (1979). Adaptive Estimation Algorithms (Convergence, Optimality, Stability), *Automat. Remote Control*, No. 3, 71–84.

Polyak, B. T. and Ya. Z. Tsypkin (1980a). Optimal Pseudogradient Adaptive Algorithms, *Automat. Remote Control*, No. 8, 74–84.

Polyak, B. T. and Ya. Z. Tsypkin (1980b). Robust Psuedogradient Adaptive Algorithms, *Automat. Remote Control*, No. 10, 91–97.

Robbins, H. and S. Monro (1951). Stochastic Approximation Methods, *Ann. Math. Stat.*, **22**, 400–407.

Rubinstein, R. Y. (1976). A Monte Carlo method for estimating the gradient in a stochastic network, Technion, HAIFA, Israel (unpublished manuscript).

Rubinstein, R. Y. (1981). *Simulation and the Monte Carlo Methods*, Wiley, New York.

Rubinstein, R. Y. and J. Kreimer (1985). Stochastic optimization via stochastic approximation, Technion, Haifa, Israel to be published in *Math. Comp. Simul.*

Rubinstein, *R. Y.* (1986) Sensitivity analysis and performance extrapolation for computer simulation models. Manuscript, Harvard University, Division of Applied Sciences.

Saks, I. (1958). Asymptotic Distribution of Stochastic Approximation Procedures. *Ann. Math. Stat.*, **20**, 373–405.

Sielkin, R. L. (1973). Stopping rules for Stochastic Approximation Procedures, *Zeitschrift Wahrscheinlichkeitstheorie*, **39**, 135–150.

Sivazlian, B. D. and L. E. Stanfel (1975). *Analysis of Systems in Operations Research, Prentice-Hall, Englewood Cliffs, NJ.*

Stroup, D. F. and M. I. Braun (1982). On a New Stopping Rule for Stochastic Approximation, *Zeitschrift Wahrscheinlichkeitstheorie*, **60**, 535–554.

Tsypkin, Ya. Z. (1970). *Adaptation and Learning in Automatic Systems*, Academic Press, New York.

Tsypkin, Ya. Z. (1979). Adaptive Optimization Algorithms When There Is a Priori Uncertainty, *Automat. Remote Control*, **6**, 94–108.

Tsypkin, Ya. Z. (1982a). Optimal Performance Criteria in Identification Problems, *Automat. Remote Control*, **11**, 5–24.

Tsypkin, Ya. Z. (1982b). Optimal Parameter Estimation Algorithms in Identification Problems, *Automat. Remote Control*, **12**, 9–23.

Tukey, I. W. (1960). A Survey of Sampling from Contaminated Distributions. Contributions to Probability and Statistics, edited by I. Olkin, Stanford University Press, Stanford, CA.

Venter, I. M. (1967). An Extension of the Robbins Monro Procedure. *Ann. Math. Stat.*, **38**, 181–190.

Wasan, M. T. (1979). *Stochastic Approximation*, Cambridge University Press, New York.

Zang, I. (1980). A Smoothing-out Technique for min max Optimization, *Math. Programming*, **19**, 61–77.

Zang, I. and A. Tishler (1982). An Absolute Deviations Curve-Fitting Algorithm for Nonlinear Models, *Studies Manage. Sci.*, **19**, 81–103.

CHAPTER 4

Perturbation Analysis for Sensitivity and Optimization of Complex Queueing Networks

This chapter surveys the perturbation analysis technique for sensitivity and of optimization discrete event dynamic systems developed by the Harvard team of Y. C. Ho, R. Suri, X. R. Cao, G. W. Diehl, J. W. Dille, and M. Zazanis. The material is based mainly on Ho et al. (1985), some parts (especially Sections 4.2–4.4) reproduced almost verbatim.

4.1 INTRODUCTION: DEDS AND THE PERTURBATION ANALYSIS APPROACH

Many important modern large-scale interconnected systems can be modeled as *discrete-event dynamic systems* (*DEDSs*). Examples of such systems are communication networks [Tenenbaum (1981)], computer systems [Sauer and Chandy (1981)], production assembly lines [Buzacott (1976)], flexible manufacturing systems [Suri and Hildebrant (1984)], traffic systems, and transportation networks [Newell (1980)]. These systems are driven by the occurrence of discrete events, and their dynamics involve state changes only at some points in time. Due to the complex interactions of such discrete events over time, studying the performance and sensitivity of DEDSs and optimizing their parameters are important tasks.

The conventional approaches for analyzing the performance of DEDSs are mainly analytic, simulation, and benchmarking.

The analytic approach is associated with the modeling DEDSs as Markovian queueing networks. Various assumptions (some of them unrealistic) have to be imposed in order for the models to be tractable analytically. For instance, where any of the factors such as nonexponential service times, multiple classes of customers, limited buffer sizes, transient behavior, and real-time scheduling or routing is significant, the analytical solution is not feasible.

In such a case simulation or benchmarking must be resorted to. By simulation, we mean computer-based discrete-event simulation [see Fishman (1978) and Rubinstein (1981)], which involves writing a computer program to mimic the detailed operation of the system and then performing a Monte Carlo experiment. Note that for n parameters each iteration step requires n new simulations [see, e.g., the finite difference estimate (3.2.20)] while optimizing DEDS. This method can therefore be computationally very demanding.

A third possibility is benchmarking, which involves constructing the actual system, leaving some scope for "tuning" or changing certain parameters, and then subjecting it to trial work loads. A well-known example of this is in the installation of a new computer operating system when the various control parameters of the system are varied by trial and error to best suit the characteristics of the users at the particular site. While optimizing systems performance, benchmarking involves a large number of experiments and is therefore also time consuming.

Recently an efficient alternative to discrete-event simulation and benchmarking has been proposed. This alternative is called *perturbation analysis (P/A) of DEDS*. It enables the sensitivity of a performance measure (gradients, Hessians, etc.) to be calculated while observing only a *single* sample path of the system.† It therefore offers computational savings for computer simulations, and it also has the ability to be applied directly on actual systems.

The perturbation analysis approach to DEDS was introduced by Ho, Eyler, and Chien (1979) to solve a design problem in production lines, that is, a set of queues in tandem. Extensions and formalizations to determine throughput sensitivity (gradient) w.r.t. (with respect to) various parameters, for tandem queues, and for certain other networks were given in Eyler (1980), Cassandras (1982), Ho, Eyler, and Chien (1983), and Ho and Cassandras (1983). The method was generalized for throughput

†Note that in Section 3.6 we have already used a single sample path for estimating the sensitivities of the model (6.1). Model (6.1) and its examples (reliability and stochastic network) can be called DESS (discrete event *static* systems) to distinguish then from DEDS (discrete event *dynamic* systems). Clearly, DEDS generalize DESS in the sense that the sample path of DESS has a single busy period only.

sensitivity w.r.t. service time parameters to Markovian queueing networks (general probability service distributions, blocking, etc.) [see Ho et al. (1984) and Dille and Ho (1985)]. Sensitivity of the sojourn time in GI/G/1 queue was studied by Zazanis and Suri (1985), and for general queueing network, by Cao and Ho (1985). Some foundations of P/A approach for a rather general class of discrete event systems can be found in Suri (1984), Cao (1985) and Rubinstein and Szidarovszky (1985, 1985a).

Before presenting the main idea of P/A, we define the performance measure and some related characteristics. Denote by $\psi(\theta, w)$ the performance measure of DEDS, where θ is a decision parameter and w the outcome of the random process. In a queueing network, for example, ψ, θ, and w may be, respectively, the throughput, mean service time of a server, and actual values (realizations) of the service time observed during simulation.

Let $l(v) = E(\psi(\theta, w))$ be the expected steady-state performance of the system. When an analytic expression is not avaliable for $l(\theta)$ (because of the complexity of DEDS), Monte Carlo simulation must be applied. As an estimate of $l(\theta)$, consider

$$\bar{l}_N(\theta) = \frac{1}{N} \sum_{k=1}^{N} \psi(\theta, w_k) ,$$

where N is the sample size (number of independent replications, regenerative cycles, batches, etc.), w_k are i.i.d. r.v.'s (independent identically distributed random vectors), and $\psi(\theta, w_k)$ is called the *sample function* corresponding to w_k. Clearly, in this case the gradient, say $\partial l/\partial\theta = \partial E(\psi(\theta, w))/\partial\theta$, of $\psi(\theta)$ is likewise not available. As estimates of $\partial l/\partial\theta$ we consider the following ones:

$$\left(\frac{\partial l}{\partial \theta}\right)_{N1} = \frac{1}{N\Delta\theta} \sum_{k=1}^{N} (\psi(\theta + \Delta\theta, w_k) - \psi(\theta, w_k)) , \tag{4.1.1}$$

$$\left(\frac{\partial l}{\partial \theta}\right)^{s}_{N1} = \frac{1}{N\Delta\theta} \sum_{k=1}^{N} \left(\psi\left(\theta + \frac{\Delta\theta}{2}, w_k\right) - \psi\left(\theta - \frac{\Delta\theta}{2}, w_k\right)\right), \tag{4.1.2}$$

$$\left(\frac{\partial l}{\partial \theta}\right)_{N2} = \lim_{\Delta\theta\to 0} \frac{1}{N\Delta\theta} \sum_{k=1}^{N} (\psi(\theta + \Delta\theta, w_k) - \psi(\theta, w_k)) , \tag{4.1.3}$$

$$\left(\frac{\partial l}{\partial \theta}\right)_{N3} = \frac{1}{N\Delta\theta} \sum_{k=1}^{N} \psi(\theta + \Delta\theta, w_k) - \sum_{k=N+1}^{2N} \psi(\theta, w_k) , \tag{4.1.4}$$

and

$$\left(\frac{\partial l}{\partial \theta}\right)^s_{N3} = \frac{1}{N\Delta\theta} \sum_{k=1}^{N} \psi\left(\theta + \frac{\Delta\theta}{2}, w_k\right) - \sum_{k=N+1}^{2N} \psi\left(\theta - \frac{\Delta\theta}{2}, w_k\right). \quad (4.1.5)$$

Note that we use superscript s in (4.1.2) and (4.1.5) to indicate symmetric finite differences. Note also that for each k in estimates (4.1.4) and (4.1.5) we use two *different* independent random vectors (numbers), namely w_k and w_{k+N} for $\theta + \Delta\theta$ and θ (and for $\theta + \Delta\theta/2$ and $\theta - \Delta\theta/2$), respectively, while estimates (4.1.1) (and (4.1.2)) use *common random numbers* (CRN) for both $\theta + \Delta\theta$ and θ (and for $\theta + \Delta\theta/2$ and $\theta - \Delta\theta/2$). We call the estimates (4.1.1)–(4.1.2), (4.1.3), (4.1.4)–(4.1.5) the CRN, the *infinitesimal* PA, and the *crude Monte Carlo* (CM) estimates respectively.

Similarly we can define the sample finite difference estimates for estimating the Hessian of $\psi(\theta, w)$.

Under some conditions [see Cao (1985) and Rubinstein and Szidarovszky (1985, 1985a)], the sample derivative (4.1.3) is an unbiased and strongly consistent estimate of the derivative of the expected value of $\psi(\theta, w)$, that is,

$$E\left\{\frac{\partial}{\partial \theta}\left[\psi(\theta, w)\right\}\right. = \frac{\partial}{\partial \theta}\{E\psi(\theta, w)\}\,.$$

Observe that the last equality means that the operators expectation and differentiation are interchangeable.

To clarify some of P/A concepts, consider the following simple example. Let the r.v. X (say, service time in a queueing system) be distributed $F(x; \theta)$, where θ is a parameter (say, mean service time). Also, let X be generated by the inverse transform method, namely,

$$X = F_\theta^{-1}(U) = \inf\{x \colon F(x; \theta) \geq U\}\,, \quad (4.1.6)$$

where $U \sim U(0, 1)$ is a random number. Clearly, generating a stream of random numbers $U_1, \ldots, U_n$ and applying (4.1.6), we obtain the sequence

$$X_i = F_\theta^{-1}(U_i)\,, \qquad i = 1, \ldots, N\,, \quad (4.1.7)$$

called the *nominal sample path*.

Next, we replace the parameter θ by $\theta + \Delta\theta$, that is, consider $F(x; \theta + \Delta\theta)$ instead of $F(x; \theta)$. Using the *same* stream of random numbers $U_1, \ldots, U_N$ and again applying (4.1.6) for $F_{\theta+\Delta\theta}^{-1}(\cdot)$, we now obtain the sequence

$$(X_i + \Delta X_i) = F_{\theta+\Delta\theta}^{-1}(U_i)\,, \qquad i = 1, \ldots, N\,, \quad (4.1.8)$$

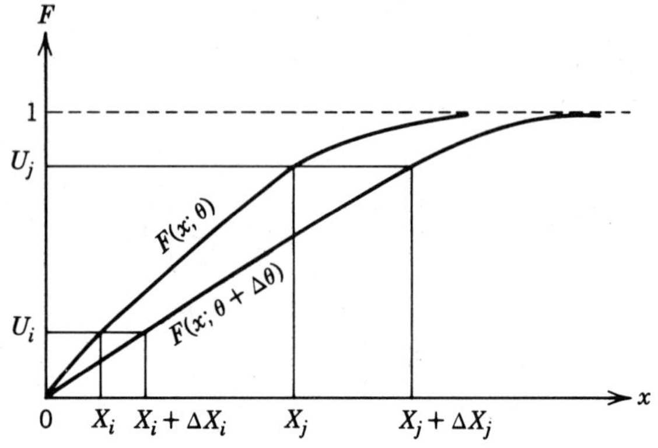

Fig. 1.

called the *perturbed sample path* (see Fig. 1). As an example of the above, let $X \sim \exp(\theta)$; that is,

$$F(x; \theta) = 1 - e^{-x/\theta} .$$

In this case the nominal sample path is

$$X_i = F_\theta^{-1}(U_i) = -\theta \ln(1 - U_i) , \qquad i = 1, \ldots, N ,$$

and its perturbed counterpart is

$$X_i + \Delta X_i = F_{\theta+\Delta\theta}^{-1}(U_i) = -(\theta + \Delta\theta) \ln(1 - U_i) , \qquad i = 1, \ldots, N .$$

In the Appendix based on Suri and Zazanis (1984), the P/A approach is applied in calculating the sensitivity of the mean waiting time of a customer in a GI/G/1 queue.

In this chapter we shall use the P/A approach to study sensitivity and optimization of a large class of closed queueing networks. More precisely, we shall deal with the sensitivity of throughput w.r.t. service time parameters and w.r.t. to the number of customers of each class. We assume that the networks have, at each server, (class-dependent) general service time distributions, first-come–first-served (FCFS) discipline, and limited buffer space so that an analytic solution is unfeasible.

This chapter is organized as follows: Section 4.2 gives the background for closed queueing networks and describes the model under consideration and some performance measures. Section 4.3 presents an overview of the perturbation analysis technique. In Section 4.4 we present simulation results illustrating the use of the P/A approach in studying the sensitivity of a broad class of closed queueing networks. Section 4.5 discusses some ideas for further research, and finally in the Appendix we apply the P/A approach to study the sensitivity of the mean waiting time of a customer in a GI/G/1 queue.

4.2 CLOSED QUEUEING NETWORKS

4.2.1 Background

A queueing network consists of a number of servers and customers. There are two ways that a queueing network can model a large-scale interconnected system. One way is to let the servers represent subsystems (e.g., individual computers in a distributed computer network) while the customers represent the interactions between them (e.g., messages or tasks being sent between the computers). A "dual" representation is to view the customers as subsystems (e.g., individual product manufacturing teams in a factory that makes several products) and servers as shared resources (e.g., tools, fixtures, pallets). The interaction between the subsystems is modeled by the customers queueing for and receiving the resources.

A subclass of these models that has been useful in practical systems is that of *closed* queueing networks. In a closed network model the number of customers of each type circulating in the network remains fixed. This is a useful model because in practice the availability of a scarce resource can limit the number of entities of a particular type in a system. As an example, for a class of automated manufacturing systems called flexible manufacturing systems, the number of fixtures for a given part type is limited, and this effectively controls the number of parts of that type in circulation [Suri and Hildebrant (1984)]. Another example is the "window" flow control protocol used in communication networks [Reiser (1979)], where a message of a given type can enter the network only when another message of that type has departed. A third example is a computer system with a fixed degree of multiprogramming, where the number of programs executing simultaneously is fixed [Buzen (1971)].

Typical design and operating parameters for these networks include decision parameters such as parameters of service time distributions, routing or work allocation to servers, number of customers of each class, buffer storage sizes, and parameters of priority rules for scheduling of service or routing.

The typical size of such queueing network models is 5–100 servers and 1–50 customer types (or classes) for models of computer systems or communication networks. Automated manufacturing systems typically have 2–40 servers and 1–20 customer classes. However, the difficulty of analyzing these networks is not immediately apparent from these (seemingly) modest numbers.

The reason these networks lead to *large* computational problems arises from the state space of such a system. In the simplest case, for a queueing network with M servers and N customers where there is only one class of customers and all service time distributions are exponential, the *state* of the network is $n = (n_1, \ldots, n_M)$, where n_i is the number of customers present at server i [Gordon and Newell (1967)]. The number of such states

$$n = \binom{M+N-1}{N}.$$

For just 10 servers and 20 customers this value exceeds 10 million. For multiple classes of customers this number grows at a faster rate, and finally, with first-come first-serve discipline the order of customers at each server also matters, and the number of states grows at a still faster rate. In the case of exponential service time distributions (the easiest to solve), the evolution of the network states is governed by a Markov chain over all the states. Thus solving for the network performance involves solving a set of L linear equations, where L is the number of states as just discussed. Finally, when the service time distributions are arbitrary, we get a hybrid (discrete/continuous) state space and an enormous number of integro-differential equations describing the system behavior. In this case even numerical solution of the equations is completely out of the question, and alternative approaches must be used.

4.2.2 Product Form Solution

For a particular class of queueing networks, however, solutions *can* be derived efficiently. This is the class of *product form* networks [Chandy and Martin (1983)]. This terminology, and the significance of these networks, can be understood by considering the single-class. M-server and N-customer closed network with exponential service times. For such a network, the steady-state probability of being in state n turns out to have the (product) form $(1/G)\Pi_{i=1}^{M} f_i(n_i)$, where the f_i are known functions and G is a normalization constant, that can be calculated according to Gordon and Newell (1967). Because of the large number of states of such networks, this was originally perceived to be a formidable task. However, Buzen (1973) derived an $O(MN)$ algorithm to calculate G. Following this, extensions allowing efficient solution of certain types of multiclass networks were also derived [see Reiser and Kobayashi (1975)]. Although these product form models are based on seemingly restrictive assumptions, the models tend to be robust to these assumptions, which often justifies their use in practical cases [(see Suri (1983)]. As a result of all these developments, closed queueing network models have been widely used for computer communication systems as well as in many other fields [see Koenigsberg (1982)].

4.2.3 The Model and its Performance

We will consider a closed queueing network with M servers, $C_1, C_2, \ldots, C_M$, and K classes of customers. Let N_k be the population of class k, B_m the buffer size at C_m (including the space at the server), $F(w; \theta_{mk})$ the cumulative distribution function of service time for class k customers at C_m, parametrized by θ_{mk}. In general, θ_{mk} could be a vector of parameters, but we consider it to be a scalar for simplicity. Extension to the vector case is straightforward. The service discipline is FCFS everywhere. The routing of customers can be arbitrarily specified. If a customer completes service at C_i and must go to C_j next, but B_j is full, then the customer must wait at C_i, which cannot serve another customer and is said to be *blocked* by C_j. We allow state-dependent routing too; that is, the customer may see that B_j is full and go to another server—the blocking condition only holds if the customer decides that it must go to C_j.

This network has the following features that make analytic solutions intractable: general service time distributions, multiple classes with FCFS discipline, and limited buffer sizes.

The *performance measure* considered here will be throughput X_k of class k, defined as follows. For each class k we define an output server $C_{o(k)}$ [that is, $o(k)$ is the index of the output server for class k]. Let R_{mk} be the number of service completions of class k at C_m. We define the termination condition for an experiment to occur when $R_{m^*k^*} = R_{\max}$ for initially specified m^*, k^*, and $R_{\max}$. Let T be the total time of the experiment at termination, then

$$X_k = \frac{R_{o(k),k}}{T} . \qquad (4.1.9)$$

Various alternative performance measures and definitions [see Dille and Ho (1985)] could be used, but the above will serve to illustrate the method. It should be noted that the method is in no way restricted to using the above definitions.

4.3 OVERVIEW OF PERTURBATION ANALYSIS TECHNIQUE

Let θ_{mk} now be a decision parameter, and suppose we are interested in the sensitivity of throughput with respect to θ_{mk}. In particular, we are interested in answering the question: If θ_{mk} had been $\theta_{mk} + \Delta\theta$, what would have been the corresponding throughput for the resulting (perturbed) path? In the *nominal sample path* each service time of class k customers at server m, say t_{mkn} for the nth such time, would have been perturbed by Δt_n. The value of Δt_n depends on $\Delta\theta$, on the nominal value t_{mkn}, and on the service time distribution and can be calculated *analytically* from the *nominal sample path*.†

Next, each generated perturbation is *propagated* along the nominal path according to the *propagation rules*, and the various perturbations are added together using superposition rules, to determine the net effect on the performance [see Ho, Cao and Cassandras (1983), Ho and Cao (1983), and the example in the Appendix]. Finally, the results can be appropriately interpreted to provide sensitivity and therefore to estimate the gradient of $l(v)$ associated with this sample path.

It is very important to note that all the above calculations can be implemented on a *single* simulation run (or experiment) with little added computational effort.‡ Thus, essentially, we can compute a whole gradient at almost the same time as performing the *single* simulation run. For n parameters we would have an *n-fold increase in efficiency* over the conventional (brute force) simulation.

We discuss now some basic perturbation analysis concepts considering two servers C_i and C_j in a general queueing network. Let the nominal sample path of the servers between times t_0 and t_3 be as in Fig. 2. The figure shows that C_i is busy for a while until a customer departs (at time t_1) after which C_i goes idle. Meanwhile, at time t_2, C_j finishes serving a customer who goes immediately to C_i, terminating C_i's idle period.

Now suppose a perturbation§ of Δt exists at C_j at time t_0. Then, as seen in Fig. 3*a*, the idle period of C_i would be terminated Δt time units later, and the perturbation of Δt would now exist at C_i. We say the perturbation has been propagated to C_i. (In general, it still exists at C_j too.) Next, suppose that a perturbation Δs existed at C_i at time t_0. Then we see that

†Recall from Section 4.1 that both the nominal $F^{-1}_{\theta_{mk}}(U_m)$ and the perturbed $F^{-1}_{\theta_{mk}+\Delta\theta}(U_m)$, $m = 1, \ldots, N$, sample paths use the same stream of random numbers $U_1, \ldots, U_N$.

‡The amount of additional memory/CPU time is usually negligible compared with the needs of the nominal simulation. Also, the P/A computations are elementary enough to be done while the simulation (or experiment) is evolving.

§The perturbation at a server is the difference in occurrence times of a specified event at that server for the nominal versus perturbed paths.

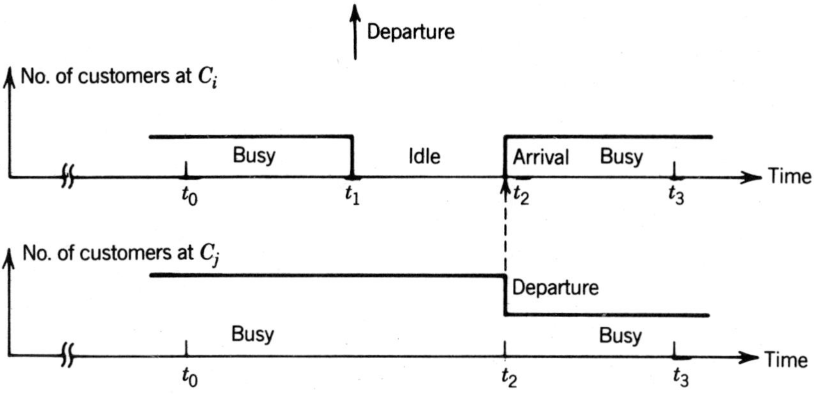

Fig. 2 Nominal sample path.

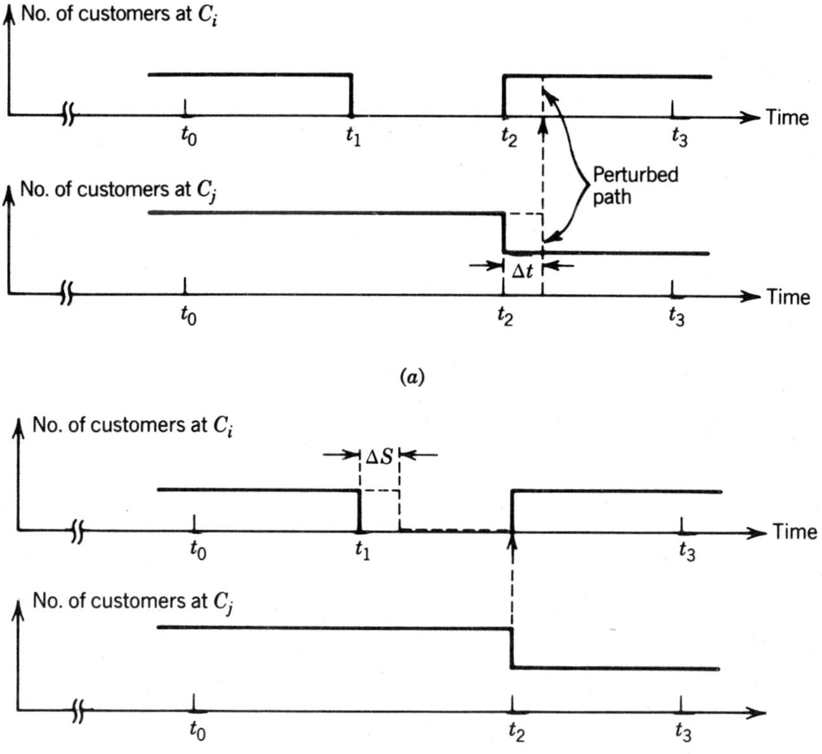

Fig. 3 (*a*) Perturbation propagation and (*b*) perturbation cancellation.

due to the idle period in this sample path, this perturbation ceases to exist after time $t_1 + \Delta s$ (Fig. 3*b*), and we say that the perturbation has been canceled.

Thus we see examples of perturbation propagation and cancellation rules. The propagation rule above could be stated as: If a customer from C_j terminates an idle period at C_i, then C_i acquires the same perturbation as C_j. Note that both perturbation generation and propagation can be *easily performed in real time*, as a system is being observed, or as a simulation is executing. Here we have presented just the simplest rules, and in more general situations the rules become more elaborate. Some of the rules can be derived formally using analytic representations [see Ho and Cao (1983) and Suri (1984)], while others are based on assumptions that have been verified only through extensive experimental results [see Section 4.4 and Ho, Cao, and Cassandras (1983)].

It is important to distinguish between *infinitesimal* and *finite* P/A. Infinitesimal P/A is associated with direct differentiation of the sample path function $\psi(\theta, w)$ w.r.t. θ (say, differentiation of the throughput w.r.t. the mean service time) and then using the infinitesimal P/A estimates [e.g., estimate (4.1.3)]. Finite P/A is associated with finite-difference P/A estimates and with finite P/A rules [see Ho, Cao, and Cassandras (1983)]. Clearly, finite difference P/A estimates must be applied to such parameters as buffer size or number of customers in a closed queueing network.

4.4. QUEUEING NETWORK AND EXPERIMENTAL RESULTS

4.4.1 Queueing Network

We consider a queueing network with nine servers and three classes of customers, which means that there are 27 *service time* distribution parameters, that is,

$$\begin{bmatrix} \theta_{1,1} & \cdots & \theta_{1,K} \\ \vdots & \vdots\ \vdots & \vdots \\ \theta_{M,1} & \cdots & \theta_{M,K} \end{bmatrix} \quad \text{for } M = 9,\ K = 3\,,$$

then there are nine *buffer size* parameters, that is, $B_1, \ldots, B_M$ for $M = 9$; and finally, three populations for the *customer classes*, $N_1, \ldots, N_K$ for $K = 3$. The network in our motivating example therefore has a total of *39 design parameters*.

The throughputs of all three classes are measured at C_1, that is, (see (4.1.9)) $o(k) = 1$ for all k. The *performance criterion* might be a given

Table 1 Conventional Results from an Experiment[a] on the Complex Network

Throughputs	X_k for all classes $k = 1, 2, 3$
Utilizations	u_m of all servers $m = 1, \ldots, 9$
Blocked proportions[b]	v_m of all servers $m = 1, \ldots, 9$
Idle proportions[c]	i_m of all servers $m = 1, \ldots, 9$
Mean total queue lengths	Q_m at all servers $m = 1, \ldots, 9$
Mean queue lengths by class[d]	q_{mk} at all servers $m = 1, \ldots, 9$; for all classes $k = 1, 2, 3$

[a]A simulation or an experiment on the actual system.
[b]Proportion of time the server was blocked by another server.
[c]Proportion of time the server was idle. ($u_m + v_m + i_m = 1$ for each m.)
[d]Mean number of customers of class k present at server m.

Table 2 Additional Results Regarding X_1 Obtained with Perturbation Analysis[a]

	Class k			Change in Buffer at m (ΔB_m)	
	1	2	3	$\Delta B_m = +1$	$\Delta B_m = -1$
Server m					
1	×	×	×	×	×
2		$\left\{\left(\frac{dX_1}{d\theta_{mk}}\right)\right\}$		×	×
3				×	×
4		Sensitivity		Δx_1 due	
5		of X_1		to ΔB_m	
6		to θ_{mk}		×	×
7				×	×
8	×	×	×	×	×
9	×	×	×	×	×
Change in Number of Class k (ΔN_k)		Δx_1 due to ΔN_k		(51 additional values)	
$\Delta N_k = +1$	×	×	×		
$\Delta N_k = -1$					

[a]Similar results are obtained for X_2 and X_3.

function, say, ψ of the throughputs of all the classes, that is,

$$\text{Performance} = \psi(X_1, \ldots, X_K).$$

If ψ is differentiable in some of its arguments, then $\partial\psi/\partial\theta_{mk}$ can be expressed in terms of $\partial\psi/\partial X_j$ and $\partial X_j/\partial\theta_{mk}$. The estimate of $\partial\psi/\partial\theta_{mk}$ can be used further in stochastic approximation, say, to solve optimization problems associated with ψ. (This might be an interesting topic for further research.) Here, however, we are concerned only with efficient evaluation of $\partial X_j/\theta_{mk}$, that is, with studying the sensitivity of throughput. Other sensitivities that might be of interest (but are not considered here) are utilizations, mean queue lengths, blocked times, and so on (see Table 1).

With P/A techniques incorporated into the experiment, we can obtain a lot of additional information (see Table 2). It can be seen that 51 additional values could be derived by P/A. Similar tables could be obtained for X_2 and X_3. Thus we can essentially obtain the results for 153 additional experiments just by observing one experiment. Real-world experiments are not just expensive but are sometimes impossible to repeat; this clearly illustrates the usefulness of the P/A approach.

4.4.2 Experimental Results

We now present a number of experimental results for the network with 39 design parameters as described in Section 4.4.1. Suppose we are given a nominal design for this network. The aim is to evaluate the effect on the throughputs of all three classes of the 39 parameters. Data for the nominal design is displayed in Table 3 and also in more graphic form in Fig. 4 (the notation used is as in Section 4.4.1).

The service time for a customer of class k at server m is taken from a uniform distribution over the range $[0.5\theta_{mk}, 1.5\theta_{mk}]$ so that the mean service time of class k at server m is θ_{mk}. Also specified are the routings for each class. Class 1 has a cyclic routing pattern (Fig. 5). Class 2 has a radial routing pattern, which brings it back to the central node after each visit to a peripheral node (Fig. 6)—the precise routing probabilities are given in Table 4. Finally, class 3 has a rather general routing pattern; Fig. 7 shows the routing of class 3 on departure from server 2; as can be seen from the detailed routing matrix (Table 5), a similar routing pattern holds for departures from all other peripheral servers, while departures from the central server have equal probability of going to the peripheral servers. This example will now be motivated by showing that it can represent two practical systems.

Table 3 Parameters for Nominal Design of Network[a]

	Values of θ_{mk} Class k			Values of B_m
	1	2	3	
Server m				
1	3	20	5	3
2	1	20	15	3
3	2	20	15	3
4	3	25	5	3
5	1	2	5	5
6	2	25	10	3
7	3	30	5	4
8	1	50	5	2
9	2	30	10	2
Values of N_k	8	4	4	(39 parameters)

[a]Symbols: θ_{mk} = mean service time at server m for class k; B_m = buffer size at server m; N_k = population of class k.

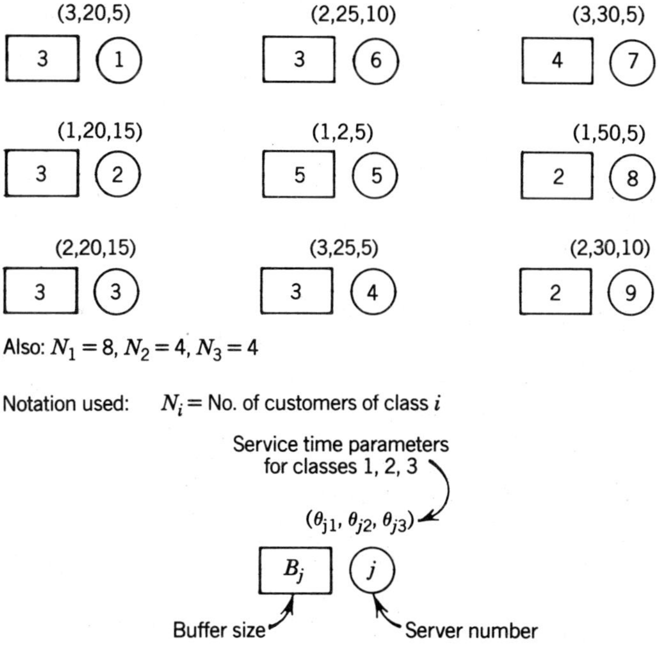

Fig. 4 Network data for nominal design.

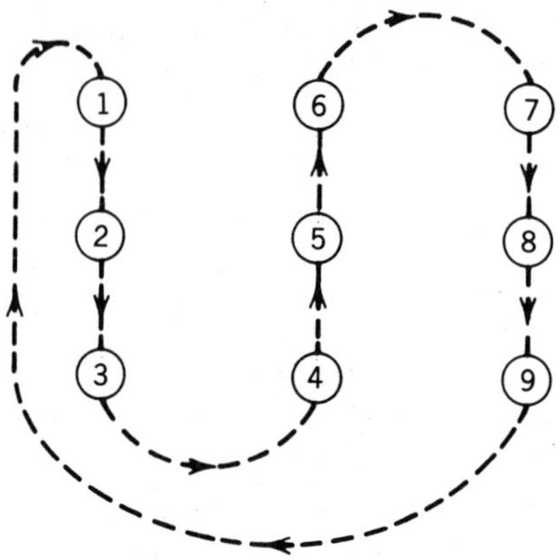

Fig. 5 Routing for class 1 customers (numbers refer to servers, as in Fig. 4).

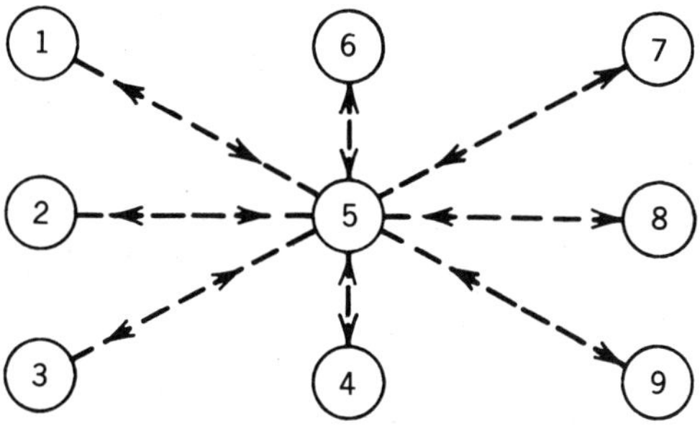

Fig. 6 Routing for class 2 customers (numbers refer to servers, as in Fig. 4).

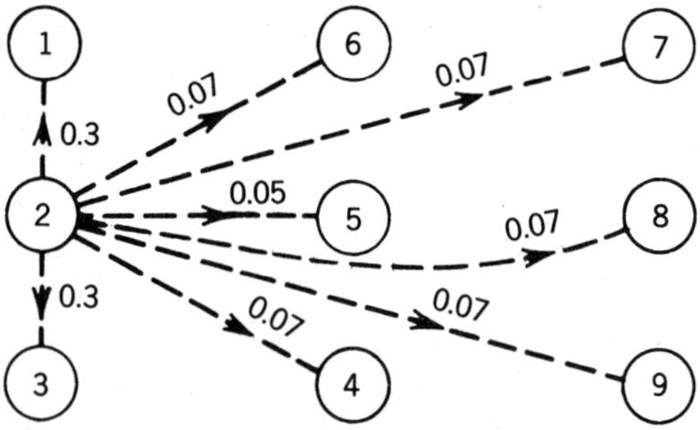

Fig. 7 Routing for class 3 customers (numbers in circles refer to servers, as in Fig. 4; numbers on dotted lines are routing probabilities).

Table 4 Routing Probability for Class 2

From	To								
	S_1	S_2	S_3	S_4	S_5	S_6	S_7	S_8	S_9
S_1					1				
S_2		(zero)			1		(zero)		
S_3					1				
S_4					1				
S_5	0.125	0.125	0.125	0.125	0	0.125	0.125	0.125	0.125
S_6					1				
S_7		(zero)			1		(zero)		
S_8					1				
S_9					1				

Table 5 Routing Probability for Class 3

From	To								
	S_1	S_2	S_3	S_4	S_5	S_6	S_7	S_8	S_9
S_1	0	0.3	0.07	0.07	0.05	0.3	0.07	0.07	0.07
S_2	0.3	0	0.3	0.07	0.05	0.07	0.07	0.07	0.07
S_3	0.07	0.3	0	0.3	0.05	0.07	0.07	0.07	0.07
S_4	0.07	0.07	0.3	0	0.05	0.07	0.07	0.07	0.3
S_5	0.125	0.125	0.125	0.125	0	0.125	0.125	0.125	0.125
S_6	0.3	0.07	0.07	0.07	0.05	0	0.3	0.07	0.07
S_7	0.07	0.07	0.07	0.07	0.05	0.3	0	0.3	0.07
S_8	0.07	0.07	0.07	0.07	0.05	0.07	0.3	0	0.3
S_9	0.07	0.07	0.07	0.3	0.05	0.07	0.07	0.3	0

(a) Communication Network Interpretation In Fig. 4 let the servers be the nodes of a store and forward communication network, with limited buffer storage at each node. The network is used for a distributed decision-making process. The customers are messages. Class 1 corresponds to a "polling" protocol message that cycles through the nodes to make sure that the network is still connected. Class 2 could be communication from peripheral decision makers to a central decision maker. Class 3 corresponds to distributed communication, with greater communication among neighboring nodes and less between distant nodes.

(b) Flexible Manufacturing System Interpretation Now, in Fig. 4 let the servers correspond to processing stations in an automated system that manufactures a variety of semiconductor chips. Class 1 corresponds to chips that follow the most regular processing cycle, which takes them through all the stations in turn. Class 2 consists of chips that must return to a central inspection station between each operation. Class 3 consists of miscellaneous processing of assorted chips, which takes place in a more random way.

An experiment (henceforth referred to as the "nominal experiment") was performed on the nominal design. This experiment involved performing 10 independent replications on a discrete-event simulation. (A "replication" has all the same system parameters and initial conditions, only the seed for the random number generator is changed.) Each replication consisted of a *run-in* time until the total number of services (over all servers) reached 10,000—during this time no statistics were accumulated. Then statistics were kept, and the run was terminated after an additional 50,000 services.

Typical statistics obtained from such an experiment by averaging over the 10 replications would be those in Table 6. However, such a table does not provide much insight into the effect of various parameters on these statistics. For example, we note first of all that *all* of the servers are utilized less than half the time. So there is no obvious "bottleneck server" that a system manager would see right away as being the server to be speeded up. Second, we see that servers 3, 8, and 9 are among the highest utilized servers. Should we invest money in speeding all of them up to get better network performance? Or might the money be better invested at one of those servers? It would be useful to be able to quantify this trade-off. The point we are making is that the statistics are useful in that they highlight some symptoms, but they do not indicate where the remedies lie.

During this same experiment P/A techniques were used to generate the sensitivity coefficients in Table 7. These predict the change in throughput for class 1 due to a change in mean service time for class k at station m (for

Table 6 Conventional Results for Simulations[a] with Nominal Parameter Values

Class k Throughput	X_k	1 (3.822×10^{-2}),	2 (0.727×10^{-2}),	3 (1.622×10^{-2}),
Server m	Utilization u_m	Blocked Proportion[b] v_m	Idle Proportion i_m	Mean Queue Q_m
1	0.34	0.30	0.36	1.36
2	0.43	0.25	0.32	1.52
3	0.47	0.20	0.33	1.46
4	0.38	0.24	0.38	1.26
5	0.19	0.69	0.12	2.96
6	0.42	0.39	0.19	1.96
7	0.41	0.48	0.11	2.89
8	0.48	0.30	0.22	1.41
9	0.46	0.24	0.30	1.18

[a]Average over 10 replications.
[b]See definitions of terms in Table 1.

Table 7 Sensitivity Values[a] Predicted by Perturbation Analysis[b] for Network Example

	Sensitivity value ($dX_1/d\theta_{mk}$) class k		
$\times 10^{-5}$	1	2	3
Server m			
1	−8.07	−7.61	−5.18
2	−6.85	−7.81	−9.62
3	−9.39	−7.11	−9.83
4	−7.78	−6.79	−4.53
5	−3.56	−18.20	−2.03
6	−12.32	−11.06	−11.96
7	−14.75	−11.46	−10.92
8	−11.85	−20.07	−12.43
9	−17.26	−14.37	−14.79

[a]These values are for X_1 (throughput of class 1) only. Similar tables are obtained for X_2 and X_3.
[b]Average over 10 replications.

all m, k)—the prediction is for the case where the mean of the uniform distribution shifts but the range remains constant. A similar table could be obtained for classes 2 and 3. Table 7 helps us to answer the question posed above. Suppose we can (at a price) speed up the processing of class 2 at servers 3, 8, or 9. The table shows us that our money would be spent 3 times more effectively at server 8 instead of server 3, and 1.5 times more effectively at server 8 instead of server 9 (see the entries in column 2 and rows 3, 8, and 9). These are large and significant ratios given that the utilizations of the servers (Table 6) hardly differed at all.

In Table 8 we predict the effects of decreasing the population of a

Table 8 Throughput Predictions for Decrease in Number of Customers of Each Class

	Throughput Values for Each Class		
$\times 10^{-2}$	X_1	X_2	X_3
Nominal experiment	3.81	0.719	1.63
Predicted for $\Delta N_1 = -1$	3.57	0.771	1.74
Predicted for $\Delta N_2 = -1$	4.49	0.635	1.91
Predicted for $\Delta N_3 = -1$	4.22	0.796	1.35

particular class by 1. The throughput of each class is shown for the case where N_k is decreased by unity (for each k), while the population for the other classes remains the same as in the nominal design. This table was also generated from one experiment (in this case a set of five independent replications). A similar computation could generate a table for the case where the population of any one class were increased by unity. These tables illustrate how P/A can derive additional useful information from an experiment performed at only one set of parameter values.

The additional CPU time used by the P/A algorithms is usually negligible compared with that required for the nominal experiment itself. For instance, each sensitivity coefficient in Table 7 added about 1% to the total CPU time required, as compared with the brute force approach, which would add 100% (a new experiment) for each coefficient required.

4.4.3 Accuracy of Method

In this section we consider the accuracy of the P/A predictions for our complex network example.

The results of various P/A predictions are compared with new experiments in Tables 9 and 10. Note that the P/A predictions are all made from one experiment, but to test these, a large number of experiments were required.

Table 9 verifies the accuracy of perturbation analysis. Since we must compare it with a brute force finite difference approximation, there is both experimental error (from the random nature of the experiments) and difference approximation error. Still, the second to the last row of the table (average values) shows good agreement between the two approaches. The first 10 rows of the table are even more interesting. They show in detail the data for each of the 10 replications. Suppose the original experiment had consisted of only one replication. Then the brute force gradient could have taken on any of the values in the last column of these 10 rows. The P/A predicted gradient would have been any of the 10 values in the preceding column. Note that while the brute force values range from +12 to −40 ($\times 10^{-5}$), the P/A values all lie between −19.3 and 20.7 ($\times 10^{-5}$). (The +12 value means that the brute force prediction actually gave the wrong sign in one case.) The variability in the two methods is quantified in the last row of the table, which gives the standard deviation of the gradient predictions over the 10 replications—0.45×10^{-5} for P/A and 15.3×10^{-5} for the brute force method, which is a considerable *decrease* in the variability of the predictions across each replication.

In Table 10 we test one of the finite perturbation analysis teshniques. Here we compare nine throughput predictions with new experiments

Table 9 Test of Accuracy of Gradient Predictions

Replication Number	Throughput of Class 1 (X_1)		Gradient ($dX_1/d\theta_{8,2}$)	
	Nominal Experiment	Perturbed[a] Experiment	Predicted by Perturbation Analysis[b]	Brute Force Method[c]
1	3.761×10^{-2}	3.861×10^{-2}	-19.3×10^{-5}	-40.0×10^{-5}
2	3.857×10^{-2}	3.827×10^{-2}	-19.5×10^{-5}	$+12.0 \times 10^{-5}$
3	3.822×10^{-2}	3.891×10^{-2}	-20.0×10^{-5}	-27.6×10^{-5}
4	3.801×10^{-2}	3.888×10^{-2}	-20.3×10^{-5}	-34.8×10^{-5}
5	3.779×10^{-2}	3.851×10^{-2}	-20.5×10^{-5}	-28.8×10^{-5}
6	3.835×10^{-2}	3.894×10^{-2}	-19.7×10^{-5}	-23.6×10^{-5}
7	3.808×10^{-2}	3.858×10^{-2}	-20.6×10^{-5}	-20.0×10^{-5}
8	3.874×10^{-2}	3.886×10^{-2}	-20.0×10^{-5}	-4.8×10^{-5}
9	3.830×10^{-2}	3.859×10^{-2}	-20.7×10^{-5}	-11.6×10^{-5}
10	3.853×10^{-2}	3.892×10^{-2}	-20.4×10^{-5}	-15.6×10^{-5}
Average[d]	3.822×10^{-2}	3.871×10^{-2}	-20.1×10^{-5}	-19.6×10^{-5}
Standard Deviation[d]	0.0336×10^{-2}	0.0227×10^{-2}	0.45×10^{-5}	15.3×10^{-5}

[a] The perturbed experiment was identical to the nominal except that $\theta_{8,2} = 47.5$ (a decrease of 5% in mean service time).
[b] All perturbation analysis predictions were derived from the nominal experiment.
[c] This involves taking the finite difference approximation using the perturbed and nominal experiments.
[d] Derived over the 10 replications.

Table 10 Comparison of Predicted Versus Observed Values of Throughput for Changes in Number of Customers in Each Class

	Throughput Values for Each Class		
$\times 10^{-2}$	X_1	X_2	X_3
For $\Delta N_1 = -1$			
Predicted[a]	3.57	0.771	1.74
Observed[b]	3.46	0.765	1.68
For $\Delta N_2 = -1$			
Predicted	4.49	0.635	1.91
Observed	4.45	0.629	1.84
For $\Delta N_3 = -1$			
Predicted	4.22	0.796	1.35
Observed	4.17	0.775	1.31

[a]All predictions are from the same original nominal experiment.
[b]Each set of observed values required a new experiment to be performed.

performed at the changed parameter values. The change involves removing one customer of a specified class. While the predictions are not as accurate as in Table 9, the error in predicted throughput is still under 5% for all nine values—which is often acceptable for practical applications. For more experimental results and accuracy of P/A see Dille and Ho (1985).

4.5 SOME IDEAS FOR FURTHER RESEARCH

Many interesting questions, both theoretical and practical, arise out of this new approach to DEDS:

We suggest the following directions for future research.

1. Find convergence conditions and speed of convergence to $\partial l / \partial v$ of the estimates of type (4.1.1)–(4.1.3) and higher-order sample derivatives [see Rubinstein and Szidarovszky (1985)] under different assumptions on the sample path $\psi(\theta, w)$.

2. Find expressions for the sensitivities of the type (4.1.1)–(4.1.3) (and higher-order derivatives) for the conventional performance measures

(throughput, sojourn time, utilization, etc.) with respect to such parameters as mean service time, routing probabilities, buffer size, and so on, in closed (open) queueing networks.

3. Find the accuracy of the finite P/A rules [zero-, first-, second-, and higher-order rules; see Ho, Cao, and Casandras (1983)] to predict (extrapolate) the performance measures of queueing networks while perturbing vector θ to $\theta + \Delta\theta$.

4. Find confidence intervals for the gradients and Hessians of $l(\theta)$ with the aid of P/A estimates, while using the

(i) regenerative method [Fishman (1978)],
(ii) method of batch means [Law and Kelton (1982)],
(iii) spectral method [Heidelberger and Welch (1981)].

Note that confidence intervals for the sensitivities, while using independent replications, are given in Rubinstein (1986).

5. Construct convergence iterative procedure of stochastic approximation type while using stochastic gradients and stochastic Hessians [see Zazanis and Suri (1985) and Rubinstein (1986)] to optimize both the unconstrained and constrained optimization problems associated with the single sample path $\psi(\theta, w)$. A useful paper on this subject is written by Y. Wardi (1985).

6. Infinitezimal P/A approach assumes generation of the sample path while using the inverse transform method. Unfortunately, this method [see Rubinstein (1981)] is not applicable in many practical situations, especially where random vectors with dependent components must be generated. Efficient use of the P/A approach without recourse on the inverse transform method (say, with the aid of the acceptance rejection) seems to us an interesting problem.

7. Variance reduction such as common and antithetic variates and control variates (see Chapters 1 and 2) may be useful to improve the accuracy of the estimates derived by the P/A approach.

APPENDIX: PERTURBATION ANALYSIS FOR GI/G/1 QUEUE

The material of this appendix is based on Suri and Zazanis (1984).

Consider a GI/G/1 queue with performance measure $E[W(\lambda, v)]$ – (steady state value of the mean time spent in the system by a customer), where λ is the interarrival rate and $v = 1/\mu$ is the mean service time. We shall apply the P/A approach on estimating the sensitivity of $E(W)$ with respect to v. Let $X(v)$ be the value of the service time for a particular customer, which is assumed to be generated by the inverse transform method that is $X(v) = F_v^{-1}(U)$.

The sequence $X_i(v) = F_v^{-1}(U_i)$, $i = 1, \ldots, N$, so generated is called [see equation (4.1.7)] the *nominal sample path*. The *perturbed sample path* (with parameter $v + \Delta v$ instead of v) can be calculated analytically from $F_{v+\Delta v}^{-1}(U_i)$, $i = 1, \ldots, n$ [see equation (4.1.8)]. Note that in both nominal and perturbed sample paths we use the same random number U_i for the ith customer.

It is well known [see Rubinstein (1981)] that if Y is distributed $F(y)$, then the r.v. $F(Y)$ is distributed $U(0,1)$. Now, since both $F_v^{-1}(U_i) = X_i(v)$ – (the nominal service time of the ith customer) and $F_{v+\Delta v}^{-1}(U_i) = X_i(v + \Delta v)$ (the perturbed service time of the ith customer) use the same random number U_i, we can write

$$F[X_i(v + \Delta v); v + \Delta v] = F[X_i(v); v] = U_i \,. \tag{A.1}$$

From the implicit function theorem we have

$$\frac{dx}{dv} = -\frac{\partial F/\partial v}{\partial F/\partial x} \tag{A.2}$$

[for $x \sim \exp(v)$, we have $x = -v \ln(1 - U)$ and $dx/dv = x/v = -\ln(1 - u)$]. If we change v to $v + \partial v$, then the service time of C_j, where C_j denotes the jth customer, would be changed by the infinitesimal amount

$$dx_j = -\frac{\partial F/\partial v}{\partial F/\partial x}\, dv \tag{A.3}$$

[for $x \sim \exp(v)$, we have $dx_j = -\ln(1 - u_j)\, dv$].

Now let us see how such change would affect the system time of all customers assuming that the order of events in the nominal and perturbed paths remains the same. In other words, we assume that if C_j initiates a busy period† in the nominal path, then, and only then, C_j initiate a busy period in the perturbed path. Such an assumption is easy to implement and is typical of infinitesimal perturbation analysis. Formally we state:

Assumption (A) [Suri and Zazanis (1984)] The set of indices (j) of customers (C_j) that initiate busy periods are the same for the nominal and perturbed paths.

†A busy period is a period between successive times when the system is empty.

Assume now that customer C_{k+1} initiates a busy period. Then by Assumption (A) it follows that in the perturbed path C_{k+1} would also find the system busy. Suppose n customers $C_{k+1}, \ldots, C_{k+n}$ are served during this busy period. Then in the perturbed path, C_{k+1} would spend in the system an additional time dx_{k+1} [see (A.3)]; C_{k+2} would spend in the system an additional time $dx_{k+1} + dx_{k+2}$ (an additional dx_{k+1} waiting for C_{k+1} and an additional dx_{k+2} for its own service time). In general,

$$\text{Additional time for } C_{k+i} = \sum_{j=1}^{i} dx_{k+j} \qquad (1 \leq i \leq n) . \tag{A.4}$$

Assume that N customers passed through the queue; then $E[W(\lambda, v)]$ can be estimated as

$$\bar{W}(\lambda, v, N) = \frac{1}{N} \sum_{i=1}^{N} W_i , \tag{A.5}$$

where W_i is the system time (time from arrival to departure) of customer C_i.

Now suppose that M busy periods were elapsed. Taking (A.5) into account, we obtain that the effect of dv on $\bar{W}$ is

$$d\bar{W} = \frac{1}{N} \sum_{m=1}^{M} \sum_{i=1}^{n_m} \sum_{j=1}^{i} dx_{k_m+j} . \tag{A.6}$$

Here n_m is the number of customers served during the mth busy period, $k_m + 1$ is the index of the customer initiating this busy period, and $N = \Sigma_{m=1}^{M} n_m$. Dividing by dv, we get

$$\frac{d\bar{W}}{dv} = \frac{1}{N} \sum_{m=1}^{M} \sum_{i=1}^{n_m} \sum_{j=1}^{i} \frac{dx_{k_m+j}}{dv} , \tag{A.7}$$

which can be used to estimate $dE(W)/dv$. Note that $(dx_{k_m+j})/dv$ and therefore $d\bar{W}/dv$ were determined from the *nominal path only*, that is, without observing the perturbed path. A similar argument can be used to derive $d\bar{W}/d\lambda$ [Suri and Zazanis (1984)].

For the M/G/1 Suri and Zazanis (1984) proved that $d\bar{W}/dv$ is a strongly consistent (and hence asymptotically unbiased) estimate of $dE(W)/dv$. They also proved that $d\bar{W}/d\lambda$ is a strongly consistent estimate for $dE(W)/d\lambda$. The extension of consistency results to a GI/G/1 queue can be found in Zazanis and Suri (1985).

Thus by performing a single simulation on the GI/G/1 queue, we can estimate simultaneously both the mean waiting time $E(W)$ and its derivative $dE(W)/dv$. Extension of this method for the case of a *vector* of parameters $\mathbf{v} = (v_1, \ldots, v_p)$ is straightforward.

Suri and Zazanis (1984) estimated $dE(W)/dv$ and $dE(W)/d\lambda$ in M/M/1, M/G/1, and GI/G/1 queues using the P/A approach and showed that their estimates $d\bar{W}/dv$ and $d\bar{W}/d\lambda$ are much more accurate than the brute force Monte Carlo estimates. They also considered the optimization problem (3.6.21) of Section 3.6, namely,

$$\min_{\nu} g(\nu) = \min_{\nu} \{C_1 \nu + C_2 E[W(\nu)]\}$$

for M/M/1 and GI/G/1 queues, where $\mu = \nu^{-1}$, and applied the following Robbins-Monro procedure:

$$v_{i+1} = v_i - a_i \frac{d\bar{W}_i}{dv_i}, \tag{A.8}$$

where $d\bar{W}_i/dv_i$ was calculated according to (A.7) and a_i was chosen according to Kesten's (1958) adaptive procedure.

The number of customers, L, simulated at each iteration was fixed. Suri and Zazanis implemented algorithm (A.8) for $L = 5$ and $L = 10$. In both cases they demonstrated fast convergence of (A.8) to the neighborhood of the optimal solution v^* and $g(v^*)$; more precisely, for both $L = 5$ and $L = 10$, v_i hits the 1% neighborhood of v^* and the 0.01% neighborhood of $g(v^*)$ after simulating less than 1000 customers.

REFERENCES

Buzacott, J. A. (1976). The Production Capacity of Job Shops with Limited Storage Space, *Int. J. Prod. Res.*, **14**, No. 5, 597–605.

Buzen, J. P. (1971). Queueing Network Models of Multiprogramming, Ph.D. Thesis, Division of Engineering and Applied Physics, Harvard University, Cambridge, MA.

Buzen, J. P. (1973). Computational Algorithms for Closed Queueing Networks with Exponential Servers, *C. ACM*, **16**, No. 9, 527–531.

Cao, X. R. (1985). Convergence of Parameter Sensitivity Estimates in a Stochastic Experiment, *IEEE Automat. Control*, **30**, 845–843.

Cao, X. and Y. C. Ho (1985). Sensitivity Analysis of Sojourn Time in Queueing Networks, Harvard University, Cambridge, MA.

Cassandras, C. (1982). Sample Path Analysis of Discrete Event Dynamic Systems, Ph.D. Thesis, Division of Applied Sciences, Harvard University, Cambridge, MA.

Chandy, K. M. and A. J. Martin (1983). A Characterization of Product Form Queueing Networks, *J. ACM*, **30**, No. 2, 286–299.

Dille, J. and Y. Ho (1985). Statistical Validation of Finite Perturbation Analysis Algorithms in Closed Queueing Network, Harvard University, Cambridge, MA.

Eyler, M. A. (1980). A New Approach to the Productivity Study of Transfer Lines, Ph.D. Thesis, Division of Applied Sciences, Harvard University, Cambridge, MA.

Fishman, G. S. (1978). *Principles of Discrete Event Simulation*, Wiley, New York.

Gordon, W. J. and G. F. Newell (1967). Closed Queueing Systems with Exponential Servers, *Op. Res.*, **15**, No. 2, 254–265.

Heidelberger, P. and P. D. Welch (1981). A Spectral Method for Confidence Interval Generation and Run Length Control in Simulations, *Comm. ACM*, **24**, 233–245.

Ho, Y. C. and X. Cao (1983). Perturbation Analysis and Optimization of Queueing Networks, *J. Optim. Theory Applic.*, **40**, No. 4, 559–582.

Ho. Y. C. and C. Cassandras (1983). A New Approach to the Analysis of Discrete Event Dynamic Systems, *Automatica*, **19**, No. 2, 149–167.

Ho, Y. C., M. A. Eyler, and T. T. Chien (1979). A Gradient Technique for General Buffer Storage Design in a Serial Production Line, *Int. J. Prod. Res.*, **17**, No. 6, 557–580.

Ho, Y. C., M. A. Eyler, and T. T. Chien (1983). A New Approach to Determine Parameter Sensitivities of Transfer Lines, *Manage. Sci.*, **29**, No. 6, 700–714.

Ho, Y. C., X. Cao, and C. Cassandras (1983). Infinitesimal and Finite Perturbation Analysis for Queueing Networks, *Automatica*, **19**, No. 4, 439–445.

Ho, Y. C., R. Suri, X. R. Cao, G. W. Diele, J. W. Dille, and M. Zazanis (1984). Optimization of Large Multiclass (Non-Product Form) Queueing Networks Using Perturbation Analysis, *Large Scale Systems*, **7**, 165–180.

Kesten, H. (1958). Accelerated Stochastic Approximation, *Ann. Math. Stat.*, **29**, 41–49.

Koenigsberg, E. (1982). Twenty Five Years of Cyclic Queues and Closed Queue Networks: A Review, *J. Opl. Res. Soc.*, **33**, 605–619.

Law, A. M. and W. D. Kelton (1982). *Simulation Modeling and Analysis*, McGraw-Hill, New York.

Newell, G. F. (1980). *Traffic Flow on Transportation Networks*, MIT Press, Cambridge, MA.

Reiser, M. (1979). A Queueing Network Analysis of Computer Communication Networks with Window Flow Control, *IEEE Trans. Communic.*, *COM-27*, **8**, 1199–1209.

Reiser, M. and H. Kobayashi (1975). Queueing Networks with Multiple Closed Chains: Theory and Computational Algorithms, *IBM J. Res. Dev.*, **19**, 283–294.

Rubinstein, R. Y. (1981). *Simulation and the Monte-Carlo Methods*, Wiley, New York.

Rubinstein, R. Y. (1986). Confidence Intervals for Sensitivity Measures of Computer Simulation Models: A Perturbation Analysis Approach, Harvard University, manuscript.

Rubinstein, R. Y. and G. Samorodnitsky (1985). Antithetic and Common Random in Simulation of Stochastic Systems, *J. Stat. Comp. Simulation*, **22**, 161–180.

Rubinstein, R. Y. and F. Szidarovszky (1985). Convergence of Perturbation Analysis Estimates for Queueing Networks: A General Approach, Harvard University, manuscript.

Rubinstein, R. Y. and F. Szidarovszky (1985a). Convergence of Perturbation Analysis Estimates for Discontinuous Sample Function: A General Approach, Harvard University, manuscript.

Sauer, C. H. and K. M. Chandy (1981). *Computer Systems Performance Modeling*, Prentice-Hall, Englewood Cliffs, NJ.

Suri, R. (1983). Robustness of Queueing Network Formulae, *J. ACM*, **30**, No. 3, 564–594.

Suri, R. (1984). Infinitesimal Perturbation Analysis for General Discrete Event Systems, Harvard University, manuscript.

Suri, R. and J. W. Dille (1984). On-Line Optimization of Flexible Manufacturing Systems Using Perturbation Analysis, *Proc. ORSA/TIMS Conference on FMS*, Ann Arbor, MI, August 1984.

Suri, R. and X. Cao (1983). The Phantom Customer and Marked Customer Methods for Optimization of Closed Queueing Networks with Blocking and General Service Times, *ACM Performance Evaluation Rev.*, **12**, 243–256.

Suri, R. and R. R. Hildebrant (1984). Modeling Flexible Manufacturing Systems Using Mean Value Analysis, *J. Manufacturing Systems*, **3**, No. 1, 27–38.

Suri, R. and M. A. Zazanis (1984). Perturbation Analysis Gives a Strongly Consistent Estimates for the M/G/1 Queue, Harvard University, manuscript.

Tanenbaum, A. S. (1981). *Computer Networks*, Prentice-Hall, Englewood Cliffs, NJ.

Wardi, Y. (1985). Simulation-based network optimization algorithms with adaptive precision, School of Electrical Engineering, Georgia Institute of Technology, Atlanta, GA.

Zazanis, M. and R. Suri (1985). Estimating second derivatives of performance measures for GI/G/1 queues from a single sample path, Manuscript, Harvard University.

CHAPTER 5

Monte Carlo Optimization

5.1 INTRODUCTION

In this chapter we shall show how Monte Carlo methods can be successfully applied in solving complex optimization problems, both deterministic and stochastic.

Under the *deterministic optimization problem* we understand the following. Given a function g from R^n to R^1 and D a subset of R^n, we seek a point $x^* \in D$ that minimizes g on D or at least satisfactorily approximates the minimum of g on D. Mathematically it can be written as

$$\text{(P1)} \qquad \min_{x \in D \subset R^n} g(x) = g(x^*) = g^* .$$

For example, D can be given as the constraints of nonlinear programming, that is,

$$D = \{x \colon g_i(x) \leq 0,\ i = 1, \ldots, M\} , \qquad (5.1.1)$$

where $g_i(x)$ is a known function.

When $g(x)$ and D have some attractive properties, for instance, $g(x)$ is a differentiable convex function and D a convex† set, then problem (P1) can be solved by conventional mathematical programming methods [see Avriel (1976)].

†A function $g(x)$ is convex if for any two points x and y in D the following inequality holds:

$$f(\lambda x + (1-\lambda)y) \leq \lambda f(x) + (1-\lambda)f(y) , \qquad 0 \leq \lambda \leq 1$$

A set $D \subset R^n$ is convex if for any two points x and y in D the point $\lambda x + (1-\lambda)y$ is in D, $0 \leq \lambda \leq 1$. Note that $\{[x + (1-\lambda)y] \colon 0 \leq \lambda \leq 1\}$ is the line segment joining x and y. Hence D is convex if the line segment between any two points in D also belongs to D.

If, however, either g or D or both are nonconvex,† then the conventional methods of mathematical programming cannot be used, and we have to resort to numerical methods, particularly to Monte Carlo methods. Under the *stochastic optimization problem* we understand the following:

$$\text{(P2)} \quad \min_{x \in D \subset R^n} \left\{ g(x) = \int \psi(x, y) f_Y(y, x)\, dt = E\psi(x, Y) = g(x^*) = g^* \right\},$$

where x and Y are n- and k-dimensional vectors, respectively, $\psi(x, Y)$ is a stochastic function, x^* is the unique optimal point, $f_Y(y, x)$ is a p.d.f.(probability density function) depending on a vector of parameters x and D in a given region not necessarily convex. For example, D can be given as in (5.1.1) or, more generally, as

$$D = \{x\colon g_i(x) = E_Y\{\psi_i(x, Y)\} \leq 0\,,\ i = 1, \ldots, M\}\,. \qquad (5.1.2)$$

Problem (P2) with D as in (5.1.2) is called a *nonlinear stochastic programming problem.*

For example, the chance constraints

$$P\left\{\sum_{j=1}^{q} a_{ij}(Y)x_j \leq b_i(Y)\right\} \geq p_i\,, \qquad i = 1, \ldots, M\,,$$

of the *linear stochastic programming* are of type (5.1.2). To see this, define $\psi_i(x, y)$ as

$$\psi_i(x, Y) = \begin{cases} p_i\,, & \text{if } \sum_{j=1}^{q} a_{ij}(Y)x_j \leq b_i(Y)\,, \\ 1 - p_i\,, & \text{otherwise}\,. \end{cases} \qquad (5.1.3)$$

Consider the following two cases associated with problem (P2).

(i) The p.d.f. $f_Y(y, x)$ of the random vector Y is *known*, and $g(x) = E_Y[\psi(x, Y)]$ {and $g_i(x) = E_Y[\psi_i(x, Y)]$, $i = 1, \ldots, M$} can be evaluated explicitly. In this case problem (P2) reduces to (P1).

(ii) The p.d.f. $f_Y(y, x)$ is *unknown* (say, only a sample from it is available) or $g(x) = E_Y\{\psi(x, Y)\}$ *cannot be evaluated analytically* even with known $f_Y(y, x)$. In this case there is no way of solving problem (P2) analytically, and we have to resort to Monte Carlo optimization methods.

†Problems where g or D or both are nonconvex are often called *nonconvex*, *multiextremal*, or *global* optimization problems.

A wide range of applied problems can be neither formulated nor solved within the framework of the deterministic optimization problem, that is, belonging to the second case. As an example, consider the problem of computing the probability of an event.

$$\left\{\sum_{j=1}^{q} a_{ij}(Y)x_j \le b_j(Y)\right\} \quad \text{for different } x \ (x \in D)\,.$$

The exact probability of this event is rarely available, even with known p.d.f. $f_Y(\cdot)$. Another situation when the exact values of $g(x)$ is not available is in the optimization of complex stochastic systems (reliability, queueing, etc.) in which the output parameters and the sensitivities (gradients, Hessians, etc.) are estimated through Monte Carlo simulation (see Section 3.6 and chapter 4). In both examples, for each x only statistical estimates of $g(x)$ and their associated sensitivities are available. They can be incorporated in iterative (adaptive) procedures (see Chapter 3) to find the optimal solution x^* of (P2).

Before proceeding with problems (P1) and (P2), let us explain what we mean by *local and global extrema*.

Consider a real-valued function g with domain D in R^n. The function g is said to have a *local minimum* at point $x^* \in D$ if there exists a real number $\delta > 0$ such that $g(x) \ge g(x^*)$ for all $x \in D$ satisfying $\|x - x^*\| < \delta$. We define a *local maximum* in a similar way, but in the sense that inequality $g(x) \ge g(x^*)$ is reversed.

We say that the function g has a *global* (*absolute*) *minimum* at $x^* \in D$ if $g(x) \ge g(x^*)$ holds for every $x \in D$. A similar definition holds for a *global maximum*. A global minimum at x^* implies that $g(x)$ achieves at x^* its smallest value among all other points in D. A local minimum, on the other hand, only guarantees that the value of $g(x)$ is a minimum with respect to other points nearby, specifically in a δ region of x^*. Thus a function may have many local minima, each with a different value of $g(x)$, say, $g(x_j^0)$, $j = 1, \ldots, k$. It is clear that every global maximum (minimum) is also a local maximum (minimum); however, the converse of this statement is, in general, not true. If $g(x)$ is a convex function in R^n and $D \subset R^n$ is a convex set, then every local minimum of g at $x \in D$ is also a global minimum of g over D [see Avriel (1976)].

This chapter is devoted to Monte Carlo (stochastic) optimization.† More specifically, Section 5.2 deals with Monte Carlo methods for global optimization, and in Section 5.3 we present an adaptive algorithm based on

†For nice surveys on global optimization, where both deterministic and stochastic approaches are considered, see Archetti and Schoen (1984) and Timmer (1984).

smoothed functionals and prove that it converges to the optimal solution of (P2). In Appendices 1–4 we present some mathematical background related to Monte Carlo optimization.

5.2 MONTE CARLO METHODS FOR GLOBAL OPTIMIZATION

Most algorithms for global optimization contain a random number generator associated with generation points in the region D and therefore are related to the Monte Carlo method. We consider several of them.

Brooks (1958) proposed, for estimating g^* in (P1), a *crude* (*nonadaptive*) *random search algorithm* that can be written as follows.

Algorithm G1-1 **(*Nonadaptive Random Search Algorithm*)**

1. Generate a sequence of N random vectors $X_1, \ldots, X_N$ uniformly distributed over D.
2. Compute $Y_k = g(X_k)$, $k = 1, \ldots, N$.
3. Estimate g^* by

$$M_N = \min(Y_1, \ldots, Y_N).$$

Intuitively, it is clear that if a large number of vectors is generated at random over any set of finite measure D, then the probability that any particular subset of D with positive measure contains at least one such vector tends to unity as the number of generated vectors tends to infinity.

Algorithm G1-1 was also studied by Rubinstein and Weissman (1977). Unlike Brooks (1958), they assumed that the sequence $X_1, \ldots, X_N$ (see step 1 of G1-1) can be generated from any (not necessarily uniform) p.d.f. $f_X(x)$ and such that $f_X(x) > 0$, $x \in D$. Our further discussion of Algorithm G1-1 is based mainly on Rubinstein and Weissman's (1977) paper.

Let P be a probability measure defined on B, the Borel σ field of D, so that (D, B, P) is a probability space.

Let $g^{-1}(a, b) = \{x \in D\colon a < g(x) \leq b\}$ and let $F(y) = P\{Y_i \leq y\}$. Then

$$F(y) = P\{g(X_i) \leq y\} = P\{g^{-1}(-\infty, y)\},$$

where $Y_i = g(X_i)$.

Proposition 1 Suppose P assigns a positive probability to every neighborhood of x^*, and suppose g is continuous at x^*. Then

$$\lim_{N\to\infty} M_N = g^* \quad \text{a.s.} \tag{5.2.1}$$

Proof The proof is simple and is based on the Borel-Cantelli lemma. Indeed, it is clear that $F(g^*) = 0$, and for each $\delta > 0$ we have

$$F(g^* + \delta) = P\{g^* < g(X_i) \le g^* + \delta\} > 0 \quad \text{by our assumption.}$$

Let $A_N(\delta)$ be the event $\{M_N \ge g^* + \delta\}$. Then $P\{A_N(\delta)\} = [1 - F(g^* + \delta)]^N$, and

$$\sum_{N=1}^{\infty} P\{A_N(\delta)\} = \frac{1 - F(g^* + \delta)}{F(g^* + \delta)} < \infty .$$

By the Borel-Cantelli lemma $P\{M_N > g^* + \delta$ infinitely often$\} = 0$ for all $\delta > 0$ and thus (5.2.1) holds. Q.E.D.

The choice of P, and consequently the resulting F, depends on our prior knowledge of x^*. If it is known that a certain region is more likely to contain x^*, then it would be more efficient to assign a higher probability to that region. If nothing is known a priori about x^*, a uniform distribution over D can be assumed.

To guarantee (5.2.1), the exact choice of P is immaterial. However, the rate of convergence is determined by the properties of F $(=P \circ g^{-1})$. For example, according to Gnedenko (1943), if there exists a constant $\alpha > 0$ such that

$$\lim_{\delta \downarrow 0} \frac{F(g^* + c\delta)}{F(g^* + \delta)} = c^{\alpha} \qquad \forall c > 0 , \tag{5.2.2}$$

then

$$\lim_{N \downarrow \infty} P\left\{\frac{M_N - g^*}{a_N} \le x\right\} = 1 - \exp\{-x^{\alpha}\} \qquad (x \ge 0) , \tag{5.2.3}$$

with a_N determined by $F(g^* + a_N) = 1/N$. Therefore, in this case M_N is asymptotically distributed Weibull. Some more properties of M_N are listed below.

(a) Geometric Distribution Let N_δ be the first N for which $M_N < g^* + \delta$. Then N_δ is a geometric r.v., that is,

$$P\{N_\delta = k\} = F(g^* + \delta)[1 - F(g^* + \delta)]^k \qquad (k = 1, 2, \ldots) \tag{5.2.4}$$

Table 1 Dependence of $k(\alpha, 1-\alpha)$ on α

α	0.500	0.200	0.100	0.050	0.020	0.010	0.005	0.002	0.001
$k(\alpha, 1-\alpha)$	1	8	22	58	194	458	1057	3104	6903

Consequently, it is well known that

$$EN_\delta = \frac{1}{F(g^* + \delta)} \equiv n_\delta ,$$

and

$$P\{N_\delta \leq k\} = 1 - [1 - F(g^* + \delta)]^k \equiv P_{\delta,k} .$$

It is clear that $\eta_\delta \to \infty$ as $\delta \downarrow 0$ and thus $F_{\delta[\eta_\delta]} = 1 - (1 - 1/\eta_\delta)^{[\eta_\delta]} = 1 - e^{-1} = 0.63$ (here $[\eta]$ is the integer part of η). Hence $\eta_\delta = EN_\delta$ is approximately 63% of N_δ, the number of trails necessary to make $M_N > g^* - \delta$ ($\delta > 0$ small).

Let $\alpha = F(g^* + \delta)$. Then $P_{\delta,k} = 1 - (1-\alpha)^k$. For every given pair (α, β) the smallest k for which $P_{\delta,k} \geq \beta$ is $k(\alpha, \beta) = \ln(1-\beta)/\ln(1-\alpha)$.

Table 1 shows some values of $k(\alpha, 1-\alpha)$ as a function of α.

(b) Lack of Memory It is well known that equation (5.2.4) implies

$$P\{N_\delta > k + m | N_\delta > m\} = P\{N_\delta > k\} . \tag{5.2.5}$$

In terms of N_M, we thus have

$$P\{M_{k+m} > g^* + \delta | M_m > g^* + \delta\} = P\{M_k > g^* + \delta\}$$

because the events $\{N_\delta > k\}$ and $\{M_k > g^* + \delta\}$ are identical. It follows that given m successive failures (to enter $\{y: y < g^* + \delta\}$) the conditional distribution of the number of trials necessary for the first success equals its unconditional distribution. In particular, we have

$$E(N_\delta | M_n > g^* + \delta) = m + EN_\delta . \tag{5.2.6}$$

We consider now some asymptotical properties of M_n.

(c) Poisson Approximation If (5.2.2) or (5.2.3) hold, then $Z_{\delta,N}$, the number of Y_i ($i = 1, 2, \ldots, N$) for which $Y_i < g^* + \delta$ is asymptotically Poisson distributed. More precisely, for fixed N and $\delta > 0$, $Z_{\delta,N}$ is a binomial r.v. with parameters N and $p = F(g^* + \delta)$. When (5.2.2) holds, by substituting $\delta = a_N$ in (5.2.2), we obtain $NF(g^* + ca_N) \to c^\alpha$, which implies that $Z_{ca_N,N}$ converges in distribution to a Poisson r.v. with parameter c^α.

Let $Y_1, \ldots, Y_N$ be a sample from any c.d.f. F. Denote again $M_N = \min(Y_1, \ldots, Y_N)$. Gnedenko (1943) showed that there are only three nondegenerate limiting distributions of M_N and gave necessary and sufficient conditions for convergence to each of the three. In our case, since we assume that a finite minimum exists, only two of the three limit distributions are possible, namely, the Weibull [given in (5.2.3)] and the Gumbel distribution $1 - \exp(-\exp x)$. Thus, when Gnedenko's condition (5.2.2) holds, we may approximate the distribution of M_N with the Weibull distribution, namely, with

$$F_N(x) = \begin{cases} 1 - e^{-[(x-g^*)/a_N]^\alpha} & \text{if } x \geq g^* , \\ 0 & \text{otherwise} , \end{cases} \tag{5.2.7}$$

where g^*, a_N, and α are called, respectively, the location, scale, and shape parameters.

We shall discuss now the conditions under which the limiting distribution of M_N [the sample minimum of problem (P1)] is Weibull.

Archetti, Betro, and Steffe (1975) showed that if

(a) $D \subset R^n$ is a compact set and $g(x)$ is a real bounded Lebesgue measurable function, with the minimum achieved at the interior point of D;

(b) the distribution function of $Y = g(X)$ is defined as

$$\begin{aligned} F(y) &= P(Y \leq y) = P\{g(X) \leq y\} \\ &= \text{meas}\{(x\colon x \in D,\ g(x) \leq y)\}/\text{meas}(D) , \end{aligned}$$

where meas denotes the Lebesgue measure in R^n; and

(c) $$\lim_{t \downarrow 0} \frac{h(t)}{t^\alpha} = H > 0 \tag{5.2.8}$$

for some $\alpha > 0$ and some $H > 0$, where

$$h(t) = \text{meas}\{x\colon x \in D,\ g(x) > g^* + t\} , \qquad t \geq 0 ,$$

g^* denotes the minimum (infinum) of g in D, then the Gnedenko condition (5.2.2) is satisfied for $F(y)$. They indicated that a wide class of functions g satisfies condition (5.2.8) and found point estimators for the unknown parameters a_n and α and the global minimum g^*.

De Haan (1981) considered Algorithm G1-1 as well and found point estimates and confidence intervals for g^*. Let $Y_{[1]}, Y_{[2]}, \ldots, Y_{[N]}$ be the order statistics of $Y_1, Y_2, \ldots, Y_N$, that is, $Y_{[1]} \leq Y_{[2]} \leq \cdots \leq Y_{(N)}$ (note that $M_N \equiv Y_{[1]}$). De Haan showed that if $X_1, \ldots, X_N$ is a sample from the uniform distribution over D and condition (5.2.2) is satisfied, then

(a) there exists a sequence of positive constants $\{a_N\}$ such that $a_N^{-1}(Y_{[1]} - g^*)$ has a nondegenerate limit distribution with distribution function as per (5.2.3).

(b) $$\left(\log \frac{Y_{[k]} - Y_{[k]}}{Y_{[3]} - Y_{[2]}}\right) \Big/ \log k$$

converges to $1/\alpha$ in distribution as $k \to \infty$, $N \to \infty$, $k/N \to 0$.

(c) the $1 - p$ confidence interval for g^* (for known α) is

$$\left\{ \frac{Y_{[1]} - (Y_{[2]} - Y_{[1]})}{(1-p)^{-1/\alpha} - 1}, Y_{[1]} \right\}.$$

Finally, Patel and Smith (1984) applied Algorithm G1-1 to the problem of minimizing a continuously differentiable concave function under linear constraints. They showed that if $X_1, \ldots, X_N$ is uniformly distributed over D, then M_N is asymptotically Weibull distributed with a shape parameter equal to n, where n is the dimension of the region D. More precisely

$$\lim_{N \to \infty} \left(\frac{M_N - g^*}{a_N} \leq x \right) = 1 - e^{-x^n} (x \geq 0).$$

Suppose now that $g(x)$ is measured with an error, that is, instead of $g(x_i)$, we observe $g(x_i, W_i) = g(x_i) + W_i$, where W_i are i.i.d. r.v.'s with a common c.d.f. H. Assume also that X and W are independent. In this case, for the following modification of G1-1 can be used.

Algorithm G1-2 (Modification of G1-1)

1. Generate $X_1, \ldots, X_N$ from any p.d.f. $f_X(x)$ [$f_X(x) > 0$, $\forall x \in D$].
2. Compute $Y_k = g(X_k, W_k) = g(X_k) + W_k$, $k = 1, \ldots, N$.
3. Estimate g^* by

$$M_N = \min(Y_1, \ldots, Y_N). \tag{5.2.9}$$

Let $W_* = \inf\{u\colon H(u) > 0\} \geq -\infty$ be the left end of the support of H. The following proposition is due to Rubinstein and Weissman (1977).

Proposition 2 Under the conditions of Proposition 1,

$$\lim_{N\to\infty} M_N = g^* + W_* \quad \text{a.s.} \tag{5.2.10}$$

Proof Let $E_N = \min_{1\le j\le N} W_i$. We say that $\{E_N\}$ is stable if there exists a sequence of constants $\{\eta_N\}$ such that for all $\delta > 0$

$$\lim_{N\to\infty} P\{|E_N - \eta_N| > \delta\} = 0 . \tag{5.2.11}$$

We consider the following three cases:

(a) $W_* > -\infty$. In this case our estimate for g^* is $M_N - W_*$, and we certainly have

$$\lim_{N\to\infty} (M_N - W_*) = g^* \quad \text{a.s.} \tag{5.2.12}$$

(b) $W_* = -\infty$, but $\{E_N\}$ is stable. In this case equation (5.2.11) implies

$$\lim_{N\to\infty} (M_N - \eta_N) = g^* \quad \text{in probability}, \tag{5.2.13}$$

and η_N is determined by $H(\eta_N) = 1/N$. A necessary and sufficient condition for (b) to hold is [see Gnedenko (1943)]

$$\lim_{u\to-\infty} \frac{H(u-\delta)}{H(u)} = 0 , \qquad \forall\delta > 0 . \tag{5.2.14}$$

(c) $W_* = -\infty$, but $\{E_N\}$ is not stable. Here we have by (5.2.10) $M_N \to -\infty$ a.s.

Therefore, for cases (a) and (b), M_N converges asymptotically (a.s. and in probability, respectively) to g^*, and in case (c), $M_N \to -\infty$ a.s. Q.E.D.

The following two examples are related to cases (b) and (c).

1. If W_i are normally distributed with mean 0 and variance σ^2, then (5.2.14) holds and $\{E_N\}$ is stable with $\eta_N = -c(2\log)^{1/2}$.
2. Suppose that the W_i have a two-sided Weibull distribution, that is,

$$h(x) = \begin{cases} 1/2e^{-|x|^\alpha} , & x < 0 , \\ 1 - 1/2e^{-x^\alpha} , & x \ge 0 . \end{cases}$$

Then it readily follows that $\{E_N\}$ is not stable for $\alpha \le 1$ but is stable for $\alpha > 1$ with $\eta_N = [\frac{1}{2} \log N]^{1/\alpha}$.

A natural extension of the crude random search algorithm G1-1 is the so-called *multistart algorithm*, which has been often used in practice in global optimization. According to this algorithm, we select at random a series of different starting points X_{0j}, $j = 1, \ldots, N$, and then use a local minimization algorithm from each of them.

Algorithm G1-3 (Multistart Algorithm)

1. Generate a sample $X_{01}, \ldots, X_{0N}$ from any p.d.f. $f_{X_0}(x) > 0$, $x \in D$ (usually the vectors X_{0k}, $k = 1, \ldots, N$, are uniformly distributed over D).
2. Consider $X_{01}, \ldots, X_{0N}$ as the starting points, and then apply N times a local optimization algorithm (gradient, Newton-Raphson, etc.) and find the local extrema $x_1^*, \ldots, x_N^*$ of $g(x)$ associated with $X_{01}, \ldots, X_{0N}$.
3. Estimate x^* by

$$\min(x_1^*, \ldots, x_N^*)$$

Let D_j be a subregion of D such that a local optimization algorithm starting at any $X_{01} \in D_j$ converges to the jth local minimum. We shall call D_j *the region of attraction* of the jth local minimum. Assume the $X_{01}, \ldots, X_{0N}$ are uniformly distributed over D and that the number of local minima is finite: Then the probability that at least one X_{0j}, $j = 1, \ldots, N$, falls in the region of attraction, say D_j^*, of the global minimum equals

$$P_N = 1 - \left(1 - \frac{m(D_j^*)}{m(D)}\right)^N , \tag{5.2.15}$$

where $m(D)$ is the measure of D. The disadvantage of the multistart algorithm is that the same local minimum may be found several times.

A substantial improvement over the multistart method can be obtained by using the so-called *clustering version of multistart*. Timmer (1984) has written an interesting and useful monograph on this subject.

Another quite interesting approach to the global optimization problem was suggested by Chichinadze (1967), who introduced a probability function $P(v)$ as the probability of $g(x) < v$, that is, if $m(V)$ is the measure of the level set

$$V = \{x: g(x) < v\} ,$$

then

$$P(v) = \frac{m(V)}{m(D)} .$$

The function $P(v)$ is, of course, not available, but if we calculate $g(x)$ at N points distributed at random over D and count the number M of these points for which $g(x) < v$, then M/N approximates $P(v)$. It is not difficult to see that the global maximum corresponds to $P(v) = 1$ and the global minimum of $P(v) = 0$. To find the solution $P(v) = 1$, Chichinadze suggested approximating $P(v)$ by a linear combination of a set of polynomial functions $P_i(v)$, $i, \ldots, k$, that is,

$$P(v) = \sum_{i=1}^{k} \lambda_i P_i(v) .$$

The range of v was divided at the points v_j, $j = 1, \ldots, s$, and the optimal values of λ_i were determined by minimizing

$$\sum_{j=1}^{S} W_j \left(\frac{M_j}{N} - \sum_{i=1}^{k} \lambda_i P_i(v_j) \right)^2 ,$$

where M_j is the number of points for which $g(x) < v_j$ and $W_j > 0$, $j = 1, \ldots, s$. The root v^* of $P(v) = 1$ was then determined to obtain an estimate of the global maximum of $g(x)$.

In the context of both the deterministic and stochastic optimization problems, considerable attention has been paid recently to the *adaptive random search algorithms*. There exists extensive literature dealing with the convergence and the speed of convergence of random search algorithms [see Anderson (1958), Baba, Shoman, and Sawaragi (1977), Boender, Rinnoy Kan, Stougie, and Timmer (1982), Devroy (1976, 1979), Ermoliev (1969, 1976), Ermakov and Zhiglyavskii (1984), Gurin, Demarsky, and Merkulov (1968), Guseva (1973), Katkovnik (1976), Katkovnik and Kulchitsky (1976), Matyas (1965), Patel, Smith and Zabinsky (1985), Rastrigin (1968), Rastrigin and Rubinstein (1969), Rubinstein (1969, 1981), Rubinstein and Karnovsky (1978), Solis and Wets (1981), Strongin (1972), Yudin (1966)]. We shall discuss some of them.

Matyas (1965) was the first to prove the convergence in probability of an adaptive random search algorithm to the global minimum of a continuous function $g(x)$, $x \in R^n$. His algorithm is as follows:

$$x_{i+1} = \begin{cases} x_i + Y_i & \text{if } g(x_i + Y_i) < g(x_i) - \varepsilon,\ \varepsilon > 0, \\ x_i & \text{otherwise,} \end{cases} \tag{5.2.16}$$

where Y_i are independent normally distributed random vectors with zero mean and covariance matrix Σ, that is, $Y_i \sim N(0, \Sigma)$. According to this algorithm, a step is made from the point x_i in the direction of Y_i only if $g(x_i + Y_i) < g(x_i) - \varepsilon_i$.

Baba et al. (1977) applied a modification of Matyas' algorithm (5.2.16) to the constrained minimization problem (P1) assuming $D = \{x: g_i(x) \leq 0, i = 1, \ldots, m, x \in R^n\}$ and proved its convergence. The algorithm follows.

***Algorithm G1-4* [*Baba et al.* (*1977*)]**

1. Find $x_1 \in D$.
2. Compute $g(x_1)$.
3. $i \leftarrow 1$.
4. Generate Y_i from $N(0, \Sigma)$.
5. If $x_i + Y_i \not\in D$, set $x_{i+1} = x$ and go to step 4.
6. Compute $g(x_i + Y_i)$.
7. $$x_{i+1} = \begin{cases} x_i + y_i, & \text{if } g(x_i + Y_i) < g(x_i), \\ x_i, & \text{otherwise}. \end{cases}$$
8. $i \leftarrow i + 1$.
9. Go to step 4.

According to this algorithm, a step is made from the point x_i in the direction Y_i only if $x_i + Y_i \in D$ and $g(x_i + Y_i) < g(x_i)$. Intuitively, it is not difficult to understand the reason for convergence of G1-4 to the global minimum x^*. Assume, for simplicity, that $g(x)$, $x \in [a, b]$, is as in Fig. 1, that is, it has two local minima at points x_1^* and x_2^*, respectively, and $x_2^* = x^*$ is the global one. Assume for simplicity that $Y \sim N(0, 1)$ and that

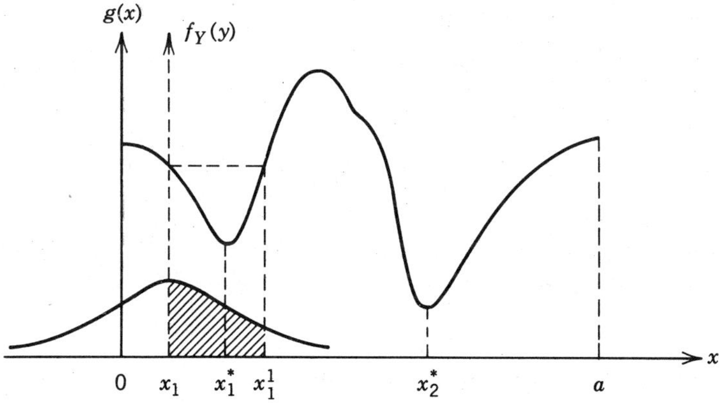

Fig. 1.

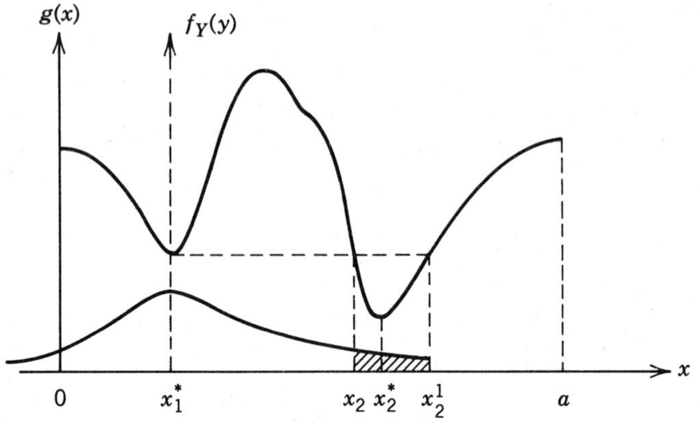

Fig. 2.

we start the algorithm G1-4 from the point x_1. It is readily seen from Fig. 1 that with probability $P_1 = \Phi(x_1 - x_1^1) - \frac{1}{2}$ (see the shaded area), the next point chosen according to G1-4 will be in the direction of x_1^*. Here Φ denotes the standard normal distribution.

Consider now the following two cases:

(i) The point x_1^* has been reached. In this case (see the shaded area in Fig. 2) there still exists a positive probability

$$P_2 = \phi(x_2^1 - x_1^*) - \Phi(x_2 - x_1^*)$$

for the algorithm G1-4 to leave the point x_1^* and therefore to move toward the global minimum x_2^*.

(ii) The point x_2^* has been reached. In this case (see Fig. 3) there is no such Y that $g(x_2^* + Y) < g(x_2^*)$ and therefore the point x_2^* will be never left.

In Appendix 2 we introduce several other random search algorithms and discuss their efficiency on linear and quadratic functions.

Gurin et al. (1968) and later Solis and Wets (1981) proposed several rather general adaptive random search algorithms and showed that many known random search algorithms can be fitted into the framework of their algorithms. We follow Solis and Wets (1981).

Let (D, B, P_i) be the sample space and P_i, $i = 1, 2, \ldots$, the probability measure with B the Borel σ field of D. Then their algorithm can be written as follows.

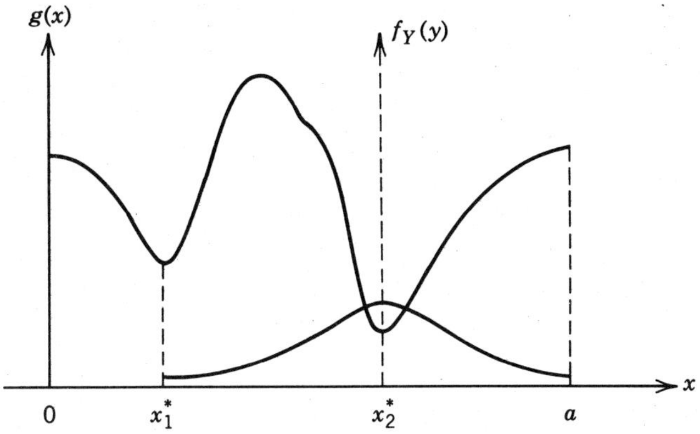

Fig. 3.

***Algorithm G1-5* [*Solis and Wets* (*1981*)]**

1. Find x_i in D and set $i = 1$.
2. Generate Y_i from the sample space (D, B, P_i).
3. Compute $X_{i+1} = \varphi(x_i, Y_i)$, where φ is the map with domain $D \times R^n$ and range D.
4. Choose P_{i+1} and set $i \leftarrow i + 1$.
5. Go to step 2.

The probability measures P_i in G1-5 generally depends on the past history and particularly on $x_1, \ldots, x_{i-1}$ generated by the preceding iteration (that is, P_i is chosen adaptively). Assuming that the conditions

(i) $g[\varphi(x, Y)] \leq g(x)$ and if $Y \in D$, then $g[\varphi(x, Y)] \leq g(Y)$ (5.2.17)

and

(ii) for any Borel subset A of D with $v(A) > 0$ where v is the Lebesgue measure†

$$\prod_{i=0}^{\infty} [1 - P_i(A)] = 0 \tag{5.2.18}$$

are satisfied, Solis and Wets proved the following theorem.

Theorem 1 [Solis and Wets (1981)] Suppose that g is a measurable function from R^n to R^1, D is a measurable subset of R^n and condition

†Condition (5.2.18) means that given any subset A in D with positive measure defined on the Borel field B, the probability of missing A by at least one Y_i, $i = 1, 2, \ldots$, must be zero.

(5.2.17) and (5.2.18) are satisfied. Then

$$\lim_{i\uparrow\infty} P(x_i \in R_{\varepsilon,M}) = 1 .$$

Here $P(x_i \in R_{\varepsilon,M})$ is the probability, that at step i, the point x_i generated by Algorithm G1-5 will be in $R_{\varepsilon,M}$, where $R_{\varepsilon,M}$ is the *optimality region* defined as

$$R_{\varepsilon,M} = \begin{cases} (x \in D\colon g(x) < \alpha + \varepsilon) & \text{if } \alpha \text{ is finite ,} \\ (x \in D\colon g(x) < M) & \text{if } \alpha = -\infty , \end{cases}$$

where $\varepsilon > 0$, $M < 0$, and α is the *essential infimum* of g defined as

$$\alpha = \inf\{t\colon v[x \in D\colon g(x) < t] > 0\} .$$

Proof From (5.2.17) it follows that if either x_i or Y_i are in $R_{\varepsilon,M}$, then $x_{i'} \in R_{\varepsilon,M}$ for all $i' \geq i+1$. Thus

$$P[x_i \in R_{\varepsilon,M}] = 1 - P[x_i \in D \backslash R_{\varepsilon,M}] \geq 1 - \prod_{l=0}^{i} [1 - P_l(R_{\varepsilon,M})] .$$

Hence

$$1 \geq \lim_{i\uparrow\infty} P[x_i \in R_{\varepsilon,M}] \geq 1 - \lim \prod_{l=0}^{i} [1 - P_l(R_{\varepsilon,M})] = 1 ,$$

where the last inequality follows from (5.2.18). Q.E.D.

Solis and Wets (1981) tested several random search algorithms on various unconstrained and constrained optimization models with different probability measures P_i and different mappings $\varphi(x_i, Y_i)$ and showed their feasibility of finding the global minimum.

Yudin (1966) suggested the algorithm

$$\frac{dx}{dt} = -\alpha V g(x) + \Xi(t) \tag{5.2.19}$$

for finding x^* in (P1). Here $\Xi(t)$ is an n-dimensional vector with components being Gaussian white noises with zero mean and correlation matrix

$$E[\Xi_i(t)\Xi_j(\tau + t)] = \sigma_{ij}\delta(\tau) , \qquad i, j = 1, \ldots, n ,$$

and $\delta(\tau)$ is the generalized Dirac function.

Yudin showed that $x(t)$ in (5.2.19) represents a Markov process of a diffusion type and that the p.d.f. $f(x, t)$ of the transition from any initial point $x_0 \in D$ to point x at time t satisfies the Kolmogorov-Fokker-Planck equation, that is,

$$\frac{\partial f(x, t)}{\partial t} - \alpha \sum_{j=1}^{n} \frac{\partial}{\partial x_j}\left(f \frac{\partial g}{\partial x_j}\right) = \frac{1}{2} \sum_{i,j} \sigma_{ij} \frac{\partial^2 f}{\partial x_i \, \partial x_j} \tag{5.2.20}$$

with the condition

$$\int_D f(x, t) \, dx = 1 . \tag{5.2.21}$$

Assuming in addition that $\sigma_{ij} = \sigma \gamma_{ij}$, where γ_{ij} is the Kronecker function, that is,

$$\gamma_{ij} = \begin{cases} 1, & i = j, \\ 0, & i \neq j. \end{cases}$$

Yudin showed that as $t \to \infty$, $f(x, t)$ has a limiting steady-state distribution given by

$$f(x) = \frac{\exp[(-2\alpha g(x)/\sigma^2)]}{\int_D \exp[-2\alpha g(x_i)/\sigma^2] \, dx} \tag{5.2.22}$$

and that the coordinates x_i^*, $i = 1, \ldots, n$, of the global minimum x^* are given by

$$x_1^* = \lim_{\sigma \to 0} \frac{\int_D x_i \exp[-2\alpha g(x)/\sigma^2] \, dx}{\int_D \exp[-2\alpha g(x)/\sigma^2] \, dx} . \tag{5.2.23}$$

It follows from (5.2.22) that the density $f(x)$ achieves its maximum at x^*. Consequently, when $t \to \infty$ and $\sigma \to 0$, we have $f(x) \to \delta(x - x^*)$.

In the particular case where $\sigma = 0$, formula (5.2.19) presents the conventional gradient algorithm that, depending on the starting point (see Algorithm G1-3), converges to one of the local minima of $g(x)$. For more references on convergence and speed of convergence of the procedure (5.2.19) see Gidas (1985a, b) and Geman and Hwang (1985).

It is of interest to note that Meerkov (1972) and Pincus (1968) arrived at formula (5.2.23) from a different direction, namely by using the classical Laplace formula for certain integrals. The following theorem is due to Pincus (1968).

Theorem 2 [Pincus (1968)] Let $g(x)$ be a real-valued, continuous function over a closed bounded domain $D \in R^n$. Further let there be a unique point $x^* \in D$ at which $\min g(x)$ is attained. Then the coordinate x_i^*, $i = 1, \ldots, n$, of x^*, $x \in D$, are given by

$$x_i^* = \lim_{\lambda \to \infty} \frac{\int_D x_i \exp[-\lambda g(x)]\, dx}{\int_D \exp[-\lambda g(x)]\, dx} . \tag{5.2.24}$$

Clearly, when $\lambda = 2\alpha/\sigma^2$, formulas (5.2.23) and (5.2.24) coincide; the main difference between them, however, is that in the former convergence to x^* is probabilistic whereas in the latter it is deterministic.

Pincus (1970) proposed a Monte Carlo algorithm for estimating x^* in (5.2.24).

For more references on stochastic method for global optimization see Archetti and Schoen (1984), Devroy (1976), (1979), Ermoliev (1976), (1983) Rinnooy Kan and Timmer (1983, 1985a, b), and Rubinstein (1969, 1981, 1983).

Remark It is of interest to note that the problem of finding the global minimum of $g(x)$ is equivalent to that of finding the mode for association with its density function. Indeed, if $g(x) > 0$, $x \in D$, then $\psi_1(x) = c^{-1}g(x)$, where $c^{-1} = [\int g(x)\, dx]^{-1}$ is a density function. If $g(x)$ is unrestricted in sign but bounded, that is, $g(x) \leq L$, then $\psi_2(x) = c^{-1}[g(x) + L]$, where $c^{-1} = [\int (g(x) + L)\, dx]^{-1}$ is again a density function.

Recently a new Monte Carlo algorithm, called simulated annealing,† was proposed by Kirkpatrick, Gelatt, and Vecchi (1983) for solving complex multiextremal optimization problems with discrete space. Their algorithm essentially presents a modification of the Metropolis, Rosenbluth, Rosenbluth, Teller, and Teller (1953) algorithm introduced for estimating the mean value of certain quantities in statistical mechanics at a given temperature. Note that in the Kirkpatrick et al. (1983) annealing algorithm the temperature of the system decreases to zero and the algorithm generates a nonstationary Markov chain with the state space being the domain of the cost function to be minimized. Applications of

†Annealing is a process of cooling metals (physical systems) with the aim of obtaining states with minimal energy.

simulated annealing can be found in combinatorial optimization in computer design problems and machine learning. We present here only the main idea of simulated annealing following the tutorial of Hajek (1985a).

Let $g(x)$ be defined on a finite set D. Assume that for each state d in D there exists a set $N(d)$ and an associated transition probability matrix $P_{dd'}$ such that $P_{dd'} > 0$ if and only if $d' \in N(d)$. Given any initial state, say $X_0 \in N(d)$, the proceeding states $X_1, X_2, \ldots,$ are updated according to the following adaptive (annealing) algorithm.

$$X_{k+1} = \begin{cases} Y_k, & \text{with probability } p_k, \\ X_k, & \text{with probability } 1 - p_k, \end{cases}$$

where Y_k is chosen from $N(d)$ with probability

$$P(Y_k = d' \mid X_k = d) = P_{dd'},$$

$$p_k = \exp\{-[g(Y_k) - g(X_k)]^+/\sigma_k\}$$

and σ_k, $k = 1, 2, \ldots,$ is a sequence (called a temperature schedule) of strictly positive numbers, such that $\sigma_1 \geq \sigma_2 \geq \cdots$ and $\sigma_k \to 0$ as $k \to \infty$.

It is not difficult to see that the sequence $X_1, X_2, \ldots,$ produced so far presents a nonhomogeneous discrete-time Markov chain. The probability p_k to choose Y_k (next potential state) increases with σ_k (the higher the "temperature" is, the more likely is that a "hill climbing" move is accepted) and decreases with $[g(Y_k) - g(X_k)]^+$. The shape of the function g, the initial temperature, the rate of decrease of the temperature, and the number of iterations are all important parameters affecting the convergence and speed of convergence of the algorithm [see Mitra, Romeo, and Sangiovanni-Vincentelli (1985)].

Let D^* denote the set of states in D where g achieves the global minimum. Hajek (1985a) discusses conditions under which

$$\lim_{k \to \infty} P(X_k \in D^*) = 1$$

and therefore the annealing algorithm converges to the global minimum. For more references on simulated annealing see Gelfand and Mitter (1985), Gidas (1985a, 1985b); Hajek (1985b), and Mitra, Romeo, and Sangiovanni-Vincentelli (1985).

In all the above Monte Carlo algorithms we assumed availability of sampling from a distribution function associated with some probability space (D, B, P). Generally, however, sampling over D, even with nice probability measures, is not a simple problem. Let us discuss this issue in

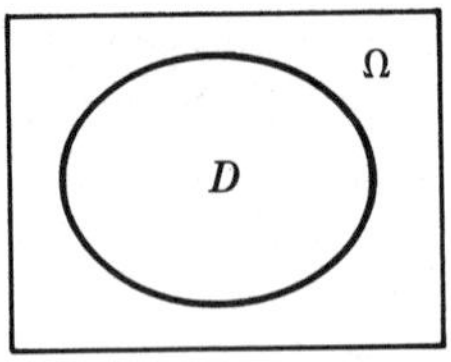

Fig. 4.

more details, assuming further that a uniform sample over D is desired. We shall distinguish the following two cases.

1. D is a nice region (multidimensional rectangle, simplex, hypersphere, or hyperellipsoid). In this case direct methods for generating random vectors uniformly distributed over D are available; they are presented in Appendix 3, reproduced mainly from Rubinstein (1982).

2. D is a complex region. In this case the simplest way to generate a random vector uniformly distributed over D is to insert the region D in a nice one, say Ω (see Fig. 4), than to generate a random vector uniformly in Ω and accept or reject the generated vector depending on whether it hits or misses the region D. We present below an algorithm known as the acceptance-rejection algorithm.

Acceptance-Rejection Algorithm

1. Generate a random vector **X** uniformly distributed in Ω.

2. If $X \in D$, accept **X** as a random vector uniformly distributed in D.

3. Go to step 1.

The efficiency of this algorithm depends on how well Ω covers D; it can be defined as

$$p = \frac{m(D)}{m(\Omega)}, \qquad 0 < p < 1 .$$

The number of trials needed before a successful one is found has a geometric distribution

$$P_X(x) = p(1-p)^x , \qquad x = 1, 2 ,$$

with the expectation equal $1/p$.

For this algorithm to be of practical interest, the following criteria must be satisfied:

1. It should be easy to generate a random vector in Ω.

2. The efficiency p should be high.

For the first criterion to be met, we have to choose Ω to be a nice region (say a multidimensional rectangle, a hypersphere, etc.) and such that direct methods (see Appendix 3) for generating uniformly distributed points would be available.

For the second criterion to be met, the volume of Ω has to be minimal and such that $D \subset \Omega$. In Appendix 4 [reproduced from Rubinstein and Samorodnitsky (1986)] we discuss the second criterion. More definitely, we assume that D is a convex region given by a system of nonlinear constraints, that is, $D = \{x\colon g_i(x) \leq 0, i, \ldots, m, x \in R^n\}$, where $g_i(x)$ are known functions, and consider the problem of its coverage with polygons and quadratic configurations of minimal volume. It is shown that the problem of optimal coverage D with an arbitrary polygon Ω can be reduced to a *convex* one of coverage with a multidimensional rectangle. If, however, rotation of the coordinate system is assumed, an additional *nonconvex* problem must be solved. It is also shown that to find the minimal covering hypersphere or hyperellipsoid, one has to solve *two convex* programming problems.

The main drawback of the acceptance-rejection method is that the number of trials in Ω needed to obtain a point in D increases explosively in the dimension n. For example, let Ω be a hypercube and the enclosing region D a hypersphere. In this case it is readily seen that the efficiency p_n equals

$$p_n = \frac{m(D)}{m(\Omega)} = \frac{\pi^{n/2}}{n2^{n-1}\Gamma(n/2)} . \tag{5.2.25}$$

For even n $(n = 2k)$ we have

$$p_k = \frac{\pi^k}{k!2^{2k}} = \frac{1}{k!}\left(\frac{\pi}{2}\right)^k 2^{-k} \quad \text{and} \quad p_k \to 0,\ k \to \infty . \tag{5.2.26}$$

The expected number of trials needed to obtain a point in D increases from 1.5 for $k = 2$ to 10^{30} for $k = 100$.

Attempts to overcome the "curse" of the dimensionality were made by Smith (1984), who proposed several *direct* methods for generating uniformly distributed (instead of acceptance-rejection) points over a bounded region D. He proved that the sample points obtained are *asymptotically* uniformly distributed over D and that the efficiency of his methods (compared with the acceptance-rejection method) increases with the dimensionality n.

Finally, we would like to point out that the problem of efficient sampling over different regions is still in its infancy. This problem is associated with many important practical applications, and we believe it will deserve more attention in the future.

5.3 SMOOTHED FUNCTIONALS FOR CONSTRAINED OPTIMIZATION PROBLEMS

In this section we consider the problem of minimization of a nonconvex function under convex constraints. We replace the original nonconvex function by an auxiliary one, called a smoothed function, which possesses some nice properties. Operating with a smoothed function and the given convex constraints, we find the minimum of the original problem.

Consider the optimization problem

$$\min_x g_0(x) = \min_x E[\psi_0(x, Y)] \tag{5.3.1}$$

subject to

$$g_j(x) = E[\psi_j(x, Y)] \leq 0 , \quad j = 1, \ldots, M , \quad x \in X \subseteq R^n , \tag{5.3.2}$$

where x and Y are n- and m-dimensional vectors, respectively. We assume that X is a convex bounded domain, $g_j(x)$ ($j = 0, 1, \ldots, M$) are bounded functions, and

$$\text{Var}[\psi_j(x, Y)] < \infty , \quad j = 0, 1, \ldots, M . \tag{5.3.3}$$

For convenience we shall deal with $E[\psi_j(x, Y)]$ of the form

$$E[\psi_j(x, Y)] = \int \psi_j(x, y) f_Y(y)\, dy , \quad j = 0, 1, \ldots, M ,$$

where $f_Y(y)$ is the probability density function.

As it was pointed out in section 5.1 that

(i) If the p.d.f. $f_Y(y)$ of the r.v. Y is *known*, and all the integrals $g_j(x) = \int \psi_j(x, y) f_Y(y)\, dy$, $j = 0, 1, \ldots, M$, can be found explicitly, then the problem (5.3.1)–(5.3.2) represent the conventional nonlinear programming problem, which can be solved either analytically or numerically by methods of nonlinear programming [see Avriel (1976)].

(ii) If at the other extreme one of the integrals in (5.3.1) and (5.3.2) cannot be found explicitly, or the p.d.f. $f_Y(y)$ is unknown, then there is no way to solve (5.3.1)–(5.3.2) analytically. In this case Monte-Carlo optimization method must be used. We shall deal here with both cases (i) and (ii).

Antonov and Katkovnik (1972) as well as Katkovnik and Kulchitsky (1972) considered the unconstrained problem (5.3.1). They introduced the function

$$\hat{g}(x, \beta) = \int_{-\infty}^{\infty} \hat{h}(v, \beta) g(x - v)\, dv = \int_{-\infty}^{\infty} \hat{h}((x - v), \beta) g(v)\, dv \quad (5.3.4)$$

and called it a *smoothed functional.*

The idea of smoothed functionals is based on the assumption that the objective function $g_0(x)$ is "ill-behaved" (it can be a multiextremal function or have a fluctuating character) (see Fig. 5) and that one could expect a better "behavior" from the smoothed function $\hat{g}(x, \beta)$.

The above authors investigated the behavior of $\hat{g}(x, \beta)$ under various assumptions on g and h and proposed several iterative procedures that converge to the optimal solution with g assumed to be a convex function.

For $\hat{g}(x, \beta)$ to have nice smoothness properties, the following assumptions on the kernel $\hat{h}(v, \beta)$ have been made in Antonov and Kotkovnik (1972):

(a) $\hat{h}(v, \beta) = \frac{1}{\beta^n} h \frac{v}{\beta} = \frac{1}{\beta^n} h\left(\frac{v_1}{\beta}, \ldots, \frac{v_n}{\beta}\right)$

is a piece-wise differentiable function with respect to v;

(b) $\lim_{\beta \to 0} \hat{h}(v, \beta) = \delta(v)$, where $\delta(v)$ is Dirac's delta function;

(c) $\lim_{\beta \to 0} \hat{g}(x, \beta) = g(x)$, if x is a point of continuity of $g(x)$;

(d) $\hat{h}(v, \beta)$ is a p.d.f.

It is obvious that the effect of smoothing depends on the parameter β: For large β the effect of smoothing is large, and vice versa. For $\beta \to 0$ we have $\hat{g}(x, \beta) \to g(x)$, that is, there is no smoothing at all.

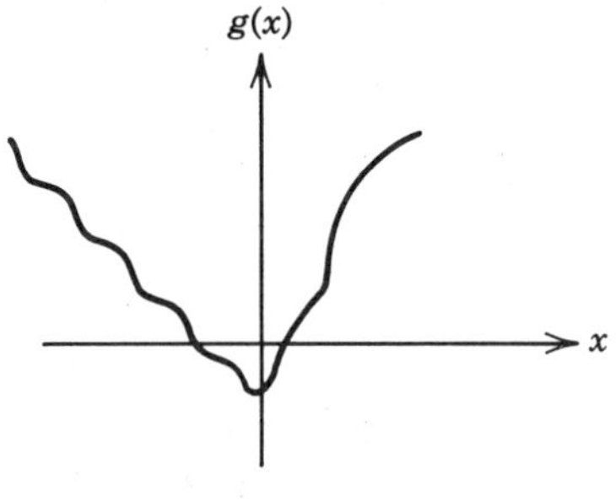

Fig. 5 An ill-behaved function.

Katkovnik (1976) suggested convergent procedures of the stochastic approximation type for finding the optimal solution of the problem (5.3.1)–(5.3.2) assuming that both the objective function $g_0(x)$ and the constraint $g_j(x) \leq 0$, $j = 1, \ldots, M$, are convex. Rubinstein (1983) proposed a convergent procedure for the unconstrained problem (5.3.1), retaining Katkovnik's smoothed functionals approach but dispensing with the convexity assumption on $g_0(x)$. Kreimer and Rubinstein (1985) considered the constrained problem [(5.3.1)–(5.3.2)] in the case where the region $D = \{x\colon g_j(x) \leq 0, j = 1, \ldots, M\}$ is convex but the objective function $g_0(x)$ is not necessarily so. They proposed an algorithm and prove it convergence almost surely to the optimal solution of (5.3.1)–(5.3.2), thereby generalizing Rubinstein's (1983) result.

The results of this section are based on Gupal (1977), Nurminski (1973a, 1973b), Rubinstein (1983), and Kreimer and Rubinstein (1985).

It is not our intention here to present simulation results based on the proposed procedure or to compare the efficiency of our procedures with that of earlier ones (such a comparison would be very difficult to establish). Our purpose is rather to apply the smoothed functional $\hat{g}(x, \beta)$ to solve the problem [(5.3.1) and (5.3.2)] under the nonconvexity assumptions about $g_0(x)$.

We believe that our discussion will promote wider use of smoothed functionals for optimization of complex nonconvex optimization problems.

Consider instead of (5.3.4) the following smoothed function:

$$\tilde{g}(x, \beta) = \frac{1}{2} \int_{-\infty}^{\infty} \hat{h}(v, \beta)[g(x - v) + g(x + v)]\, dv$$

$$= \frac{1}{2} \int_{-\infty}^{\infty} h(v)[g(x - \beta v) + g(x + \beta v)]\, dv\,. \tag{5.3.5}$$

We shall further use the following two properties of $\tilde{g}(x, \beta)$:

Property 1 If $g(x)$ is convex, then $\tilde{g}(x, \beta)$ is also convex.

Proof The proof of this property is straightforward. For $0 < \lambda < 1$

$$\lambda \hat{g}(x, \beta) + (1 - \lambda)\hat{g}(y, \lambda) - \hat{g}(\lambda x + (1 - \lambda)y, \beta)$$

$$= \int \hat{h}(v, \beta)[\lambda g(x - v) + (1 - \lambda)g(y - v) - g(\lambda x + (1 - \lambda)y - v)]\, dv\,.$$

The convexity of $g(x)$ implies

$$g(\lambda x + (1-\lambda)y - v) = g(\lambda(x-v) + (1-\lambda)(y-v))$$
$$\leq \lambda g(x-v) + (1-\lambda)g(y-v) .$$

From the above two formulas and from the fact that $h(v, \beta) \geq 0$ the proof follows immediately. Q.E.D.

Property 2 If $g(x)$ is convex, then $\tilde{g}(x, \beta) \downarrow g(x)$ as $\beta \downarrow 0$.

Proof We prove first that:
(i) $g(x) \leq \tilde{g}(x, \beta)$.
We have

$$g(x) \leq \tfrac{1}{2} g(x - \beta v) + \tfrac{1}{2} g(x + \beta v) ,$$

since $x = \frac{1}{2}(x - \beta v) + \frac{1}{2}(x + \beta v)$ and g is convex. Therefore

$$g(x) = \int_{-\infty}^{\infty} h(v) g(x) \, dv$$
$$\leq \frac{1}{2} \int_{-\infty}^{\infty} h(v)[g(x - \beta v) + g(x + \beta v)] \, dv = \tilde{g}(x, \beta) .$$

Next we prove that:
(ii) $\tilde{g}(x, \beta)$ is a nondecreasing function of β. Assume $\beta_1 < \beta_2$; then

$$x - \beta_1 v = \lambda_1(x - \beta_2 v) + (1 - \lambda_1)(x + \beta_2 v)$$

and

$$x + \beta_1 v = \lambda_2(x - \beta_2 v) + (1 - \lambda_2)(x + \beta_2 v) ,$$

where

$$\lambda_1 = \frac{\beta_1 + \beta_2}{2\beta_2} , \qquad \lambda_2 = \frac{\beta_2 - \beta_1}{2\beta_2} , \qquad (\lambda_1 + \lambda_2 = 1) .$$

Therefore it follows from the convexity of $g(x)$ that

$$g(x - \beta_1 v) \leq \lambda_1 g(x - \beta_2 v) + (1 - \lambda_1) g(x + \beta_2 v) ,$$

and

$$g(x + \beta_1 v) \leq \lambda_2 g(x - \beta_2 v) + (1 - \lambda_2) g(x + \beta_2 v) .$$

Adding the last two inequalities together, we have

$$\begin{aligned} g(x-\beta_1 v)+g(x+\beta_1 v) &\le (\lambda_1+\lambda_2)g(x-\beta_2 v) \\ &\quad +[(2-(\lambda_1+\lambda_2))g(x+\beta_2 v)] \\ &= g(x-\beta_2 v)+g(x+\beta_2 v)\,, \end{aligned}$$

from which it readily follows that

$$\tilde{g}(x, \beta_1) \le \tilde{g}(x, \beta_2)\,.$$

Now from (i), (ii), and assumption (c) on the kernel $\hat{h}(v, \beta)$, we obtain

$$\tilde{g}(x, \beta) \downarrow g(x) \quad \text{as } \beta \downarrow 0\,. \qquad \text{Q.E.D.}$$

The gradient of $\tilde{g}(x, \beta)$ in (5.3.5) equals

$$\begin{aligned} \nabla_x \tilde{g}(x, \beta) &= \frac{1}{2\beta}\int_{-\infty}^{\infty} \nabla \hat{h}_v(v, \beta)[g(x-v)-g(x+v)]\,dv \\ &= \frac{1}{2\beta}\int_{-\infty}^{\infty} \nabla h_v(v)[g(x-\beta v)-g(x+\beta v)]\,dv \end{aligned} \tag{5.3.6}$$

and is called the *smoothed gradient.*

By analogy with (5.3.6), the smoothed gradient of $\tilde{\psi}(x, Y, \beta)$ is

$$\begin{aligned} \nabla_x \tilde{\psi}(x, Y, \beta) &= \frac{1}{2}\int_{-\infty}^{\infty} \nabla_v \hat{h}(v, \beta)[\psi(x-v, Y_1)-\psi(x+v, Y_2)]\,dv \\ &= \frac{1}{2\beta}\int_{-\infty}^{\infty} \nabla_v h(v)[\psi(x-\beta v, Y_1)-\psi(x+\beta v, Y_2)]\,dv\,. \end{aligned} \tag{5.3.7}$$

Explicit expressions for the multiple integrals of type (5.3.7) are usually unavailable (especially when ψ is ill-behaved), but numerical methods—in particular the Monte Carlo method—can be used. The following Monte Carlo estimator of $\nabla_x \tilde{\psi}(x, Y, \beta)$ is proposed in Katkovnik (1976) and in Rubinstein (1983):

$$\bar{\nabla}_x \psi(x, Y, \beta, N) = \frac{1}{2N\beta}\sum_{j=1}^{N} \frac{\nabla_v h(V_j)}{\varphi_V(V_j)}[\psi(x-\beta V_j, Y_{1j})-\psi(x+\beta V_j, Y_{2j})] \tag{5.3.8}$$

and is called the *parametrical statistical gradient* (PSG). Here $\varphi_V(v)$ is a p.d.f. from which a sample of size N is taken. For $\varphi_V(v) = h(v)$, that is, $h(v)$ a p.d.f., we have

$$\bar{\nabla}_x \psi(x, Y, \beta, N) = \frac{1}{2N\beta} \sum_{j=1}^{N} \frac{\nabla_v h(V_j)}{h(V_j)} [\psi(x - \beta V_j, Y_{1j}) - \psi(x + \beta V_j, Y_{2j})], \tag{5.3.9}$$

where V_j and Y_{1j}, Y_{2j}, $j = 1, \ldots, N$ are samples from $h(v)$ and $f_Y(y)$ respectively. Table 2 presents some smoothed gradients and their estimators.

Assuming that the r.v.'s V and Y are independent and taking the expectation of $\bar{\nabla}_x \psi(x, Y, \beta, N)$ with respect to Y and V, we obtain

$$E[\bar{\nabla}_x \psi(x, Y, \beta, N)] = \frac{1}{N} E\left[\sum_{j=1}^{N} D_j \psi(x, Y_j, \beta)\right] = \nabla_x \tilde{g}(x, \beta), \tag{5.3.10}$$

where

$$D_j(x, Y_j, \beta) = \frac{1}{2\beta} \frac{\nabla_v h(V_j)}{\varphi_V(V_j)} [\psi(x - \beta V_j, Y_{1j}) - \psi(x + \beta V_j, Y_{2j})]. \tag{5.3.11}$$

That is, the PSG $\bar{\nabla}_x \psi(x, Y, \beta, N)$ is an unbiased estimator for the smoothed gradient $\nabla_x \tilde{g}(x, \beta)$. Assuming also independence of Y_j, $j = 1, \ldots, N$, we obtain the variance of the sth component of $\bar{\nabla}_x \psi(x, Y, \beta, N)$ as

$$\begin{aligned} \operatorname{Var}[\bar{\nabla}_x \psi(x, Y, \beta, N)]_s &= \frac{1}{N^2} \sum_{j=1}^{N} \operatorname{Var}[D_j(x, Y_j, \beta)]_s \\ &= N^{-1}(2\beta)^{-2} \sigma_s^2(x), \qquad s = 1, \ldots, n, \end{aligned} \tag{5.3.12}$$

where

$$\sigma_s^2 = \operatorname{Var}\left(\frac{[\nabla_v h(V)]_s}{\varphi_V(V)} [\psi(x - \beta V, Y_1) - \psi(x + \beta V, Y_2)] \right), \tag{5.3.13}$$

and

Table 2 Smoothed gradients and their estimators[a]

$\hat{h}(v, \beta)$	$h(v)$	$\hat{\nabla}_x \psi(x, Y, \beta)$	$\bar{\nabla}_x \psi(x, Y, \beta, N)$	Algorithm
$\frac{1}{\beta^n} \delta(v/\beta)$	$\delta(v)$	$\hat{\nabla}_x \psi(x, Y)$	$\frac{1}{N} \sum_{j=1}^{N} \hat{\nabla}_x \psi(x, Y_j)$	Gupal (2.11) and (2.12)
$\frac{1}{2^n \beta^n} \prod_{s=1}^{n} [U(v_s - \beta) - U(v_s + \beta)]$	$\frac{1}{2^n} \prod_{s=1}^{n} [U(v_s - 1) - U(v_s + 1)]$	$\frac{1}{2\beta} \sum_{s=1}^{n} [\psi(x + \beta l_s, Y_1) - \psi(x - \beta l_s, Y_2)] l_s$	$\frac{1}{2N\beta} \sum_{j=1}^{N} \sum_{s=1}^{n} [\psi(x + \beta l_s, Y_{ij}) - \psi(x - \beta l_s, Y_{2j})] l_s$	Kiefer-Wolfswitz (Wasan, 1969)
$\frac{1}{(2\pi)^{n/2} \beta^n} \exp\left(-\sum_{s=1}^{n} \frac{v_s^2}{2\beta^2}\right)$	$\frac{1}{(2\pi)^{n/2}} \exp\left(-\sum_{s=1}^{n} \frac{v_s^2}{2}\right)$	$\frac{1}{2\beta} E_V\{[\psi(x + \beta V, Y_1) - \psi(x - \beta V, Y_2)] V\}$	$\frac{1}{2N\beta} \sum_{j=1}^{N} [\psi(x + \beta V_j, Y_{1j}) - \psi(x - \beta V_j, Y_{2j})] V_j$	Random search (Rubinstein, 1983)
$\frac{1}{\beta^n} \left(1 - \sum_{s=1}^{n} \frac{v_s^2}{\beta^2}\right) U\left(1 - \sum_{s=1}^{n} \frac{v_s^2}{\beta^2}\right)$, $0 \le v_s \le 1$	$\left(1 - \sum_{s=1}^{n} v_s^2\right) U\left(1 - \sum_{s=1}^{n} v_s^2\right)$, $0 \le v_s \le 1$	$\frac{n}{2\beta} E_V\{[\psi(x + \beta V, Y_1) - \psi(x - \beta V, Y_2)] V\}$	$\frac{n}{2N\beta} \sum_{j=1}^{N} [\psi(x + \beta V_j, Y_{1j}) - \psi(x - \beta V_j, Y_{2j})] V_j$	Random search ((Rubinstein, 1983)

[a] Here $U(x) = \begin{cases} 1, & \text{if } x \ge 0, \\ 0, & \text{if } x < 0. \end{cases}$

$$\begin{aligned}
&E[\|\bar{\nabla}_x \psi(x, Y, \beta, N)\|^2] \\
&\quad = E\left[\sum_{s=1}^{n} [\bar{\nabla}_x \psi(x, Y, \beta, N)]_s^2\right] \\
&\quad = \sum_{s=1}^{n} E[\bar{\nabla}_x \psi(x, Y, \beta, N)]_s^2 \\
&\quad = \sum_{s=1}^{n} \{\mathrm{Var}[\bar{\nabla}_x \psi(x, Y, \beta, N)]_s + [E[\bar{\nabla}_x \psi(x, Y, \beta, N)]_s]^2\} \\
&\quad = \sum_{s=1}^{n} \mathrm{Var}[\bar{\nabla}_x \psi(x, Y, \beta, N)]_s + \|E[\bar{\nabla}_x \psi(x, Y, \beta, N)]\|^2 \\
&\quad \le \sigma^2 N^{-1}(2\beta)^{-2} + \|\nabla_x \tilde{g}(x, \beta)\|^2 . \qquad (5.3.14)
\end{aligned}$$

Here $[\nabla_v h(V)]_s$ is the sth coordinate of the vector $\nabla_v h(v)$ and

$$\sigma^2 = n \max_s \sup_{x \in X} \sigma_s^2(x) < \infty , \qquad (5.3.15)$$

which is finite in view of (5.3.3) and of the compactness of X.

The Proposed Algorithm

We first consider the unconstrained problem (5.3.1) by applying the algorithm

$$x_{i+1} = \pi_X\{x_i - a_i \bar{\nabla}_x \psi(x_i, y_i, \beta_i, N_i)\} , \qquad (5.3.16)$$

where $\bar{\nabla}_x \psi(x_i, y_i, \beta_i, N_i)$ is defined in (5.3.8), π_X is a projection operator on the set X, i.e., for every $x \in R^n$, $\pi_X(x) \in X$, and $\|x - \pi_X(x)\| = \min_{y \in X} \|x - y\|$.

Theorem [Rubinstein (1983)] Let $x \in R^k$ be a convex, closed, and bounded set and $g(x)$ be a bounded function for each $x \in X$ satisfying the following conditions:

$$\begin{gathered}
\langle (x - x^*), \nabla_x g(x, \beta_1) \rangle \ge K_1 \|x - x^*\|^2 - \gamma_i , \\
\langle \nabla_x g(x, \beta), \nabla_x g(x, \beta) \rangle \le K_2 \|x - x^*\|^2 , \\
0 < \alpha < 2K_1 K_2^{-1} , \\
\lim_{i \to \infty} \beta_i^{-2} N_i^{-1} = 0 , \qquad \lim_{i \to \infty} \gamma_i = 0 , \qquad \gamma_i > 0 , \\
E\|x_1\|^2 < \infty ,
\end{gathered}$$

where $\|x\| = (\Sigma_{k=1}^{n} x_k^2)^{1/2}$ is the norm of x and K_1 and K_2 are positive constants. Then process (5.3.16) converges in the mean square to the point x^*, that is, $\lim_{i\to\infty} E\|x_i - x^*\|^2 = 0$. If we replace condition $\lim_{i\to\infty} \beta_i^{-2} N_i^{-1} = 0$ by

$$\sum_{i=1}^{\infty} \beta_i^{-2} N_i^{-1} < \infty$$

and condition $\lim_{i\to\infty} \gamma_i = 0$, $\gamma_i > 0$, by

$$\sum_{i=1}^{\infty} \gamma_i < \infty ,$$

then (5.3.16) converges a.s. to x^*.

Consider now the constraint problem (5.3.1) and (5.3.2), which we solve again by applying algorithm (5.3.16) with

$$\bar{\nabla}_x \psi(x_i, y_i, \beta_i, N_i) = \begin{cases} \bar{\nabla}_x \psi_0(x_i, y_i, \beta_i, N_i) & \text{if } \max_{i \le j \le M} z_i^j \le 0 , \\ \bar{\nabla}_x \psi_k(x_i, y_i, \beta_i, N_i) & \text{if } z_i^k = \max_{i \le j \le M} z_i^j > 0 , \end{cases} \tag{5.3.17}$$

$$z_{i+1}^j = z_i^j + \rho_i(\theta_i^j - z_i^j) , \qquad j = 1, \ldots, M , \tag{5.3.18}$$

$$E[\theta_i^j | x_0, z_0, \ldots, x_i, z_i] = g_j(x_i) , \qquad j = 1, \ldots, M , \tag{5.3.19}$$

and ρ_i, a_i are sequences defined in (5.3.24) below.

(a) For example,

$$\theta_i^j = \psi_j(x_i, Y_i) , \qquad j = 1, \ldots, M .$$

(b) If θ_i^j are unbounded, then the values z_i^j are normalized as

$$z_{i+1}^j = \begin{cases} Z_i^j + \rho_i(\theta_i^j - z_i^j) = \tilde{z}_{i+1}^j & \text{if } |\tilde{z}_{i+1}^j| \le C , \\ C \operatorname{sign}(\tilde{z}_{i+1}^j) & \text{otherwise} , \end{cases} \tag{5.3.20}$$

where C is a sufficiently large number.

Theorem [Kreimer and Rubinstein (1985)] Let $X \subset \mathbb{R}^n$ be a convex closed and bounded set, $g_j(x)$ $(j = 1, \ldots, M)$ be continuous and convex functions, and $g_0(x)$ be a function bounded for each x and satisfying following conditions (examples of functions satisfying (5.3.21) and (5.3.22) are given in the following section):

$$\langle (x - x^*), \nabla_x \tilde{g}_0(x, \beta_i) \rangle \geq \gamma(\|x - x^*\|) + 0(\beta_i) \tag{5.3.21}$$

$$\|\nabla_x \tilde{g}_0(x, \beta)\| \leq K_2 , \qquad x \in X , \tag{5.3.22}$$

where x^* is a solution of the problem (5.3.1)–(5.3.2), $\langle \ , \ \rangle$ denotes a scalar product, and $\gamma(t)$ is defined in (5.3.24).

Assume further that the set

$$D = \{x: g_j(x) \leq 0, \ j = 1, \ldots, M\}$$

satisfies Slater's condition, that is,

$$\{x: g_j(x) < 0, \ j = 1, \ldots, M\} \neq \emptyset , \tag{5.3.23}$$

and

(i) $a_i > 0 , \qquad \rho_i > 0 ;$

(ii) $\sum_{i=0}^{\infty} a_i = \infty , \qquad \sum_{i=0}^{\infty} \rho_i = \infty ;$

(iii) $\sum_{i=0}^{\infty} a_i^2 < \infty , \qquad \sum_{i=0}^{\infty} \rho_i^2 < \infty ;$

(iv) $\dfrac{a_i}{\rho_i} \downarrow 0$ as $i \to \infty ;$

(v) $\beta_i^{-2} N_i^{-1} < C < \infty , \qquad i = 0, 1, \ldots ;$ (5.3.24)

(vi) $\gamma(t)$ is a continuous function,

$$\gamma(t) > 0 \text{ for } t > 0 \ [\text{e.g.}, \gamma(t) = K_1 t^r, \ r > 0];$$

(vii) $E[|\theta_i^j|^2] \leq C < \infty \quad (j = 1, \ldots, M);$

(viii) $E[(z_0^j - g_j(x_0))^2] < \infty \quad (j = 1, \ldots, M).$

Then $\lim_{i \to \infty} x_i = x^*$ almost surely.

Examples

We shall now show that conditions (5.3.21) and (5.3.22) are satisfied for the so-called G-convex functions. We present an example of a function satisfying (5.3.21) and (5.3.22) that is neither convex nor concave in the neighborhood of x^*.

Definition [Avriel and Zang (1974)] A function f defined on a convex set $X \subset \mathbb{R}^n$ is called G convex if for any $x_1 \in X$, $x_2 \in X$, $q_1 \geq 0$, and $q_2 \geq 0$ $(q_1 + q_2 = 1)$

$$f(q_1 x_1 + q_2 x_2) \leq G^{-1}[q_1 G(f(x_1)) + q_2 G(f(x_2))] ,$$

where $G(y)$, $y \in R^1$, is continuous and strictly monotone and G^{-1} is the inverse of G. If the last inequality is strict, then f is called strictly G convex.

Proposition 3 Let $f(x)$ be a continuously differentiable strictly G-convex function defined on a compact set $X \in R^n$. Then $f(x)$ satisfies conditions (5.3.21) and (5.3.22) if G is strictly increasing and continuously differentiable.

Proof Our proof is based on the following result due to Avriel and Zang (1974) [$f(x)$ is strictly G convex iff $g(x) = G(f(x))$ is strictly convex] and on the next lemma.

Lemma 1 Let the original function $g(x)$ be differentiable and satisfy

$$\langle (x - x^*), \nabla g(x) \rangle \geq \gamma(\|x - x^*\|) , \tag{5.3.21'}$$

$$\|\nabla g(x)\| \leq K_2 . \tag{5.3.22'}$$

Then the smoothed function $\tilde{g}(x, \beta)$ [see equation (5.3.5)] satisfies (5.3.21) and (5.3.22) if

$$E(\|V\|) = \int \|v\| h(v)\, dv < \infty . \tag{5.3.25}$$

Proof The gradient of $\tilde{g}(x, \beta)$ is

$$\tilde{\nabla} g(x, \beta) = \frac{1}{2} \int h(v)[\nabla g(x - \beta v) + \nabla g(x + \beta v)]\, dv .$$

It follows from (5.3.22′) and $E(\|V\|) < \infty$ that $\|\nabla \tilde{g}(x, \beta)\| \leq K_2$, that is, condition (5.3.22) holds for $\tilde{g}(x, \beta)$.

Now, making use of (5.3.25) we have

$$\begin{aligned}
&\langle x - x^*, \nabla\tilde{g}(x, \beta)\rangle \\
&= \frac{1}{2}\int_{-\infty}^{\infty} h(v)[\langle x - x^*, \nabla g(x - \beta v)\rangle + \langle x - x^*, \nabla g(x + \beta v)\rangle]\, dv \\
&= \frac{1}{2}\int_{-\infty}^{\infty} h(v)\{[\langle x - \beta v - x^*, \nabla g(x - \beta v)\rangle + \langle x + \beta v - x^*, \nabla g(x + \beta v)\rangle] \\
&\quad + \beta[\langle v, \nabla g(x - \beta v)\rangle - \langle v, \nabla g(x + \beta v)\rangle]\}\, dv \\
&\geq \frac{1}{2}\int_{-\infty}^{\infty} h(v)[\gamma(\|x - \beta v - x^*\|) + \gamma(\|x + v - x^* \vee)]\, dv - K_2\beta E[\|V\|] \\
&= \tilde{\gamma}(\|x - x^*\|, \beta) - K_2\beta E[\|V\|] = \gamma(\|x - x^*\|) + 0(\beta)\,.
\end{aligned}$$

Thus, (5.3.21) holds for $\tilde{g}(x, \beta)$. Q.E.D.

To proceed with Proposition 3 note that conditions (5.3.21) and (5.3.22) are formulated for the smoothed function $\tilde{g}(x, \beta)$ and the conditions (5.3.21′) and (5.3.22′) for the original function $g(x)$. From Lemma 1, it is clear that Proposition 3 will be proved if we prove that $f(x)$ satisfies conditions (5.3.21′) and (5.3.22′) rather than (5.3.21) and (5.3.22).

To prove that (5.3.21′) holds for $f(x)$, note that it holds for $g(x) = G(f(x))$ since it is assumed to be strictly convex. Rewriting it as

$$G'(f(x))\langle \nabla f(x), x - x^*\rangle \geq \gamma(\|x - x^*\|)$$

and bearing in mind that

$$G'(f(x)) \leq M\,,$$

where $M = \max_{x \in X} G'(f(x)) > 0$, we have

$$\langle \nabla f(x), x - x^*\rangle \geq \frac{1}{M}\gamma(\|x - x^*\|)\,.$$

Thus, condition (5.3.21′) holds for $f(x)$. Condition (5.3.22′) holds for $f(x)$ automatically since it is taken as continuously differentiable and it is defined on a compact set X. This proves Proposition 3. Q.E.D.

We present now an example of a function satisfying (5.3.21) and (5.3.22) that is neither convex nor concave in any neighborhood of the minimum point.

Let $X = [-1, 1]$ and let $g(x) = g(-x)$, where

$$g(x) = \begin{cases} 0\,, & x = 0\,, \\ \dfrac{1}{2^n} + 2\left(x - \dfrac{1}{2^n}\right), & \dfrac{5}{3}\dfrac{1}{2^{n+1}} \le x \le \dfrac{1}{2^n}\,, \\ \dfrac{1}{2^{n+1}} + \dfrac{1}{2}\left(x - \dfrac{1}{2^n}\right), & \dfrac{1}{2^{n+1}} \le x \le \dfrac{5}{3}\dfrac{1}{2^{n+1}}\,, \end{cases}$$

and $0 \le n < \infty$. Then

$$\nabla g(x) = \begin{cases} 2 \text{ or } \frac{1}{2}\,, & x > 0\,, \\ -2 \text{ or } -\frac{1}{2}\,, & x < 0\,. \end{cases}$$

Clearly, $g(x)$ is neither convex nor concave in any neighborhood of $x^* = 0$ and conditions (5.3.21) and (5.3.22) are satisfied.

APPENDIX 1: EFFICIENCY OF RANDOM SEARCH ALGORITHMS

In this appendix† we introduce several random search algorithms and discuss their efficiency. The algorithms under consideration are

(a) Algorithm RS-1 (the random search double-trial algorithm),
(b) Algorithm RS-2 (acceptance-rejection random search algorithm),
(c) Algorithm RS-3 (optimum trial random search algorithm), and
(d) Algorithm RS-4 (statistical gradient random search algorithm).

Algorithms RS-1 through RS-4 are presented here for the deterministic problem (P1) and for $D = R^n$. Their presentations for the stochastic problem (P2) are similar and left as an exercise for the reader.

Algorithm RS-1

According to this algorithm, we approximate the point x^* step by step. If on the ith iteration ($i = 1, 2, \ldots$) we have reached the point x_i, then the next point x_{i+1} (see Fig. 6) is chosen as

$$x_{i+1} = x_i - \frac{\alpha_i}{2\beta_i}\,[g(x_i + \beta_i \Xi_i) - g(x_i - \beta_i \Xi_i)]\Xi_i\,, \qquad \text{(A1.1)}$$

where Ξ_i, $i = 1, 2, \ldots,$ is a sequence of random vectors continuously

† The material of this Appendix is taken from Rubinstein and Samorodnitsky (1985).

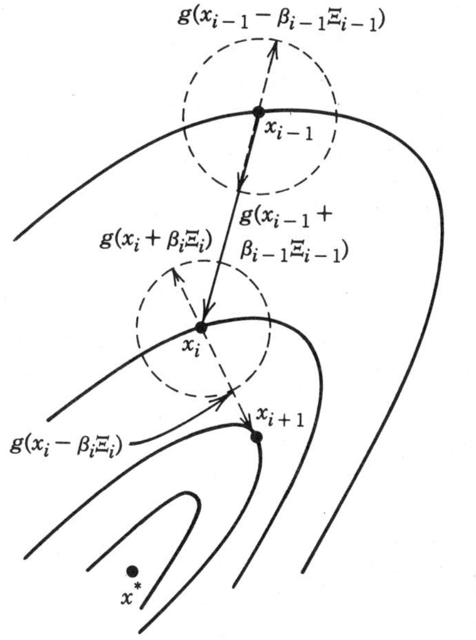

Fig. 6 Graphical representation of the random search double-trial algorithm RS-1.

distributed over some subregion in R^n. (Often Ξ_i, $i = 1, \ldots, n$, are assumed to be independent uniformly distributed on the unit n-dimensional sphere.) Here $\alpha_i > 0$ and $\beta_i > 0$ are step parameters. It follows from (A1.1) that, depending on† $\operatorname{sign}[g(x_i + \beta_i\Xi_i) - g(x_i - \beta_i\Xi_i)]$, the direction from x_i to x_{i+1} will coincide either with Ξ_i or with its opposite, that is, $-\Xi_i$.

We shall consider also the following modification of (A1.1), suggested by Rastrigin (1968):

$$x_{i+1} = x_i - \alpha_i \operatorname{sign}[g(x_i + \beta_i\Xi_i) - g(x_i - \beta_i\Xi_i)]\Xi_i \,. \qquad \text{(A1.2)}$$

Clearly the step size in (A1.2) equals $\Delta x_i = x_{i+1} - x_i = \pm\alpha_i\Xi_i$.

Algorithm RS-2

The algorithm can be written as

$$x_{i+1} = \begin{cases} x_i & \text{if } g(x_i + \beta_i\Xi_i^{(K)}) - g(x_i) \geq 0\,, \qquad K = 1, \ldots, T-1\,, \\ x_i - \dfrac{\alpha_i}{\beta_i}[g(x_i + \beta_i\Xi_i^{(T)}) - g(x_i)]\Xi_i^{(T)} & \text{if } g(x_i + \beta_i\Xi_i^{(T)}) - g(x_i) < 0\,. \end{cases}$$

(A1.3)

† $\operatorname{sign} z = \begin{cases} 1, & z > 0\,, \\ -1, & z \leq 0\,. \end{cases}$

According to this algorithm, we generate for each i a sequence of independent random vectors $\Xi_i^{(K)}$, $K = 1, \ldots, T$ (T is a random variable called the stopping time), and consider the increments $g(x_i + \beta_i \Xi_i^{(K)}) - g(x_i)$. If $g(x_i + \beta_i \Xi_i^{(K)}) - g(x_i) \geq 0$ for some index K, no iteration is made, that is, $x_{i+1} = x_i$. This is the same as returning from point $x_i + \beta_i \Xi_i^{(K)}$ to point x_i (unsuccessful step). In this case we proceed in generating $\Xi_i^{(K+1)}$, $\Xi_i^{(K+2)}, \ldots$ until we have some index T such that $g(x_i + \beta_i \Xi_i^{(T)}) - g(x_i) < 0$ (successful step), and the iteration is made along the vector $\Xi_i^{(T)}$, that is according to

$$x_{i+1} = x_i - \frac{\alpha_i}{\beta_i} [g(x_i + \beta_i \Xi_i^{(T)}) - g(x_i)] \Xi_i^{(T)} .$$

By analogy with (A1.2), we introduce the following modified version of Algorithm RS-2:

$$x_{i+1} = \begin{cases} x_i & \text{if } g(x_i + \beta_i \Xi_i^{(K)}) - g(x_i) \geq 0 , \quad K = 1, \ldots, T-1 , \\ x_i + \alpha_i \Xi_i^{(T)} & \text{if } g(x_i + \beta_i \Xi_i^{(T)}) - g(x_i) < 0 . \end{cases} \tag{A1.4}$$

Algorithm (A.1.4) will be presented for simplicity in the form where both indices K and T are surpressed, that is,

$$x_{i+1} = \begin{cases} x_i & \text{if } g(x_i + \beta_i \Xi_i) \geq g(x_i) , \\ x_i + \alpha_i \Xi_i & \text{otherwise.} \end{cases} \tag{A1.5}$$

The number of trials that have to be performed in (A1.5) up to the first successful step has a geometric distribution with the parameter

$$P_i = \Pr\{(x_i + \beta_i \Xi_i) - g(x_i) < 0\} . \tag{A1.6}$$

The mean number of such trials clearly equals $1/P_i$, and the mean number of measurements of $g(x_i)$ up to the first success is

$$E(N_i) = 1 + \frac{1}{P_i} . \tag{A1.7}$$

Algorithm RS-3

Each iteration of this algorithm comprises the following steps:

1. Generate a sequence $\Xi_{i1}, \ldots, \Xi_{iN}$ of independent random vectors (say, uniformly distributed on the unit hypersphere).

2. Find $Y_{ik} = g(x_i + \beta_{ik}) - g(x_i)$, $k = 1, \ldots, N$.
3. Set $Y^0_{iN} = \min(Y_{i1}, \ldots, Y_{iN})$.
4. The next point x_{i+1} is chosen as

$$x_{i+1} = x_i - \alpha_i \beta_i Y^0_{iN} \Xi^0_{iN}, \qquad \alpha_i > 0,\ \beta_i > 0 .$$

According to this algorithm, at each iteration we generate a sequence of independent random vectors $\Xi_{i1}, \ldots, \Xi_{iN}$. The next point x_{i+1} is chosen in the direction Ξ^0_{iN} if the greatest decrease of $g(x)$ that is the vector Ξ^0_{iN} corresponds to the random trial optimal among those available.

Remark It is proved in Rubinstein and Karnovsky (1978) that under some regularity condition the vector Ξ^0_{iN} converges almost surely to the direction of the greatest decrease as $N \to \infty$.

Algorithm RS-4

Each iteration of the algorithm comprises the following steps:

1. Generate a sequence of independent random vectors (say, uniformly distributed on the unit hypersphere).
2. Find $Y_{ik} = g(x_i + \beta_{ik}) - g(x_i)$, $k = 1, \ldots, N$.
3. Set

$$V_{i\Xi} = \frac{1}{N} \sum_{k=1}^{N} Y_{ik} \Xi_{ik} .$$

4. The next point x_{i+1} is chosen as

$$x_{i+1} = x_i - \alpha_i \beta_i^{-1} V_{i\Xi} , \qquad \alpha_1 > 0,\ \beta_i > 0 .$$

Thus, x_{i+1} is chosen in the direction $V_{i\Xi}$, which is obtained by averaging the vectors Ξ_{ik}, $k = 1, \ldots, N$, weighted with their corresponding increments Y_{ik}. In the particular case where

$$N = n \quad \text{and} \quad \Xi_{ik} = e_k = \underbrace{0, \ldots, 0, 1}_{k}, 0, \ldots, 0, \qquad k = 1, \ldots, n ,$$

we obtain the finite difference gradient algorithm

$$x_{i+1} = x_i + \alpha_i \tilde{\nabla} g(x_i) ,$$

where

$$\tilde{\nabla} g(x) = \left(\frac{\tilde{\partial} g(x)}{\partial x_i}, \ldots, \frac{\tilde{\partial} g(x)}{\partial x_n} \right)$$

$$= \left(\frac{g(x_1 + \beta_1, x_2, \ldots, x_n) - g(x)}{\beta_1}, \ldots, \frac{g(x_1, \ldots, x_n + \beta_n) - g(x)}{\beta_n} \right).$$

We leave for the reader to prove that if Ξ_{ik} are independent and uniformly distributed on the surface of the unit hypersphere and $g(x)$ is a linear function, then for any i the direction of $V_{i\Xi}$, on the average, coincides with that of the gradient of $g(x)$. This is the reason why the algorithm is called *statistical gradient algorithm* (see Rastrigin (1968)).

We shall now turn to the efficiency of the random search algorithms. There are different criteria for comparing algorithms. Usually, algorithms are compared according to their local and global properties. Local properties are associated with a single iteration, global properties with multiple iterations. Before algorithms can be compared, they must be set under identical conditions by defining some test functions (linear, quadratic, multiextremal, etc.) for which the extremum is sought and the initial point x_1 at which the search originates.

When comparing algorithms according to their global properties, the following criteria may be used: Find an index k that corresponds to the best algorithm among S algorithms available, such that

(i) $$\min_{s=1,\ldots,S} E(\|x_i^{(s)} - x^*\|) = E(\|x_i^{(k)} - x^*\|),$$

when the number of iteration i is given;

(ii) $$\max_{s=1,\ldots,S} \Pr(\|x_i^{(s)} - x^*\| \le \varepsilon_x) = \Pr(\|x_i^{(k)} - x^*\| \le \varepsilon_x),$$

where i and ε_x are given;

(iii) $$\max_{s=1,\ldots,S} \Pr(|g(x_i^{(s)}) - g(x^*)| \le \varepsilon_g) = \Pr(|g(x_i^{(k)}) - g(x^*)| \le \varepsilon_g),$$

where i and ε_g are given;

(iv) $$\min_{s=1,\ldots,S} i^{(s)} = i^{(k)},$$

subject to

$$E(x_i^{(s)}) \in [R_x = \{x: \|x - x^*\| \le \varepsilon_x\}]$$

or

$$E(x_i^{(s)}) \in [R_g = \{x: \|g(x) - g(x^*)\| \le \varepsilon_g\}];$$ and

(v) $\min_{s=1,\dots,S} i^{(s)} = i^{(k)}$,

subject to

$$E(x_i^{(s)}) \in [R_x = \{x: \|x - x^*\| < \varepsilon_x\}]$$

and

$$\mathrm{Var}\, x_i^{(s)} \le c,\ c > 0\,.$$

It is readily seen that the first three criteria are associated with finding the best algorithm when the number of iteration i is given; the last two are associated with finding the best algorithm that hits, at the minimum number of iterations, a given region R_x or R_g containing the extremum point x^*.

When comparing algorithms according to their local properties, we assume that some point, say, x_i is given and we are allowed a single step (iteration). Let x_{i+1} be the point reached after this single iteration and $\Delta g_i = g(x_{i+1}) - g(x_i)$ the increment in the value of g. We now define the following two local characteristics of the random search algorithms:

$$C_n = \frac{E(-\Delta g_i)}{E(N_i)}\,, \tag{A1.8}$$

$$D_n = \frac{C_n}{\sqrt{\mathrm{Var}(\Delta g_i)}}\,, \tag{A1.9}$$

where N_i is the number of observations (measurements) of $g(x)$ at the ith step.

Generally, comparison of algorithms according to their integral properties entails difficulties. Attempts to overcome these difficulties have been made by Rastrigin (1968) and Lawrence and Steiglitz (1972). Rastrigin showed that with linear g, $D = R^n$, $n > 2$, Algorithm RS-2 is faster than a sequential coordinate-wise minimization algorithm. On the other hand, Lawrence and Steiglitz (1972) showed experimentally that near a local minimum a coordinate-wise algorithm is faster than RS-2 but if $D \subset R^n$, then it is often trapped on the boundary of D.

Assume that we have two random search algorithm. Let the efficiency associated with first one be $C_n^{(1)}$ and that associated with the second be $C_n^{(2)}$. We say that the first algorithm is more efficient than the second according to criterion C_n if $C_n^{(1)}/C_n^{(2)} > 1$, and similarly according to criterion D_n.

In Rubinstein (1981) it is shown that for linear and quadratic functions Algorithm RS-1 is more efficient than RS-3 according to the criterion c_n for $n > 1$ and that they have approximately the same efficiency according to the criterion D_n.

In the rest of this Appendix we compare the efficiency of the random search algorithms RS-1 and RS-2 [versions (A1.2) and (A1.5)] on linear and quadratic models for both deterministic and stochastic optimization problems, assuming that $\alpha_i = 1$ and Ξ_i, $i = 1, 2, \ldots$, are independent vectors uniformly distributed on the surface of the n-dimensional unit hypersphere.

We show that for the linear model Algorithm RS-1 is more efficient that Algorithm RS-2 for both deterministic and stochastic cases. For the quadratic model there exists a distance r^* from the extremum x^* such that, for $r > r^*$, RS-1 is more efficient than RS-2 and vice versa for $r \leq r^*$.

As a further research it would be interesting to find the efficiencies of Algorithms RS-3 and RS-4 and then to compare all four of them, that is, RS-1 through RS-4.

A. Linear Model, Deterministic Optimization Problem

For the linear model we assume that the function $g(x)$ is approximately linear, which is equivalent to stating that in Taylor's expansion

$$\begin{aligned} g(x_{i+1}) &= g(x_i + \Delta x_i) = g(x_i) + \langle \Delta x_i, \nabla g(x_i) \rangle + o(\Delta x_i) \\ &= g(x_i) + \|\Delta x_i\| \cdot \|g(x_i)\| \cos \varphi_i + o(\Delta x_i) \end{aligned} \tag{A1.10}$$

Therefore, at each iteration made by the random search algorithms, we approximate $g(x)$ linearly on the interval Δx_i. We assume further, without loss of generality, that $\|\nabla g(x_i)\| = 1$, $i = 1, 2, \ldots$.

(a) Algorithm RS-1 Substituting (A1.2) in (A1.10), we obtain

$$g(x_{i+1}) = g(x_i + \Delta x_i) = g(x_i) - \alpha_i \|\nabla g(x_i)\| \cdot \|\Xi_i\| \cdot \cos \varphi_i + o(x_i) \tag{A1.11}$$

where

$$\cos \varphi_i = \frac{\langle \nabla g(x_i), (\text{sign } \Delta g_i) \Xi_i \rangle}{\|\nabla g(x_i)\| \cdot \|\Xi_i\|} .$$

Taking into account that $\|\Xi_i\| = 1$, $\|\nabla g(x_i)\| = 1$, and $\alpha_i = 1$, we can rewrite (A1.11) as

$$g(x_{i+1}) = g(x_i) - \cos \varphi_i + o(\Delta x_i) .$$

It is shown in Appendix 2 that under the assumption that Ξ_i is uniformly distributed on the surface of the unit hypersphere and the direction of the gradient corresponds to $\varphi_i = 0$, the p.d.f. of the random angle is

$$h_n(\varphi_i) = B_n \sin^{n-2} \varphi_i, \qquad -\tfrac{1}{2}\pi \leq \varphi_i \leq \tfrac{1}{2}\pi , \tag{A1.12}$$

where

$$B_n = \frac{\Gamma(n/2)}{\sqrt{\pi}\Gamma(n-1/2)} \tag{A1.13}$$

and Γ denotes the gamma function. Further, we shall omit the subscript i in φ_i, since the distribution of the latter does not depend on it.

Taking into account that for Algorithm RS-1 we need two observations of $g(x)$ at points $g(x + \beta\Xi)$ and $g(x - \beta\Xi)$, respectively, the efficiency C_n [see (A1.8)] is

$$C_n = \frac{E[g(x_i) - g(x_{i+1})]}{2} = \frac{E(\cos \varphi)}{2} . \tag{A1.14}$$

The expected value and the variance of $\cos \varphi$ are, respectively,

$$\begin{aligned} E(\cos \varphi) &= \int_{-\pi/2}^{\pi/2} \cos \varphi h_n(\varphi)\, d\varphi = 2B_n \int_0^{\pi/2} \cos \varphi \sin^{n-2} \varphi\, d\varphi \\ &= \frac{2B_n}{n-1} \end{aligned} \tag{A1.15}$$

$$\begin{aligned} \operatorname{Var}(\cos \varphi) &= E(\cos^2 \varphi) - [E(\cos \varphi)]^2 \\ &= 2B_n \int_0^{\pi/2} \cos^2 \varphi \sin^{n-2} \varphi\, d\varphi - \left(\frac{2B_n}{n-1}\right)^2 \\ &= \frac{1}{n} - \left(\frac{2B_n}{n-1}\right)^2 . \end{aligned} \tag{A1.16}$$

Substituting (A1.15) into (A1.14), we obtain

$$C_n = \frac{B_n}{n-1} \tag{A1.17}$$

and the relationships

$$C_{n+1} = \frac{1}{2\pi n C_n} \qquad C_{n+2} = \frac{n}{n+1} C_n$$

are also readily verified.

For the criterion D_n [see (A1.9)] we have

$$D_n = \frac{C_n}{\sigma} = \frac{B_n}{(n-1)\left[\frac{1}{n} - \left(\frac{2B_n}{n-1}\right)^2\right]^{1/2}} = \frac{1}{\left[\frac{(n-1)^2}{nB_n^2} - 4\right]^{1/2}} \tag{A1.18}$$

Table 3 and Fig. 7 present the efficiencies C_n and D_n as a function of n, from which it follows that both C_n and D_n decrease with n. Taking into account that for large n (see (A-3)) $B_n \approx [(n-1)/2\pi]^{1/2}$ we have, respectively,

Table 3 Efficiency and σ^2 as Functions of n for Algorithm RS-1

n	C_n	σ^2	$D_n = C_n/\sigma$
2	0.3183	0.095	1.0327
3	0.25	0.083	0.8677
4	0.2122	0.069	0.8027
5	0.1875	0.059	0.7694
6	0.1697	0.051	0.7482
7	0.1562	0.045	0.7344
8	0.1454	0.040	0.7233
9	0.1366	0.036	0.7156
10	0.1292	0.033	0.7090
11	0.1229	0.030	0.7040
12	0.1174	0.028	0.6990

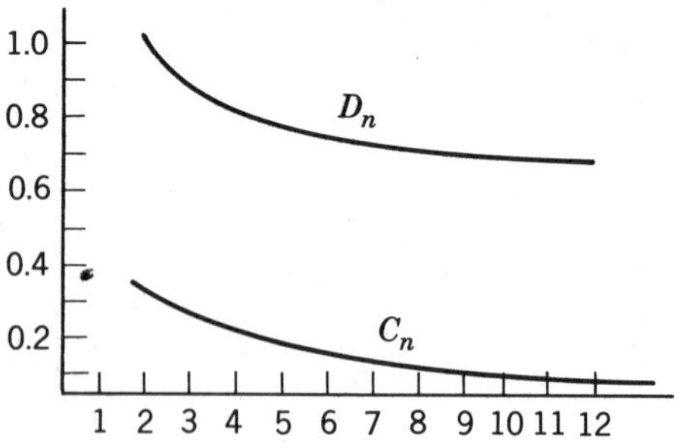

Fig. 7 Efficiencies C_n and D_n as functions of n for Algorithm RS-1.

$$C_n = \frac{1}{\sqrt{2\pi(n-1)}}, \tag{A1.19}$$

$$D_n = \left[\left(\frac{2\pi(n-1)}{n} - 4\right)^{1/2}\right]^{-1} \approx \frac{1}{\sqrt{2\pi - 4}} \approx 0.6618. \tag{A1.20}$$

(b) Algorithm RS-2 From (A1.6) and (A1.7) it follows that for a linear function and for Ξ uniformly distributed on the surface of the n-dimensional hypersphere, $P_i = P = \frac{1}{2}$ and $E(N_i) = E(N) = 3$.

It is readily seen that formulas (A1.11)–(A1.13) are also valid for Algorithm RS-2. This means that the expected value $E(\cos\varphi)$ and variance $\mathrm{var}(\cos\varphi) = \sigma^2$ [see (A1.15) and (A1.16)] are the same for both RS-1 and RS-2. However, the former is 1.5 times more efficient than the latter according to both C_n and D_n, since it requires exactly two measurements of $g(x)$ for each iteration, as against an average of three for RS-2.

B. Linear Model, Stochastic Optimization Problem

We assume that the noise is additive, that is,

$$\psi(x, W) = g(x) + W, \quad \text{where } W \sim N(0, \sigma^2).$$

(a) Algorithm RS-1 For stochastic optimization problem (P2) Algorithm RS-1 can be written by analogy with (A1.2) as

$$x_{i+1} = x_i - \alpha_i \,\mathrm{sign}[\psi(x_i + \beta_i \Xi_i, W_i^{(1)}) - \psi(x_i - \beta_i \Xi_i, W_i^{(2)})]\Xi_i. \tag{A1.21}$$

It follows from (A.21) that the step is made in the direction of decreasing $g(x)$, subject to occurrence of the event

$$\{\mathrm{sign}[g(x + \beta\Xi) + W_1 - g(x - \beta\Xi) - W_2]$$
$$= \mathrm{sign}[g(x + \beta\Xi) - g(x - \beta\Xi)]\}, \tag{A1.22}$$

where W_1 and W_2 are the realizations of the noise observed as points $x + \beta\Xi$ and $x - \beta\Xi$, respectively. The probability of this event is

$$P_1 = \frac{1}{2}\left[1 + \Phi\left(\frac{|\Delta g_1|}{2\sigma}\right)\right], \tag{A1.23}$$

where

$$\Phi(x) = \frac{1}{\sqrt{\pi}} \int_{-\infty}^{x} e^{-t^2}\, dt \tag{A1.24}$$

and

$$\Delta g_1 = g(x + \beta\Xi) - g(x - \beta\Xi) .$$

Taking into account that $\alpha = 1$ and $\|\nabla g(x)\| = 1$, we have $\Delta g_1 = 2\cos\varphi$ and therefore

$$P_1 = \frac{1}{2}\left[1 + \Phi\left(\left|\frac{\cos\varphi}{\sigma}\right|\right)\right] . \tag{A1.25}$$

Introducing

$$Y_1 = \begin{cases} \cos\varphi \text{ with probability } P_1 , & -\tfrac{1}{2}\pi \le \varphi \le \tfrac{1}{2}\pi , \\ -\cos\varphi \text{ with probability } 1 - P_1 , & -\tfrac{1}{2}\pi \le \varphi \le \tfrac{1}{2}\pi , \end{cases} \tag{A1.26}$$

we obtain for the expected value of Y_1

$$\begin{aligned} E(Y_1) &= 2B_n \int_0^{\pi/2} P_1 \cos\varphi \sin^{n-2}\varphi \, d\varphi \\ &\quad - 2B_n \int_0^{-\pi/2} (1 - P_1) \cos\varphi \sin^{n-2}\varphi \, d\varphi \\ &= 2B_n \int_0^{\pi/2} \Phi \frac{|\cos\varphi|}{\sigma} \cos\varphi \sin^{n-2}\varphi \, d\varphi . \end{aligned} \tag{A1.27}$$

The efficiency C_n is

$$C_n = \frac{E(Y_1)}{2} = B_n \int_0^{\pi/2} \Phi \frac{|\cos\varphi|}{\sigma} \cos\varphi \sin^{n-2}\varphi \, d\varphi . \tag{A1.28}$$

The variance of Y_1 is

$$\operatorname{Var}(Y_1) = E(Y_1^2) - E^2(Y_1) ,$$

where

$$\begin{aligned} E(Y_1^2) &= 2B_n \int_0^{\pi/2} P_1 \cos^2\varphi \sin^{n-2}\varphi \, d\varphi \\ &\quad + 2B_n \int_0^{\pi/2} (1 - P_1) \cos^2\varphi \sin^{n-2}\varphi \, d\varphi \\ &= 2B_n\left(\frac{1}{2B_n} - \frac{1}{2B_{n+2}}\right) = \frac{1}{n} . \end{aligned} \tag{A1.29}$$

Finally

$$\mathrm{Var}(Y_1) = \frac{1}{n} - \left(2B_n \int_0^{\pi/2} \Phi \, \frac{|\cos \varphi|}{\sigma} \cos \varphi \sin^{n-2} \varphi \, d\varphi\right)^2 , \quad \text{(A1.30)}$$

and the efficiency D_n is

$$D_n = \frac{C_n}{\sqrt{\mathrm{Var}(Y_1)}} = \frac{B_n \int_0^{\pi/2} \Phi(|\cos \varphi|/\sigma) \cos \varphi \sin^{n-2} \varphi \, d\varphi}{\left(1/n - \left(2B_n \int_0^{\pi/2} \Phi(|\cos \varphi|/\sigma) \cos \varphi \sin^{n-2} \varphi \, d\varphi\right)^2\right)^{1/2}} .$$

(A1.31)

For small x the integral $\Phi(x)$ can be approximated as $\Phi(x) \approx 2/\sqrt{\pi}x$, or, equivalently, as

$$\Phi\left(\frac{|\cos \varphi|}{\sigma}\right) \approx \frac{2}{\sqrt{\pi}} |\cos \varphi| . \quad \text{(A1.32)}$$

Substituting (A1.32) in (A1.28) and (A1.31), we obtain for large σ respectively,

$$C_n = \frac{1}{n\sigma\sqrt{\pi}} , \quad \text{(A1.33)}$$

$$D_n = \frac{1}{(n\sigma^2\pi - 4)^{1/2}} . \quad \text{(A1.34)}$$

Comparing the efficiencies C_n and D_n of RS-1 for the stochastic and deterministic optimization problems, we may readily conclude that for the stochastic model both C_n and D_n have been significantly reduced.

(b) Algorithm RS-2 For the stochastic optimization problem Algorithm RS-2 can be written, in analogy with (A1.4), as

$$x_i + 1 = \begin{cases} x_i , & \text{if } \psi(x_i + \beta_i \Xi_i^{(K)}, W_i^{(K)}) - \psi(x_i, W_i^{(0)}) \geq 0 , \\ & K = 1, \ldots , T-1 , \\ x_i + \alpha_i \Xi_i^{(T)} , & \text{if } \psi(x_i + \beta_i \Xi_i^{(T)}, W_i^{(T)} - \psi(x_i, W_i^{(0)}) < 0 . \end{cases}$$

(A1.35)

In this case we again generate for each i a sequence of independent random vectors $\Xi_i^{(K)}$, $K = 1, 2, \ldots$, uniformly distributed on the surface of the unit hypersphere until we find the first random index T (stopping time) for which the event

$$A_1 = \{g(x_i + \beta_i \Xi_i^{(T)}, W_i^{(T)}) - \psi(x_i, W_i^{(0)} < 0\} \tag{A1.36}$$

occurs. Assuming again $\psi(x, W) = g(x) + W$ where $W \sim N(0, \sigma^2)$, we can rewrite (A1.36) as

$$A_1 = \{g(x_i + \beta_i \Xi_i^{(T)}) + W_i^{(T)} - g(x_i) - W_i^{(0)} < 0\} . \tag{A1.37}$$

Therefore a step is performed by (A1.35) only if the event A_1 occurs. However, this step is successful, that is, is made in the direction of decreasing $g(x)$, only if the chosen $\Xi_i^{(T)}$ lies in the direction for which the event

$$A_2 = \{g(x_i + \beta_i \Xi_i^{(T)}) - g(x_i) < 0\} \tag{A1.38}$$

occurs.

It is readily seen that the conditional probability $P(A_2 \mid A_1)$ is equal to

$$P(A_2 \mid A_1) = 2 \int_{\Delta g_2 < 0} P_2(\Delta g_2) f(\Delta g_2)\, d(\Delta g_2) , \tag{A1.39}$$

where $\Delta g_2 = g(x_i + \beta_i \Xi_i^{(T)}) - g(x_i)$ and $f(\Delta g_2)$ is a p.d.f. of the random variable Δg_2. Here

$$P_2(\Delta g_2) = \frac{1}{2}\left[1 + \Phi\left(\frac{|\Delta g_2|}{2\sigma}\right)\right] . \tag{A1.40}$$

Assuming again $\alpha = 1$, $\|\nabla g(x_i)\| = 1$, omitting (without loss of generality the subscript i, and denoting $P_2 = P_2(\Delta g_2)$, we obtain $\Delta g_2 = \cos \varphi$, and

$$P_2 = \frac{1}{2}\left[1 + \Phi\left(\frac{|\cos \varphi|}{2\sigma}\right)\right] .$$

Defining, in analogy with (A1.26),

$$Y_2 = \begin{cases} \cos \varphi \text{ with probability } P_2 , & -\frac{1}{2}\pi \le \varphi \le \frac{1}{2}\pi , \\ -\cos \varphi \text{ with probability } 1 - P_2 , & -\frac{1}{2}\pi \le \varphi \le \frac{1}{2}\pi , \end{cases} \tag{A1.41}$$

we readily find that $E(Y_2)$ and $\mathrm{Var}(Y_2)$ remain almost the same as in

(A1.27) and (A1.30); the only difference being in the argument of $\Phi(\cdot)$; namely, in order to obtain $E(Y_2)$ and $\mathrm{Var}(Y_2)$, we have to replace $\Phi(|\cos\varphi|/\sigma)$ both in $E(Y_1)$ and in $\mathrm{Var}(Y_1)$ [see (A1.27)–(A1.30)] by $\Phi(|\cos\varphi|/2\sigma)$. Since $\Phi(|\cos\varphi|/\sigma) > \Phi(|\cos\varphi|/2\sigma)$, we have, directly from (A1.27) and (A1.30),

$$E(Y_2) < E(Y_1) \tag{A1.42}$$

and

$$\mathrm{Var}(Y_2) > \mathrm{Var}(Y_1)\,. \tag{A1.43}$$

Taking again into account that we need exactly two measurements of $g(x)$ for RS-1 and on the average three measurements for RS-2 and the results (A1.42) and (A1.43), we readily conclude [see also (A1.28) and (A1.31)] that for the stochastic linear model according to both C_n and D_n, RS-1 is at least 1.5 times more efficient than RS-2. In the absence of noise (deterministic linear model) we have the exact lower bound, that is,

$$C_n^{(1)} = 1.5C_n^{(2)} \quad \text{and} \quad D_n^{(1)} = 1.5D_n^{(2)}\,.$$

C. Quadratic Model, Deterministic Optimization Problem

We consider the quadratic model

$$g(x) = x^t x\,, \qquad x \in R^n\,,$$

where t is the transpose operator.

(a) Algorithm RS-1 The value of the function g at the point x_{i+1} according to this algorithm is

$$g(x_{i+1}) = g(x_i - \alpha_i\,\mathrm{sign}[g(x_i + \beta_i \Xi_i) - g(x_i - \beta_i \Xi_i)]\Xi_i)\,, \tag{A1.44}$$

which is the same as

$$x_{i+1}^t x_{i+1} = x_i^t x_i - 2\alpha|x_i|\cos\varphi_i + \alpha^2\,, \quad \text{where } -\tfrac{1}{2}\pi \le \varphi < \tfrac{1}{2}\pi\,. \tag{A1.45}$$

Assuming again, without loss of generality, that $\alpha = 1$ and denoting $x^t x = r^2$, we have (see Fig. 8)

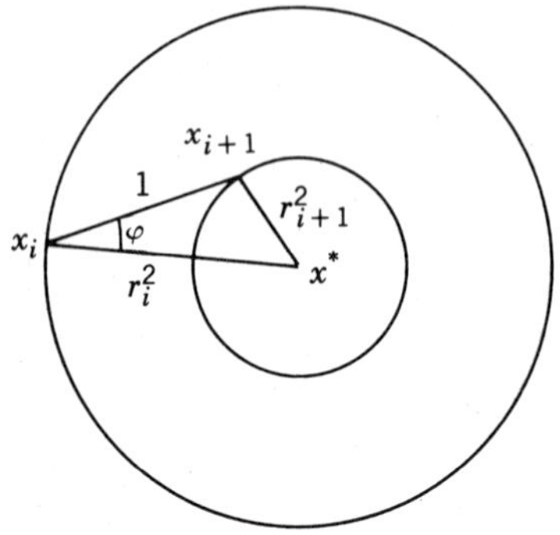

Fig. 8 Representation of an iteration for algorithm RS-1 (quadratic model).

$$r_{i+1}^2 = r_i^2 - 2r_i \cos \varphi_i + 1 , \qquad -\tfrac{1}{2}\pi \leq \varphi_0 < \tfrac{1}{2}\pi . \qquad \text{(A1.46)}$$

The average increment in the value of g after one step equals

$$\begin{aligned} E(\Delta g_1) &= E(r_i^2 - r_{i+1}^2) = E(2r \cos \varphi - 1) \\ &= \int_{-\pi/2}^{\pi/2} (2r \cos \varphi - 1) h_n(\varphi)\, d\varphi \\ &= 2B_n \int_0^{\pi/2} (2r \cos \varphi - 1) \sin^{n-2}\varphi \, d\varphi \\ &= \frac{4rB_n}{n-1} - 1 . \end{aligned} \qquad \text{(A1.47)}$$

Equating (A1.47) to zero and solving it with respect to r, we find

$$\tilde{r}_1 = \frac{n-1}{4B_n} . \qquad \text{(A1.48)}$$

The surface of the hypersphere with radius $\tilde{r}$ represents the stable zone, which means that on its way to this zone the system moves at random around x^* at the average distance of $\tilde{r}_1$.

It follows from (A1.48) that $\tilde{r}_1$ increases with n. For large n $B_n \approx [(n - 1/2\pi)]^{1/2}$ and

$$\tilde{r}_1 = [\tfrac{1}{2}\pi(n-1)]^{1/2} \approx 1.25\sqrt{n-1} . \qquad \text{(A1.49)}$$

The variance of Δg_1 is

$$\mathrm{Var}(\Delta g_1) = E(\Delta g_1^2) - [E(\Delta g_1)]^2$$

$$= 2B_n \int_0^{\pi/2} (2r\cos\varphi - 1)^2 \sin^{n-2}\varphi \, d\varphi - \left(\frac{4rB_n}{n-1} - 1\right)^2$$

$$= 4r^2\left[\frac{1}{n} - \frac{4B_n^2}{(n-1)^2}\right]. \tag{A1.50}$$

Taking again into account that the algorithm RS-1 we need exactly two measurements of $g(x)$, we have

$$C_n = \frac{2rB_n}{n-1} - \tfrac{1}{2}, \tag{A1.51}$$

and

$$D_n = \frac{2rB_n/(n-1) - 1/2}{2r[1/n - 4B_n^2/(n-1)^2]^{1/2}}. \tag{A1.52}$$

(b) Algorithm RS-2 According to Algorithm RS-2, a successful step is made if $g(x + \beta\Xi) - g(x) < 0$, which is the case only if, for a given point x, the point $x + \beta\Xi$ lies within a feasible cone (Fig. 9) with $-\varphi_0 \leq \varphi \leq \varphi_0$, $(\varphi \leq \frac{1}{2}\pi)$.

It is readily seen that, for a given step size, φ_0 increases with r $[\varphi_0 = \arc\cos(1/2r)]$. As $r \to \infty$, $\varphi_0 \to \frac{1}{2}\pi$ and we have again the linear model.

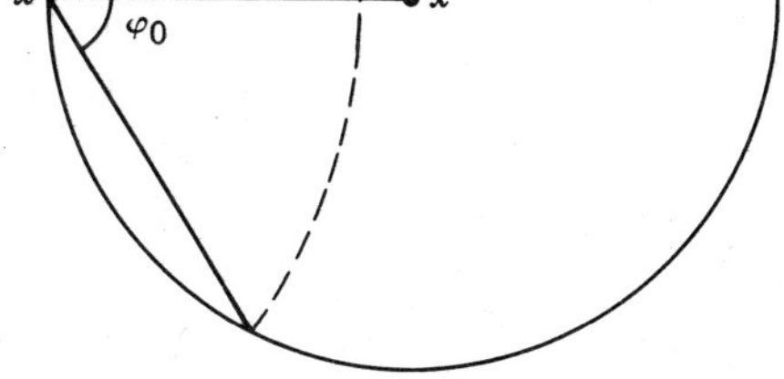

Fig. 9 Representation of feasible cone for algorithm RS-2 (quadratic model).

Since Ξ is uniformly distributed on the surface of the unit hypersphere, we have, in analogy with (A1.12),

$$h_n(\varphi) = \frac{|\sin \varphi|^{n-2}}{\int_{-\varphi_0}^{\varphi_0} |\sin \varphi|^{n-2}\, d\varphi}\,, \qquad -\varphi_0 \le \varphi \le \varphi_0\,, \tag{A1.53}$$

and, in analogy with (A1.46),

$$r_{i+1}^2 = r_i^2 - 2r_i \cos \varphi_i + 1\,, \qquad -\varphi_0 \le \varphi_i \le \varphi_0\,. \tag{A1.54}$$

The expected value of Δg_2 is

$$E(\Delta g_2) = E(r_i^2 - r_{i+1}^2) = \frac{\int_{-\varphi_0}^{\varphi_0} (2r \cos \varphi - 1)|\sin \varphi|^{n-2}\, d\varphi}{\int_{-\varphi_0}^{\varphi_0} |\sin \varphi|^{n-2}\, d\varphi}\,. \tag{A1.55}$$

The probability of a successful step is

$$P = \Pr\{g(x + \beta\Xi) - g(x) < 0\} = B_n \int_0^{\varphi_0} \sin^{n-2} \varphi\, d\varphi\,. \tag{A1.56}$$

Substituting (A1.56) into (A1.55), we have

$$\begin{aligned} E(\Delta g_2) &= \frac{B_n}{P} \int_0^{\varphi_0} (2r \cos \varphi - 1) \sin^{n-2} \varphi\, d\varphi \\ &= \frac{B_n}{P}\left(\frac{2r \sin^{n-1} \varphi_0}{n-1} - \frac{P}{B_n}\right). \end{aligned} \tag{A1.57}$$

Equating (A1.57) to zero and solving it with respect to r, we find

$$\tilde{r}_2 = \tfrac{1}{2}\,, \tag{A1.58}$$

which represents the stable zone for RS-2.

Comparing $\tilde{r}_1$ with $\tilde{r}_2$ [equations (A1.48) and (A1.58)] we see that RS-2 achieves closer proximity to x^* than RS-1.

Another interesting feature of RS-2 is that it stops exactly at the distance $\tilde{r}_2 = \frac{1}{2}$ from the extremum x^*. By contrast, RS-1 never stops and continues to move at random around the surface of the stable hypersphere with radius $\tilde{r}_1 = (n-1)/4B_n$.

Recalling that the average number of measurements of $g(x)$ for RS-1 is $E(N) = (P+1)/P$ [see (A1.7)], we obtain

$$C_n = \frac{E(\Delta g_2)}{E(N)} = \frac{2rB_n \sin^{n-1} \varphi_0}{(1+P)(n-1)} - \frac{P}{1+P}$$

$$= \left[\frac{2rB_n \sin^{n-1} \varphi_0}{P(n-1)} - 1\right] \frac{P}{1+P} . \tag{A1.59}$$

Further

$$\mathrm{Var}(\Delta g_2) = E(\Delta g_2^2) - [E(\Delta g_2)]^2$$

$$= \frac{\int_0^{\varphi_0} (2r \cos \varphi - 1)^2 \sin^{n-2} \varphi \, d\varphi}{\int_0^{\varphi_0} \sin^{n-2} \varphi \, d\varphi} - \left(\frac{2rB_n \sin^{n-1} \varphi_0}{P(n-1)} - 1\right)^2 . \tag{A1.60}$$

After some algebraic manipulations we have

$$\mathrm{Var}(\Delta g_2) = 4r^2\left(1 - \frac{n-1}{n}\frac{P_{n+2}}{P} - \frac{B_n^2 \sin^{2(n-1)} \varphi_0}{P^2(n-1)^2}\right), \tag{A1.61}$$

where $P_{n+2} = \int_0^{\varphi_0} B_{n+2} \sin^{n-2} \varphi \, d\varphi$. The efficiency D_n is

$$D_n = \frac{C_n}{\sqrt{\mathrm{Var}(g_2)}} = \frac{\dfrac{2rB_n \sin^{n-1} \varphi_0}{(1+P)(n-1)} - \dfrac{P}{1+P}}{2r\left(1 - \dfrac{n-1}{n}\dfrac{P_{n+2}}{P} - \dfrac{B_n^2 \sin^{2n-2} \varphi_0}{P^2(n-1)^2}\right)^{1/2}} . \tag{A1.62}$$

To compare the efficiencies of RS-1 and RS-2, we shall make use of the following lemma.

Lemma 2 There exists r^* that depends on n such that, for $r > r^*$, RS-1 is more efficient than RS-2 according to criterion C_n and vice versa for $r \geq r^*$.

Proof We know that, for $\tilde{r}_1 = (n-1)/4B_n$, C_n (RS-1) $= 0$ and C_n (RS-2) > 0. We also know that, for sufficiently large r, C_n (RS-1) $>$ C_n (RS-2) as the quadratic model reduces to a linear one. Thus, if r^* exists it must be greater than $\tilde{r}_1 = (n-1)/4B_n$. To prove the lemma, it suffices to show that $\Delta C = C_n$ (RS-1) $- C_n$ (RS-2) is an increasing function of r.

Indeed,

$$\Delta C = C_n\ (\text{RS-1}) - C_n\ (\text{RS-2}) = \frac{2rB_n}{n-1}\left(1 - \frac{\sin^{n-1}\varphi_0}{1+P}\right) - \frac{1}{2} + \frac{P}{1+P},$$

and after more manipulations we have

$$\frac{d(\Delta C)}{dr} > \frac{2B_n}{n-1}\frac{P}{1+P} - \frac{P}{(1+P)^2}\frac{B_n}{2r^2}.$$

Taking into account that $r > (n-1)/4B_n$, we have

$$\frac{d(\Delta C)}{dr} > \frac{2B_n}{n-1}\frac{P}{1+P} - \frac{P}{(1+P)^2} 8B_n \frac{B_n^2}{(n-1)^2}.$$

Using the inequality $B_n < (n-1)/2\pi$ we have, after some algebraic manipulations,

$$\frac{d(\Delta C)}{dr} > \frac{2B_n}{n-1}\frac{P}{1+p}\left(1 - \frac{2}{\pi(1+P)}\right) > 0. \qquad \text{Q.E.D.}$$

While in real-world optimization problems r^* is almost unobtainable, an important suggestion follows from Lemma 2, namely, to start the optimization process with RS-1 until the stable zone $\tilde{r}_1 = (n-1)/4B_n$ has been reached, and then switch to RS-2.

D. Quadratic Model, Stochastic Optimization Problems

In this section we shall compare the two algorithms according to C_n only, as the expressions for D_n are extremely complicated.

(a) Algorithm RS-1 It is readily seen that formulas (A1.21)–(A1.24), derived for the linear stochastic model, remain valid in this case.

Defining

$$V_1 = \begin{cases} 2r\cos\varphi - 1 \text{ with probability } P_1, & -\frac{1}{2}\pi \le \varphi \le \frac{1}{2}\pi, \\ -2r\cos\varphi - 1 \text{ with probability } 1 - P_1, & -\frac{1}{2}\pi \le \varphi \le \frac{1}{2}\pi. \end{cases} \tag{A1.63}$$

We readily conclude that

$$C_n = \frac{E(\Delta g)}{N} = \frac{E(V_1)}{2}, \tag{A1.64}$$

where

$$E(V_1) = 2B_n \int_0^{\pi/2} (-2r\cos\varphi - 1)[1 - P_1(\varphi)] \sin^{n-2}\varphi \, d\varphi$$

$$+ 2B_n \int_0^{\pi/2} (2r\cos\varphi - 1)P_1(\varphi) \sin^{n-2}\varphi \, d\varphi$$

$$= \frac{4rB_n}{n-1} - 1 - 8rB_n \int_0^{\pi/2} [1 - P_1(\varphi)] \sin^{n-2}\varphi \cos\varphi \, d\varphi$$

$$= 4rB_n I_n^{-1} , \tag{A1.65}$$

and

$$I_n = \int_0^{\pi/2} \Phi\left(\frac{\cos\varphi}{\sigma}\right) \cos\varphi \sin^{n-2}\varphi \, d\varphi . \tag{A1.66}$$

Comparing (A1.65) with (A1.47), we find that the efficiency of RS-1 in the situation of noise has been reduced by a factor of

$$4rB_n \int_0^{\pi/2} [1 - P_1(\varphi)] \cos\varphi \sin^{n-2}\varphi \, d\varphi .$$

Lemma 3 The stable zone of RS-1 in the situation of noise is located at least at the distance $\tilde{r}_1 = \frac{1}{4}(n\sigma\sqrt{\pi})$ from the extremum point x^*.

Proof Taking into account that

$$\Phi\left(\frac{\cos\varphi}{\sigma}\right) \leq \frac{2\cos\varphi}{\sigma\sqrt{\pi}} , \tag{A1.67}$$

we obtain

$$I_n = \int_0^{\pi/2} \Phi\left(\frac{\cos\varphi}{\sigma}\right) \cos\varphi \sin^{n-1}\varphi \, d\varphi \leq \frac{1}{\sigma\sqrt{\pi}}\left(\frac{1}{B_n} - \frac{1}{B_{n+2}}\right) . \tag{A1.68}$$

Substituting (A1.68) into (A1.65), we find

$$E(Y_1) \leq \frac{4r}{\sqrt{\pi}\sigma n} - 1 , \tag{A1.69}$$

and the result follows immediately from the solution of $E(V_1) = 0$.

Q.E.D.

It is worthwhile noting that for the deterministic quadratic model $\tilde{r}_1 = O(\sqrt{n})$ [see (A1.48) and (A1.49)] while for the stochastic one $\tilde{r}_1 \geq O(n)$ (see Lemma 3), that is, the stable zone is larger in the latter case.

(b) Algorithm RS-2 The random angle between the direction of the successful step and the antigradient is distributed according to

$$h_n(\varphi) = \begin{cases} \dfrac{1}{P} B_n \sin^{n-2} \varphi P_2(\varphi), & 0 \leq \varphi \leq \varphi_0, \\ \dfrac{1}{P} B_n \sin^{n-2} \varphi [1 - P_2(\varphi)], & \varphi_0 < \varphi \leq \pi, \end{cases} \tag{A1.70}$$

where P is the probability of a successful step equal to

$$P = \int_0^{\varphi_0} B_n \sin^{n-2} \varphi P_2(\varphi)\, d\varphi + \int_{\varphi_0}^{\pi} B_n \sin^{n-2} \varphi [1 - P_2(\varphi)]\, d\varphi . \tag{A1.71}$$

It is readily seen that

$$C_n = \frac{E(V_2)P}{1+P}, \tag{A1.72}$$

where

$$E(V_2) = \frac{1}{P} B_n \int_{\varphi_0}^{\pi} (2r \cos \varphi - 1)[1 - P_2(\varphi)] \sin^{n-2} \varphi\, d\varphi$$
$$+ \frac{1}{P} B_n \int_0^{\varphi_0} (2r \cos \varphi - 1) P_2(\varphi) \sin^{n-2} \varphi\, d\varphi . \tag{A1.73}$$

After some manipulations we find

$$C_n = \frac{1}{1+P}\left(B_n \int_{\varphi_0}^{\pi} \phi \frac{|\cos \varphi|}{2\sigma} (1 - 2r \cos \varphi) \sin^{n-2} \varphi\, d\varphi \right.$$
$$\left. - \frac{1}{2} B_n \int_0^{\pi} \phi \frac{|\cos \varphi|}{2\sigma} \sin^{n-2} \varphi\, d\varphi \right). \tag{A1.74}$$

Making use of (A1.67), we can prove, in analogy with Lemma 3, that the stable zone for RS-2 is located at least at the distance

$$\tilde{r}_2 = \frac{\sigma n \sqrt{\pi}}{2} - B_n \frac{n}{n-1} \tag{A1.75}$$

from the extremum point.

It is interesting to point out that for both the deterministic and stochastic models we have, respectively, $\tilde{r}_2 = \frac{1}{2}$ and $\tilde{r}_2 \geq O(n)$, that is, the stable zone for the latter is again much larger.

It can also be proved that Lemma 2 is also valid for the stochastic model, and that $\tilde{r}_2 < \tilde{r}_1$, that is, the stable zone for RS-2 is smaller than that for RS-1. We omit the proof of these statements, as they are similar to their counterparts for the deterministic quadratic model.

APPENDIX 2: DISTRIBUTION OF A RANDOM ANGLE IN THE *n*-DIMENSIONAL SPACE

Let Ξ be a random vector uniformly distributed over the surface of unit n-dimensional hypersphere with its center as origin and let R be any given unit vector issuing from the origin (see Fig. 10).

The p.d.f. of the random angle between Ξ and R is sought. For reasons of symmetry, we shall confine ourselves to the semisphere $0 \leq \varphi \leq \pi$. The p.d.f. is then [see Rastrigin (1968)]

$$h_n(\varphi) = \frac{\sin^{n-2} \varphi}{\int_0^{\pi} \sin^{n-2} \varphi \, d\varphi} = B_n \sin^{n-2} \varphi , \qquad 0 \leq \varphi \leq \pi , \tag{A2.1}$$

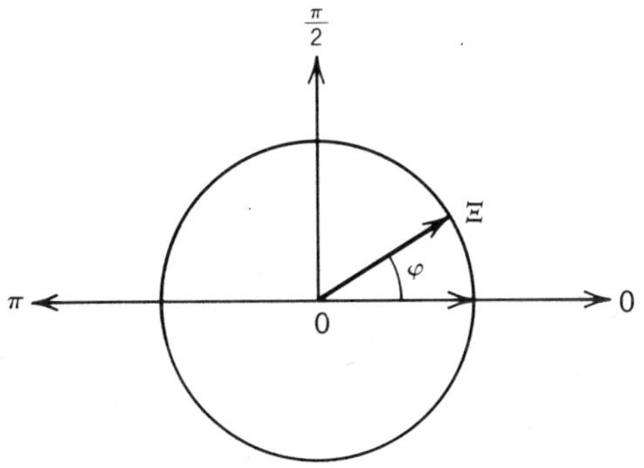

Fig. 10 Graphical representation of random vector Ξ.

where

$$B_n = \left(\int_0^\pi \sin^{n-2} \varphi \, d\varphi \right)^{-1} = \frac{\Gamma(n/2)}{\sqrt{\pi}\Gamma[(n-1)/2]} \,. \tag{A2.2}$$

For sufficiently large n we readily obtain

$$B_n \approx \left(\frac{n-1}{2\pi} \right)^{1/2} \,. \tag{A2.3}$$

The expected value of the r.v. φ is readily obtained,

$$E(\varphi) = \tfrac{1}{2}\pi \,, \tag{A2.4}$$

from which it follows that, on the average, R and Ξ are orthogonal.

It readily follows from (A2.1) that as n increases, $h_n(\varphi)$ approaches Dirac's delta function, that is,

$$h_n(\varphi) \underset{n \to \infty}{\to} \delta(\tfrac{1}{2}\pi - \varphi) \,. \tag{A2.5}$$

Figure 11 represents $h_n(\varphi)$ for different n.

For the rate of convergence of $h_n(\varphi)$ to $\delta(\cdot)$ and its relationship to generation of random polytops, see Dam, Frenk, and Telgen (1983).

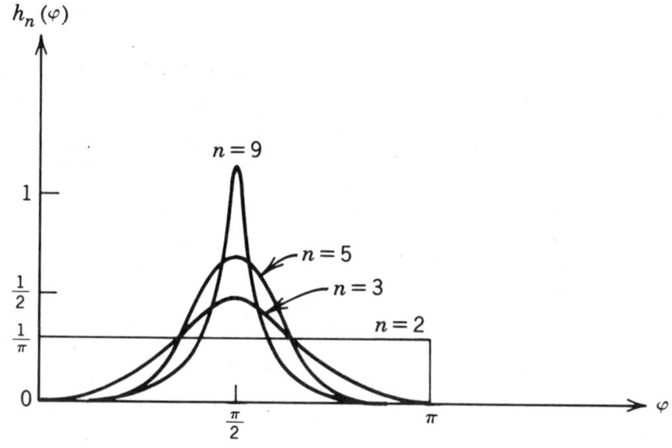

Fig. 11 Density function of φ for different n.

APPENDIX 3: GENERATING RANDOM VECTORS UNIFORMLY DISTRIBUTED INSIDE AND ON THE SURFACE OF DIFFERENT REGIONS

This appendix deals with several algorithms for generating random vectors uniformly distributed on and inside of

(a) a simplex,
(b) a hypersphere, and
(c) a hyperellipsoid.

Most of the material is taken from Rubinstein (1982).

A. Generating Random Vectors Over a Simplex

The p.d.f. of a random vector x uniformly distributed *inside* an n-dimensional simplex

$$S_n = \left\{(x_1, \ldots, x_n): x_i \geq 0,\ i = 1, \ldots, n,\ \sum_{i=1}^{n} x_i \leq 1\right\} \tag{A3.1}$$

is

$$f_X(x) = \begin{cases} n! & \text{if } x = \{x_1, \ldots, x_n\} \in S_n\,, \\ 0 & \text{otherwise}\,. \end{cases} \tag{A3.2}$$

This density is a particular case of Dirichlet density

$$f_X(x) = \frac{\Gamma(\nu_1 + \cdots + \nu_{n+1})}{\Gamma(\nu_1) \cdots \Gamma(\nu_{n+1})} x_1^{\nu_1 - 1} \cdots x_n^{\nu_n - 1} (1 - x_1 - \cdots - x_n)^{\nu_{(n+1)} - 1} \tag{A3.3}$$

with parameters $\nu_1, \ldots, \nu_n, \nu_{n+1}$ denoted $D(\nu_1, \ldots, \nu_n, \nu_{n+1})$ [see Wilks (1962)].

Indeed, for $\nu_1 = \nu_n = \nu_{n+1} = 1$, equation (A3.3) equals (A3.2). We shall denote the density (A3.2) $D(1, \ldots, 1, 1)$. It is also known (see Wilks (1962)) that if Y_i, $i = 1, \ldots, n+1$, are independent r.v.'s distributed according to gamma distributions with parameters ν_i and 1, that is,

$$f_{Y_i}(y) = \begin{cases} \dfrac{y^{\nu_i - 1} e^{-y}}{\Gamma(\nu_i)} & y > 0,\ \nu_i > 0\,, \\ 0 & \text{otherwise}\,, \end{cases} \tag{A3.4}$$

then the random vector

$$X = (X_1, \ldots, X_n) = \left(\frac{Y_1}{\sum_{i=1}^{n+1} Y_i}, \ldots, \frac{Y_n}{\sum_{i=1}^{n+1} Y_i} \right) \tag{A3.5}$$

is distributed $D(\nu_1, \ldots, \nu_n, \nu_{n+1})$. Taking into account that in the particular case where $\nu_i = 1$, $i = 1, \ldots, n+1$, $f_{Y_i}(y)$ is distributed exp(1), the algorithm for generating r.v.'s from $D(1, \ldots, 1, 1)$ can be written as follows:

Algorithm 1 **(*Exponential Approach*)**

1. Generate $n + 1$ random variables from exp(1).
2. Apply formula (A3.5) and deliver X as a vector distributed from $D(1, \ldots, 1, 1)$.

It is readily seen that if we expand the n-dimensional vector (A3.5) to the $(n + 1)$-dimensional one, namely to

$$X = (X_1, \ldots, X_{n+1}) = \left(\frac{Y_1}{\sum_{i=1}^{n+1} Y_i}, \ldots, \frac{Y_{n+1}}{\sum_{i=1}^{n+1} Y_i} \right), \tag{A3.6}$$

we obtain a uniform distribution *on* the surface of the $(n + 1)$-dimensional simplex

$$S_{n+1} = \left\{ (x_1, \ldots, x_{n+1}) \colon x_i \geq 0,\ i = 1, \ldots, n+1,\ \sum_{i=1}^{n+1} x_i = 1 \right\}$$

instead of the uniform distribution inside the n-dimensional simplex considered in (A3.5).

An alternative procedure for generating random vectors uniformly distributed *inside* an n-dimensional simplex is based on the following result. Let $Y_{(1)}, \ldots, Y_{(n)}$ be the order statistics from any continuous c.d.f. $F_Y(y)$. Then the random vector $X = (X_1, \ldots, X_n)$ with

$$\begin{aligned} X_1 &= F_Y(Y_{(1)}), \\ X_2 &= F_Y(Y_{(2)}) - F_Y(Y_{(1)}), \\ &\vdots \\ X_n &= F_Y(Y_{(n)}) - F_Y(Y_{(n-1)}) \end{aligned} \tag{A3.7}$$

is distributed $D(1, \ldots, 1, 1)$ [see Wilks (1962)]. Taking into account that the random variable $F_Y(Y)$ is distributed $U(0, 1)$ we can rewrite equation (A3.7) as

$$\begin{aligned} X_1 &= U_{(1)}, \\ X_2 &= U_{(2)} - U_{(1)}, \\ &\vdots \\ X_n &= U_{(n)} - U_{n-1}. \end{aligned} \tag{A3.8}$$

The following is an alternative algorithm for generating from $D(1, \ldots, 1, 1)$.

Algorithm 2 **(*Order Statistics Approach*)**

1. Generate n independent r.v.'s $U_1, \ldots, U_n$ from $U(0, 1)$.
2. Arrange them to be order statistics $U_{(1)}, \ldots, U_{(n)}$.
3. Apply formula (A3.8) and deliver X as a vector distributed $D(1, \ldots, 1, 1)$.

To obtain a vector uniformly distributed *on* the surface of an $(n+1)$-dimensional simplex, we expand as in (A3.6) the n-dimensional (A3.8) to the $(n+1)$-dimensional

$$\begin{aligned} X_1 &= U(1), \\ X_2 &= U_{(2)} - U_{(1)}, \\ &\vdots \\ X_n &= U_{(n)} - U_{(n-1)}, \\ X_{n+1} &= 1 - U_{(n)}. \end{aligned} \tag{A3.9}$$

Note 1 Both Algorithms 1 and 2 obtained so far are suitable for generating random vectors uniformly distributed over an n-dimensional simplex with vertices

$$x_0 = (0, 0, \ldots, 0), \qquad x_1 = (1, 0, \ldots, 0), \ldots, x_n = (0, 0, \ldots, 1).$$

Assume now that we want to generate random vectors uniformly distributed over an n-dimensional simplex with arbitrary vertices, say, $z_0, z_1, \ldots, z_n$ $(z_i = z_{i1}, \ldots, z_{in})$, $i = 0, 1, \ldots, n)$. It is readily seen that if X is uniformly distributed over the simplex (A3.11), then the linear transformation

$$Z = CX + z_0$$

with the matrix

$$C = \begin{vmatrix} z_{11} - z_{01} & z_{12} - z_{02}, \ldots, z_{1n} - z_{0n} \\ z_{21} - z_{01} & z_{22} - z_{02}, \ldots, z_{2n} - z_{0n} \\ \\ z_{n1} - z_{01} & z_{n2} - z_{02}, \ldots, z_{nn} - z_{0n} \end{vmatrix}$$

preserves uniformity, that is, the vector Z will be uniformly distributed on the simplex with vertices $z_0, z_1, \ldots, z_n$.

B. Generating Random Vectors Uniformly Distributed over an n-Dimensional Unit Hypersphere

The density function of a random vector uniformly distributed on the surface of an n-dimensional sphere of radius r is

$$f_Y(y) = \begin{cases} \dfrac{\Gamma(n/2)}{2\pi^{n/2} r^{n-1}}, & y = \left\{(y_1, \ldots, y_n); \sum_{i=1}^{n} y_i^2 = r^2\right\}, \\ 0 & \text{otherwise}. \end{cases} \tag{A3.10}$$

Formula (A3.10) follows from the fact that the surface of the n-dimensional sphere of radius r is

$$\frac{2\pi^{n/2} r^{n-1}}{\Gamma(n/2)}.$$

Before proceeding with the algorithms for generating points uniformly distributed *on* the surface and *inside* the n-dimensional sphere, let us note that any *surface* algorithm (see below) can be converted into an *interior* algorithm and vice versa. Indeed, let $Y = (Y_1, \ldots, Y_n)$ be a vector uniformly distributed *on* the surface of an n-dimensional sphere; then the vector $X = (X_1, \ldots, X_n)$ with components $X_i = U^{1/n} Y_i$, where $U \sim U(0, 1)$, is distributed uniformly *inside* this hypersphere. Conversely, let $X = (X_1, \ldots, X_n)$ be a vector uniformly distributed *inside* the n-dimensional sphere; then the vector $Y = (Y_1, \ldots, Y_n)$ with components $Y_i = (\sum_{i=1}^{n} X_i^2)^{-1/2} X_i$ is distributed uniformly *on* the surface of this n-dimensional sphere. This is so, since Y is a projection of the interior point to the surface of the sphere.

The first algorithm for generating a vector uniformly distributed *on* the surface of an n-dimensional unit sphere is based on von Neumann's acceptance-rejection method and is as follows:

Algorithm 3 (Acceptance-Rejection)

1. Generate U_i, $i = 1, \ldots, n$, uniformly on $(0, 1)$.

2. Set $X_i = 1 - 2U_i$, $i = 1, \ldots, n$ and $Y^2 = \Sigma_{i=1}^n X_i^2$.

3. If $Y^2 < 1$, then accept

$$Z = (Z_1, \ldots, Z_n)\,, \quad \text{where } Z_i = \frac{X_i}{Y}, \; i = 1, \ldots, n\,,$$

as the desired vector.

4. Go to step 1.

Clearly, if we rewrite step 3 as

3′. if $Y^2 < 1$, then accept $(Y_1, \ldots, Y_n)$ as the desired vector, then $Y = (Y_1, \ldots, Y_n)$ will be distributed uniformly *inside* the n-dimensional unit sphere.

The main advantage of the acceptance-rejection method is its simplicity. Its main disadvantage is that the number of trials (see Section 4.2) needed to obtain a point inside (on) the hypersphere increases explosively in n. For this reason it can be recommended only in low dimension ($n \geq 5$).

Another algorithm for generating a vector uniformly distributed *on* the surface of a unit hypersphere is based on the following:

Theorem Let $X_1, \ldots, X_n$ be i.i.d. r.v.'s distributed $N(0, 1)$. Then the vector

$$Y = (Y_1, \ldots, Y_n) = \left(\frac{X_1}{Z}, \ldots, \frac{X_n}{Z}\right), \tag{A3.11}$$

where $Z = (\Sigma_{i=1}^n X_i^2)^{1/2}$, is distributed uniformly on the surface of the unit hypersphere $\Sigma_{i=1}^n y_i^2 = 1$.

Proof Let us make the following transformation:

$$\begin{cases} z = v\,, \\ x_i = v y_i\,, \; i = 1, \ldots, n\,. \end{cases} \tag{A3.12}$$

The Jacobian of this transformations is $J = z^n$, and we have

$$f_{Y_1,\ldots,Y_n,V}(y_1,\ldots,y_n,v) = v^n \frac{1}{(2\pi)^{n/2}} \exp\left(-\sum_{i=1}^{n} \frac{(y_i v)^2}{2}\right)$$

$$= v^n \frac{1}{(2\pi)^{n/2}} \exp\left(-\frac{v^2}{2}\sum_{i=1}^{n} y_i^2\right). \qquad \text{(A3.13)}$$

Since

$$\sum_{i=1}^{n} y_i^2 = \sum_{i=1}^{n}\left(\frac{x_i}{z^2}\right)^2 = \frac{1}{z^2}\sum_{i=1}^{n} x_i^2 = 1\,,$$

we obtain

$$f_{Y_1,\ldots,Y_n,V}(y_1,\ldots,y_n,v) = v^n \frac{1}{(2\pi)^{n/2}} e^{-v^2/2}\,. \qquad \text{(A3.14)}$$

It follows from (A3.13) and (A3.14) that $(Y_1,\ldots,Y_n)$ and V are independent and that the vector $(Y_1,\ldots,Y_n)$ is uniformly distributed on the surface of the unit sphere $\Sigma_{i=1}^n y_i^2 = 1$. Q.E.D.

Algorithm 4 **[*Normal Approach, Muller (1959)*]**

1. Generate n random variables X_i, $i = 1,\ldots,n$, from $N(0,1)$.
2. Compute $Z = (\Sigma_{i=1}^n X_i^2)^{1/2}$.
3. Apply formula (A3.11) and deliver Y as a desired vector.

An alternative procedure due to Hicks and Weeling (1959) is as follows: Selecting a point on a 1-sphere ($x_1 = 1$ or $x_1 = -1$), determine from this a random point on a 2-sphere and so on until the required dimensionability is reached. Given a point $(x_1, x_2, \ldots, x_n)$ on an n-sphere, the point on the $(n+1)$-sphere is

$$(rx_1, rx_2, \ldots, rx_n, s\sqrt{1-r^2})\,,$$

where s is a random sign and r has the p.d.f.

$$f_r(r) = \begin{cases} k\dfrac{r^{n-1}}{\sqrt{1-r^2}} & 0 \le r \le 1,\ k = \left(\displaystyle\int_0^1 \frac{r^{n-1}}{\sqrt{1-r^2}}\,dr\right)^{-1}, \\ 0 & \text{otherwise}\,. \end{cases}$$

Consider now the polar transformation

$$
\begin{aligned}
X_1 &= r \cos \varphi_1 \cos \varphi_2 \cdots \cos \varphi_{n-2} \cos \varphi_{n-1} , \\
X_2 &= r \cos \varphi_1 \cos \varphi_2 \cdots \cos \varphi_{n-2} \sin \varphi_{n-1} , \\
X_3 &= r \cos \varphi_1 \cos \varphi_2 \cdots \sin \varphi_{n-2} , \\
&\;\;\vdots \\
X_{n-1} &= r \cos \varphi_1 \sin \varphi_2 , \\
X_n &= r \sin \varphi_1 ,
\end{aligned}
\tag{A3.15}
$$

where $-\frac{1}{2}\pi \leq \varphi_i \leq \frac{1}{2}\pi$, $i = 1, \ldots, n-2$, $0 \leq \varphi_{n-1} \leq 2\pi$, $0 \leq r \leq 1$; and r and $\varphi_1, \ldots, \varphi_{n-1}$ are independent r.v.'s. To generate points uniformly distributed *inside* the hypersphere with radius r, one can use the following algorithm:

Algorithm 5 **(*Polar Transformation*)**

1. Generate U_1 from $U(0, 1)$.
2. Set $r = U_1^{1/n}$.
3. Generate U_2 from $U(0, 1)$.
4. Set $\varphi_{n-1} = 2\pi U_2$.
5. Generate random angles φ_i, $i = 1, \ldots, n-1$, from the p.d.f.

$$f_i(x) = c_i \cos^{n-i-1} x \,, \quad \text{where } c_i^{-1} = \int_{-\pi/2}^{\pi/2} \cos^{n-i-1} x \, dx \,.$$

6. Apply formula (A3.15) and deliver X as a desired vector.

Deak (1979) showed numerically that for any n $(n > 1)$ the normal approach (Algorithm 4) is faster than both Hick's and Wheeling's (1959) algorithm and the polar transformation (Algorithm 5) is superior than the acceptance-rejection algorithm (Algorithm 3) for $n \geq 4$.

Note 2 Let $X = (X_1, \ldots, X_n)$ be a vector uniformly distributed *inside* (*on*) the n-dimensional unit hypersphere; then the vector $Y = (Y_1, \ldots, Y_n)$ with coordinates $Y_i = rX_i + b_i$, $i = 1, \ldots, n$, is distributed uniformly *inside* (*on*) the sphere with the radius r and with the center at point $b = (b_1, \ldots, b_n)$.

C. Generating Random Vectors Uniformly Distributed over the Hyperellipsoid

The equation of the hyperellipsoid with the center at origin can be represented as

$$X^{Xt}\Sigma X = r^2 , \tag{A3.16}$$

where Σ is a positive defined and symmetric $(n \times n)$ matrix, and t is the transpose operator. In the particular case where $\Sigma = 1$, we obtain the hypersphere with radius r. Since Σ is positively defined and symmetric, there exists a unique lower triangular matrix

$$C = \begin{vmatrix} c_{11} & 0 & \cdots & 0 \\ c_{21} & c_{22} & \cdots & 0 \\ \vdots & \vdots & \vdots & \vdots \\ c_{n1} & c_{n2} & \cdots & c_{nn} \end{vmatrix}$$

such that $\Sigma = CC^t$. It can be readily proven that if the vector X is uniformly distributed over (i.e., inside or on) the n-dimensional sphere with radius r, then the vector $Y = CX$ is uniformly distributed over the hyperellipsoid [equation (A3.16)]. The algorithm follows.

Algorithm *6*

1. Generate $X = (X_1, \ldots, X_n)$ uniformly distributed over the hypersphere with radius r.
2. Find the matrix C from $CC^t = \Sigma$.
3. Deliver $Y = CX$ as a desired vector.

Remark The elements c_{ij} of the matrix C can be calculated from the recursive formula [see Rubinstein (1981)]

$$c_{ij} = \frac{\sigma_{ij} - \sum_{k=1}^{j-1} c_{ik} c_{jk}}{\left(\sigma_{jj} - \sum_{k=1}^{j-1} c_{jk}^2\right)^{1/2}} ,$$

where $\Sigma_{k=1}^{0} c_{ik} c_{jk} = 0$, $1 \le j \le i \le n$, and σ_{ij} are elements of Σ.

APPENDIX 4: OPTIMAL COVERAGE OF CONVEX REGIONS

In this appendix, taken from Rubinstein and Samorodnitksy (1986), we consider the problem of optimal coverage of a region D given by the nonlinear constraints

$$D = \{x \colon g_i(x) \le 0,\ i = 1, 2, \ldots, M,\ x \in R^n\}$$

with polygons and quadratic configurations.

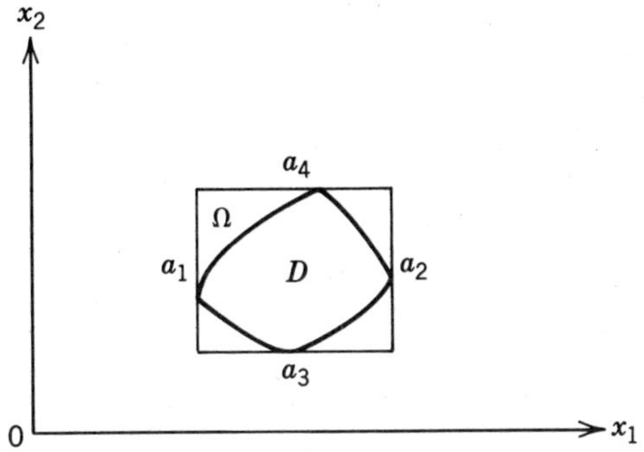

Fig. 12.

A. Coverage with Polygons

(a) Rectangles We start with the two-dimensional case, and before describing the general algorithm, we discuss the simplest case. Our objective is to cover a given region D by a rectangle Ω of minimal area.

We first consider the case when the sides of rectangle Ω are parallel to the axes x_1 and x_2 (see Fig. 12).

Clearly, in order to find such a rectangle of minimal area, we only have to determine the extreme points at the top and at the bottom (a_4 and a_3) and on the right and left (a_2 and a_1). Mathematically it can be written as

$$a_1 = \min_{(x_1,x_2)\in D} x_1 , \qquad a_2 = \max_{(x_1,x_2)\in D} x_1 ,$$
$$a_3 = \min_{(x_1,x_2)\in D} x_2 , \qquad a_4 = \max_{(x_1,x_2)\in D} x_2 . \tag{A4.1}$$

Obviously, since D is a convex region, (A4.1) represents a set of convex optimization problems. Its generalization to the n-dimensional case ($n > 2$) is straightforward; we have to solve $2n$ convex problems of type (A4.1).

Example 1 Let us take D defined by (see Fig. 13)

$$x_1 + x_2 \geq 1 , \qquad x_1^2 + x_2^2 \leq 1 .$$

Here, clearly, $a_1 = 0$, $a_2 = 1$, $a_3 = 0$, and $a_4 = 1$ and mes $\Omega = 1$.

Assume now that rotation of the coodinate system $\{x_1, x_2\}$ through an angle α is allowed. For a *fixed* α we have

$$\begin{bmatrix} x_1(\alpha) \\ x_2(\alpha) \end{bmatrix} = \begin{bmatrix} x_1 \cos\alpha + x_2 \sin\alpha \\ x_2 \cos\alpha - x_1 \sin\alpha \end{bmatrix} . \tag{A4.2}$$

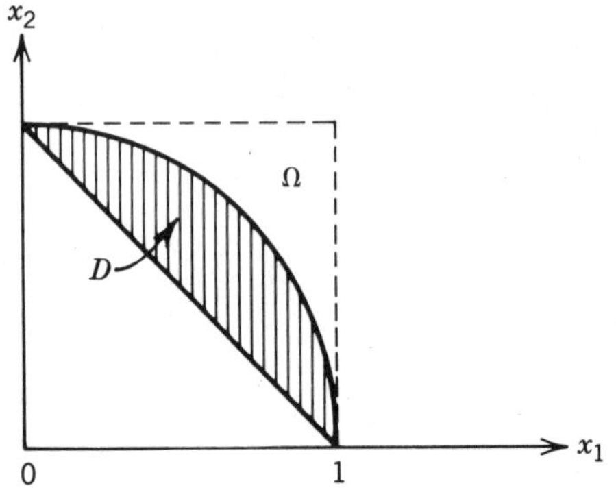

Fig. 13.

The problem of finding a minimal rectangle $\Omega(\alpha)$ in the new system of coordinates is analogous to problem (A4.1) and can be written as

$$a_1(\alpha) = \min_{(x_1,x_2)\in D} x_1(\alpha) = \min_{(x_1,x_2)\in D} (x_1 \cos \alpha + x_2 \sin \alpha) , \qquad \text{(A4.3)}$$

$$a_2(\alpha) = \max_{(x_1,x_2)\in D} x_1(\alpha) = \max_{(x_1,x_2)\in D} (x_1 \cos \alpha + x_2 \sin \alpha) , \qquad \text{(A4.4)}$$

$$a_3(\alpha) = \min_{(x_1,x_2)\in D} x_2(\alpha) = \min_{(x_1,x_2)\in D} (x_2 \cos \alpha - x_1 \sin \alpha) , \qquad \text{(A4.5)}$$

$$a_4(\alpha) = \max_{(x_1,x_2)\in D} x_2(\alpha) = \max_{(x_1,x_2)\in D} (x_2 \cos \alpha - x_1 \sin \alpha) . \qquad \text{(A4.6)}$$

The area of $\Omega(\alpha)$ is

$$\text{mes}\, \Omega(\alpha) = [a_2(\alpha) - a_1(\alpha)][a_4(\alpha) - a_3(\alpha)] .$$

To find the optimal among all possible rectangles $\Omega(\alpha)$, we have to solve one more optimization problem:

$$\min_{0\le\alpha\le\pi/2} \text{mes}\, \Omega(\alpha) = \min_{0\le\alpha\le\pi/2} [a_2(\alpha) - a_1(\alpha)][a_4(\alpha) - a_3(\alpha)] . \qquad \text{(A4.7)}$$

Example 2 Consider the same region D as in Example 1, that is, assuming additionally rotation of the coordinates (see Fig. 14). It is readily seen that

$$a_2(\alpha) = \cos \alpha , \qquad a_1(\alpha) = -\sin \alpha ,$$

$$a_3(\alpha) = \begin{cases} \sin \alpha , & \alpha \le \frac{1}{4}\pi , \\ \cos \alpha , & \alpha > \frac{1}{4}\pi , \end{cases} \qquad a_4(\alpha) = 1 .$$

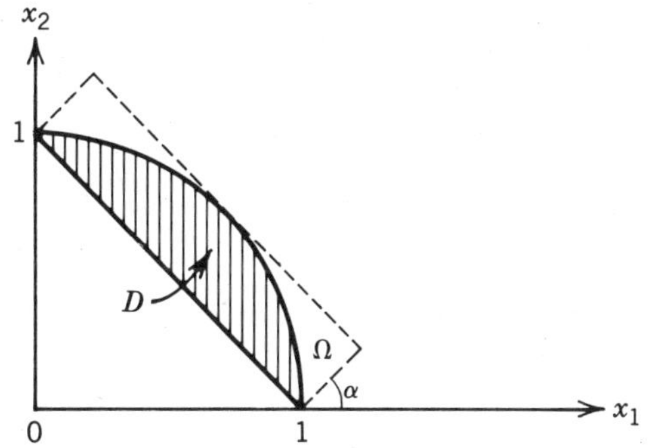

Fig. 14.

and problem (A4.7) becomes

$$\min_{0\le\alpha\le\pi/2} \operatorname{mes}\Omega(\alpha) = \min\begin{cases} \min\limits_{0\le\alpha\le\pi/4} (\cos\alpha + \sin\alpha)(1-\sin\alpha)\,, \\ \min\limits_{\pi/4<\alpha\le\pi/2} (\cos\alpha + \sin\alpha)(1-\cos\alpha)\,. \end{cases}$$

It is readily obtainable that the optimal solution is achieved when $\alpha = \pi/4$ and equals $\operatorname{mes}\Omega(\pi/4) = \sqrt{2} - 1 = 0.41$. Thus, the optimal rectangle obtained by rotating the axis has an area 0.41, which is $1/0.41 = 2.44$ times less than the optimal rectangle obtained without rotation (see Example 1).

The problem of finding a minimal volume of $\Omega(\alpha)$ in (A4.7) is generally a multiextremal one and can be solved by the methods considered in Boender et al. (1982), Dixon and Szegö (1975, 1978), Rubinstein (1981), and this chapter.

Generalization to the n-dimensional case ($n > 2$) is straightforward. In this case we have to solve $2n$ *convex* optimization problems of type (A4.3)–(A4.6) followed by a nonconvex one of type (A4.7), which can be written as

$$\min_{\alpha}\{\operatorname{mes}\Omega(\boldsymbol{\alpha})\} = \min_{\boldsymbol{\alpha}} \prod_{i=1}^{n} [a_{2i}(\boldsymbol{\alpha}) - a_{2i-1}(\boldsymbol{\alpha})]\,,$$

where $\boldsymbol{\alpha}$ denotes rotation of the axes.

Thus in the case without rotation of the coordinate system we have to solve $2n$ convex problems of type (A4.3)–(A4.6), and in the case with rotation of the coordinates we need to solve, in addition, one *multiextremal* problem of type (A4.7).

We will now suggest a method that can sometimes be useful in practice and that reduces the number of problems of type (A4.3)–(A4.6) from $2n$

to n. However, it involves construction of a new set closely associated with D. Specifically, we define a conjugate set of $D \subset R^n$ as

$$\bar{D} = \{x \in R^n: x = y - z,\ y \in D,\ z \in D\}\,. \tag{A4.8}$$

Lemma 4 $\bar{D}$ is convex if D is.

Proof Let $x, y \in \bar{D}$, $\alpha \in [0, 1]$. Then, by (A4.8), there exist z, t, v, w all from D such that

$$x = z - t\,, \tag{A4.9}$$

$$y = v - w\,. \tag{A4.10}$$

Now

$$\begin{aligned} \alpha x + (1-\alpha)y &= \alpha(z-t) + (1-\alpha)(v-w) \\ &= [\alpha z + (1-\alpha)v] - [\alpha t + (1-\alpha)w]\,. \end{aligned} \tag{A4.11}$$

By convexity of D, we have $\alpha z + (1-\alpha)v \in D$ and $\alpha t + (1-\alpha)w \in D$, which together with (A4.8) and (A4.11) yields $\alpha x + (1-\alpha)y \in \bar{D}$, and hence $\bar{D}$ is convex. Q.E.D.

For example, it is readily seen that problem (A4.7) reduces to

$$\min_{0 \le \alpha \le \pi/2} \operatorname{mes} \Omega(\alpha) = \min_{0 \le \alpha \le \pi/2} b_1(\alpha) b_2(\alpha)\,, \tag{A4.12}$$

where

$$b_1(\alpha) = \max_{x \in \bar{D}} \{x_1 \cos\alpha + x_2 \sin\alpha\}\,, \tag{A4.13}$$

$$b_2(\alpha) = \max_{x \in \bar{D}} \{x_2 \cos\alpha - x_1 \sin\alpha\}\,. \tag{A4.14}$$

Comparing (A4.13) and (A4.14) with (A4.3)–(A4.6), we find that the number of problems to be solved has been halved.

Example 3 Let D be a regular hexagon with sides of unit length and center at the origin (see Fig. 15). Clearly, $\bar{D}$ is likewise a regular hexagon concentric with the former with sides of double length. Thus, we have halved the number of optimization problems (A4.3)–(A4.6) without affecting their complexity.

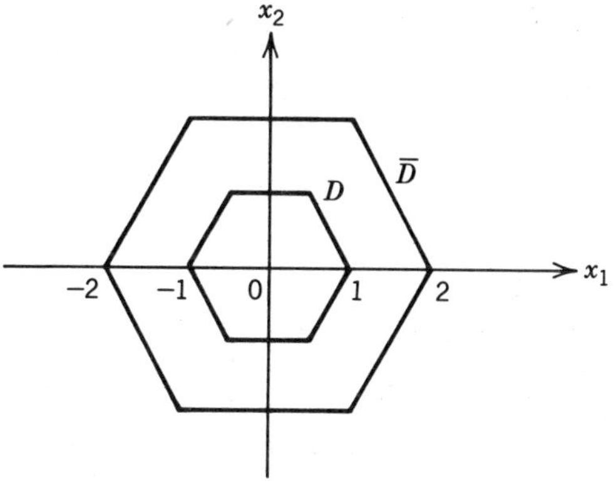

Fig. 15.

(b) Polygons (Polyhedra) We now show how to cover arbitrary convex regions D with polygons (polyhedra) of minimal volume. First, we introduce the following definitions:

Definition 1 A polygon Ω is said to be *normalized* (notation Ω_{nor}) if it is described by the system of linear inequalities

$$\Omega_{\text{nor}} = \{x\colon A_i x + b_i \leq 0,\ A_i = (a_{i1}, \ldots, a_{in}),\ i = 1, \ldots, m\}\,, \tag{A4.15}$$

where a_{ij} and b_i are given constants such that $\operatorname{mes} \Omega_{\text{nor}} = 1$.

We shall define now two types of similarity of polygons.

Definition 2 A pair of polygons

$$\begin{aligned} \Omega_1 &= \{x\colon A_i x + b_i \leq 0,\ i = 1, \ldots, m\}\,, \\ \Omega_2 &= \{x\colon C_i x + d_i \leq 0,\ i = 1, \ldots, m\}\,, \end{aligned} \tag{A4.16}$$

have similarity of type 1 or 2 if there exists rotation of the system of coordinates such that Ω_2 can be presented, respectively, as

$$\Omega_2 = \{x\colon C_i(\alpha)x + d_i(\alpha) \leq 0,\ i = 1, \ldots, m\}\,,$$

where

$$C_i(\alpha) = A_i\,,$$

and

$$\Omega_2 = \{x\colon C_i(\alpha)x + d_i(\alpha) \le 0,\ i = 1, \ldots, m\}\ ,$$

where

$$C_i(\alpha) = \frac{A_i}{k},\ d_i(\alpha) = b_i - A_i x_0 \quad \text{for some } k > 0,\ x_0 \in R^n\ .$$

For example, the polygons (see Fig. 16*a*)

$$\Omega_1 = \{(x_1, x_2)\colon x_2 \ge 0,\ x_1 - x_2 \ge 0,\ x_2 \le 1,\ x_1 \le 2\}\ ,$$
$$\Omega_2 = \{(x_1, x_2)\colon x_2 \ge 0,\ x_1 - x_2 \ge 0,\ x_2 \le 0.5,\ x_1 \le 1\}\ ,$$

have similarity of type 2, and the polygons (see Fig. 16*b*)

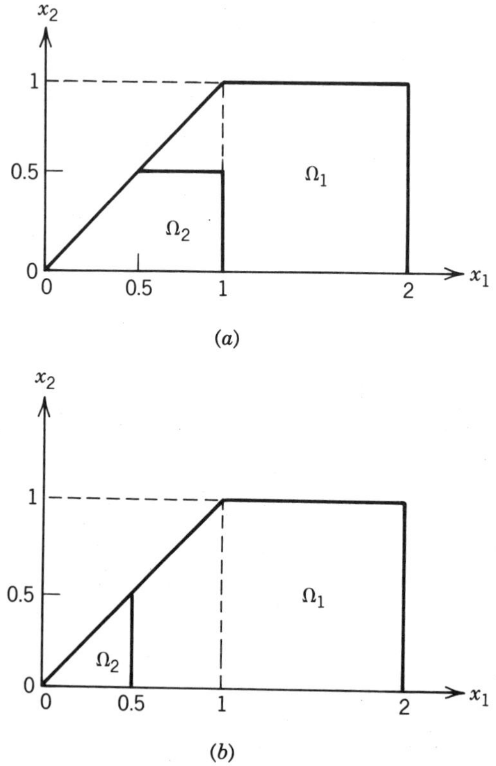

Fig. 16.

$$\Omega_1 = \{(x_1, x_2): x_2 \geq 0,\ x_1 - x_2 \geq 0,\ x_2 \leq 1,\ x_1 \leq 2\}\ ,$$

$$\Omega_2 = \{(x_1, x_2): x_2 \geq 0,\ x_1 - x_2 \geq 0,\ x_2 \leq 0.5,\ x_1 \leq 0.5\}\ ,$$

have similarity of type 1.

Remark The similarity of type 2 coincides with the conventional concept of similarity used in geometry, meaning that the shapes of Ω_1 and Ω_2 are preserved. The similarity of type 1 generally does not require preservation of the shapes of Ω_1 and Ω_2 (see Fig. 16*b*).

We shall call the coverage problem with polygons that have similarity of type 1 (type 2) to Ω_{nor} the coverage problem of type 1 (type 2). Our problem is to find a polygon that is similar in the sense of type 1 (type 2) to the normalized Ω_{nor}, covers D, and has a minimal volume. The coverage problem with polygons of type 2 is more difficult than the one with polygons of type 1 since the latter has more degrees of freedom (see Definition 2).

First, we consider the coverage problem of type 1. Let D be arbitrary and Ω be a triangle (see Fig. 17). The extension of our results to any covering polygon of type 1 will be straightforward. Let the normalized triangle Ω_{nor} be given by

$$\Omega_{nor} = \{x: A_i x + b_i \leq 0,\ A_i = (a_{i1}, a_{i2}),\ i = 1, 2, 3\}\ .$$

For a fixed $\alpha \in [0, 2\pi]$ (see Fig. 17) the area of any triangle that covers D is equal to

$$(P(\alpha)[P(\alpha) - r_1(\alpha)][P(\alpha) - r_2(\alpha)][P(\alpha) - r_3(\alpha)])^{1/2}\ , \quad \text{(A4.17)}$$

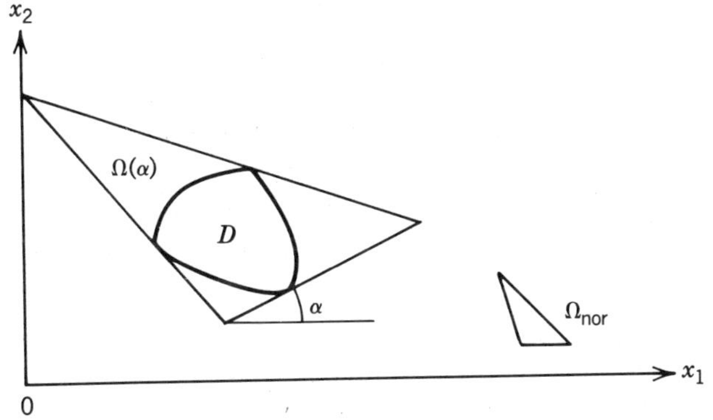

Fig. 17.

where

$$P(\alpha) = \tfrac{1}{2}(r_1(\alpha) + r_2(\alpha) + r_3(\alpha)),$$

and $r_i(\alpha)$, $i = 1, 2, 3$, are the lengths of the sides of the triangle.

To find the minimal triangle $\Omega(\alpha)$ for a fixed α, we have to solve the following three convex mathematical programming problems:

$$\max_{(x_1,x_2)\in D} \{a_{i1}(x_1 \cos\alpha + x_2 \sin\alpha) + a_{i2}(-x_1 \sin\alpha + x_2 \cos\alpha)\},$$
$$i = 1, 2, 3. \qquad \text{(A4.18)}$$

Solving (A4.18), we obtain three points, say $F_1(\alpha)$, $F_2(\alpha)$, $F_3(\alpha)$, from which we find

$$r_1^*(\alpha) = \|F_3(\alpha) - F_2(\alpha)\|,$$
$$r_2^*(\alpha) = \|F_3(\alpha) - F_1(\alpha)\|,$$
$$r_3^*(\alpha) = \|F_2(\alpha) - F_1(\alpha)\|,$$

representing the length of the sides of the minimal covering triangle $\Omega(\alpha)$.

The optimal among all possible triangles can be found by analogy to (A4.7) by solving

$$\min_{0\le\alpha\le 2\pi} \{\text{mes}\,\Omega(\alpha)\} = \min_{0\le\alpha\le 2\pi} (P(\alpha)[P(\alpha) - r_1^*(\alpha)]$$
$$\times [P(\alpha) - r_2^*(\alpha)][P(\alpha) - r_3^*(\alpha)])^{1/2}. \qquad \text{(A4.19)}$$

We shall now show that problem (A4.18) can be expressed similarly to problems (A4.3)–(A4.6), that is, to coverage by a rectangle problem. Indeed,

$$a_{i1}(x_1 \cos\alpha + x_2 \sin\alpha) + a_{i2}(-x_1 \sin\alpha + x_2 \cos\alpha)$$
$$= x_1(a_{i1} \cos\alpha - a_{i2} \sin\alpha) + x_2(a_{i1} \sin\alpha + a_{i2} \cos\alpha)$$
$$= (a_{i1}^2 + a_{i2}^2)^{1/2}[x_1 \cos(\alpha + \varphi) + x_2 \sin(\alpha + \varphi)], \qquad \text{(A4.20)}$$

where $\varphi = \arctan(a_{i2}/a_{i1})$. If follows from (A4.20) that the coefficients a_{i1} and a_{i2} represent an additional rotation of the system of coordinates.

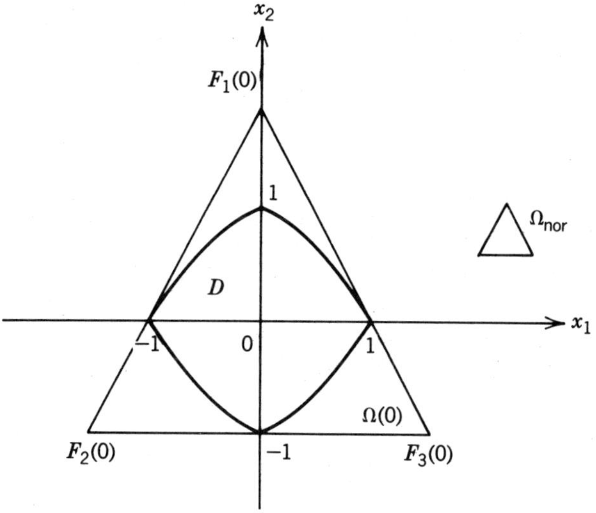

Fig. 18.

Example 4 Let D be a region defined by (see Fig. 18)

$$(x_1 - 1)^2 + (x_2 - 1)^2 \leq 5\,, \qquad (x_1 - 1)^2 + (x_2 + 1)^2 \leq 5\,,$$

$$(x_2 + 1)^2 + (x_2 - 1)^2 \leq 5\,, \qquad (x_2 + 1)^2 + (x_2 + 1)^2 \leq 5\,,$$

and Ω_{nor} an equilateral triangle.

It is readily seen from Fig. 18 that on rotating the triangle counterclockwise through a 30° angle, the side $F_1(\alpha)F_2(\alpha)$ becomes parallel to the axis x_2. Therefore it suffices to consider only $0 \leqslant \alpha \leqslant 30°$. Further, because of the symmetry of D about the line $x_2 = x_1$ we have to consider only $0 \leq \alpha \leq 15°$. Solving (A4.18), we obtain

1. $0 \leq \alpha \leq \arctan \frac{1}{2}$:

$$F_2(\alpha) = \left(\frac{\sqrt{5} - \sin(60 - \alpha)}{\cos(60 - \alpha)[\tan \alpha + \tan(60 - \alpha)]}, \right.$$
$$\left. \frac{\sqrt{5} \tan \alpha - \sin(60 - \alpha) \tan \alpha}{\cos(60 - \alpha)[\tan \alpha + \text{tg}(60 - \alpha)]} - 1 \right),$$

$$F_1(\alpha) = \left(\frac{\sqrt{5} - \sin(60 + \alpha)}{\cos(60 + \alpha)[\tan \alpha - \tan(60 + \alpha)]}, \right.$$
$$\left. \frac{\sqrt{5} \tan \alpha - \sin(60 + \alpha) \tan \alpha}{\cos(60 + \alpha)[\tan \alpha - \tan(60 + \alpha)]} - 1 \right).$$

2. $\arctan \frac{1}{2} \le \alpha \le 15°$:

$$F_2(\alpha) = \left(\frac{\sqrt{5} - \sin(60 - \alpha)}{\cos(60 - \alpha)[\tan \alpha + \tan(60 - \alpha)]}, \right.$$

$$\left. \frac{\sqrt{5} \tan \alpha - \sin(60 - \alpha) \tan \alpha}{\cos(60 - \alpha)[\tan \alpha + \tan(60 - \alpha)]} - 1 \right),$$

$$F_1(\alpha) = \left(\frac{1 + \tan(60 + \alpha)}{\tan \alpha - \tan(60 + \alpha)}, \frac{\tan \alpha - 1}{\tan \alpha - \tan(60 - \alpha)} - \frac{1}{\sqrt{3}} - 1 \right).$$

Since the triangle is equilateral, it suffices to minimize the length of its side $r_i(\alpha) = r(\alpha)$, $i = 1, 2, 3$. The minimal $r(\alpha)$ is achieved when $\alpha = 0$, and a simple calculation yields

$$r_{\text{opt}} = r_{\alpha=0} = 0.827 .$$

It is very important to note that the problem of *coverage of any region* with polygons of type 1 can [after some simple mathematical manipulations analogous to (A4.20)] be reduced to the *convex problem* of type (A4.3)–(A4.6), that is, to the *coverage by a multidimensional rectangle.*

Let us consider now the coverage problem of type 2. Assume the polygon with which we seek to cover the given configuration D to be given in normalized form [(A4.15)]. We assume also that $0 \in \Omega_{\text{nor}}$, that is, $b_i \le 0$, $i = 1, \ldots, m$. Taking an arbitrary point $x^0 \in R^n$, we define the displacement of the polygon Ω_{nor} by x^0 as

$$\Omega_{\text{norm}}^{x0} \stackrel{\text{det}}{=} \{x \in R^n : x - x^0 \in \Omega_{\text{nor}}\}$$

$$= \{x \in R^n : A_i(x - x^0) + b_i \le 0, i = 1, \ldots, m\} , \qquad \text{(A4.21)}$$

and the *contraction* or *expansion* of the polygon by a factor k $(k > 0)$ relative to x^0, as

$$\Omega_{\text{nor}}^{x0} : k \stackrel{\text{det}}{=} \{x \in R^n : A_i(x - x^0)k + b_i \le 0, \; i = 1, \ldots, m\} . \qquad \text{(A4.22)}$$

Apparently, the contraction (expansion) operation transforms the polygon into another similar one obtainable by changing the scale by the same factor along all axes, with center at x^0.

Lemma 5 If $k > 1$, then $\Omega_{\text{nor}}^{x0} : k \subseteq \Omega_{\text{nor}}^{x0}$.

Proof Let $x \in \Omega_{\text{nor}}^{x0} : k$. Then

$$A_i(x - x^0)k + b_i \leq 0 \rightarrow A_i(x - x^0) + \frac{b_i}{k} \leq 0 , \qquad i = 1, \ldots, m .$$

Since $b_i \leq 0$, we have

$$A_i(x - x^0) + b_i \leq A_i(x - x^0) + \frac{b_i}{k} \leq 0$$

and therefore $x \in \Omega^{x0}_{\text{nor}}$. Q.E.D.

We shall now present an algorithm for the coverage problem of type 2. Denote

$$C_i = \max_{x \in D} A_i x . \tag{A4.23}$$

It is obvious that for $\Omega^{x0}_{\text{nor}}: k$ to cover D, the following chain of inequalities must be valid for every $x \in D$:

$$A_i(x - x^0)k + b_i \leq 0 \rightleftarrows A_i x + \frac{b_i - A_i x^0 k}{k}$$
$$\leq 0 \rightarrow C_i + \frac{b_i - A_i x^0 k}{k} \leq 0 .$$

The last inequality can be rewritten as

$$\frac{b_i}{k} \leq A_i x^0 - C_i . \tag{A4.24}$$

Consider the following two cases:

(a) If $A_i x^0 - C_i \geq 0$, then (A4.24) is satisfied for all $k > 0$, since $b_i \leq 0$.

(b) If $A_i x^0 - C_i < 0$ for some i, then we have, from (A4.24), $k \leq b_i/(A_i x^0 - C_i)$.

Denote

$$k^*(x^0) = \min_i \frac{b_i}{A_i x^0 - C_i} .$$

Here $k^*(x^0) \geq 0$, since $b_i \leq 0$, $i = 1, \ldots, m$.

Substituting $k^*(x^0)$ in (A4.22), we find the optimal covering polygon $\Omega^{x0}_{\text{nor}}: k^*(x^0)$ with center at x^0. To find the optimal among all possible covering polygons, we must solve

$$k^*(x^*) = \sup_{x^0 \in R^n} k^*(x^0) .$$

Substituting $k^*(x^*)$ for k and x^* for x_0 in (A4.22), we obtain the optimal covering polygon Ω_{nor}: $k^*(x^*)$ of type 2.

The algorithm for finding Ω_{nor}: $k^*(x^*)$ consists of the following steps:

1. Solve $C_i = \max_{x \in D} A_i x$, $\forall i = 1, \ldots, m$.

2. Choose $x^0 \in R^n$ arbitrary.

3. Find the indices i for which

$$A_i x^0 - C_i < 0 \, . \tag{A4.25}$$

4. For all the indices defined in step 3, find

$$\min_i \frac{b_i}{A_i x^0 - C_i} = k^*(x^0) \, . \tag{A4.26}$$

5. Solve

$$k^*(x^*) = \sup_{x^0 \in R^n} k^*(x^0) \, . \tag{A4.27}$$

6. Substitute $k^*(x^*)$ for k and x^* for x^0 in (A4.22).

This algorithm enables us to find the optimal covering polygon Ω_{nor}: $k^*(x^*)$ with sides parallel to those of Ω_{nor}. Now, rotating the coordinates as described earlier, we obtain the coefficient, say, $k^*(x^*, \alpha)$, and thus the optimal polygon for each angle α. The one with minimum volume that covers D is obtainable from (A4.22) with the optimal coefficient k found from the solution of

$$\max_\alpha k^*(x^*, \alpha) \, . \tag{A4.28}$$

Note that problems (A4.27) and (A4.28) are generally *nonconvex* optimization problems. We shall show now that problem (A4.27) is equivalent to

$$k^*(x^*) = \sup_{x^0 \in S} k^*(x^0) \, , \tag{A4.29}$$

where

$$S = \{x \in R^n \colon |x - y_0| \le rl\} \, , \quad r = \sup_{x \in \Omega_{\text{nor}}} |x| \, , \tag{A4.30}$$

where $l = 1/k^*(x^0)$, x^0, y_0 being arbitrary, and such that $x^0 \in R^n$, $y_0 \in D$.

Since (A4.29) has a compact feasible set, it follows that for it [and thus for (A4.27)] the optimal solution exist.

To prove equivalence of (A4.27) and (A4.29), let us note that $x \in \Omega^x_{\text{nor}}$ since $b_i \le 0$. We have now

$$\sup_{x_1 \in \Omega^x_{\text{nor}}:\, k} |x - x_1| = \frac{r}{k} \,.$$

While covering D we shall, of course, cover $y_0 \in D$. However, covering y_0 implies the establishment of $|y_0 - x| \le r/k$. But if $|x - y_0| > rl$, then we would obtain $k < 1/l = k^*(x^0)$, which is meaningless since we already have $k^*(x^0)$.

The following example illustrates the above results for $\alpha = 0$, that is, when the sides of the optimal covering polygon are parallel to those of Ω_{nor}.

Example 5 Let $D = D_1 \cup D_2$ (Fig. 19),

$$D_1 = \{(x_1, x_2): x_1^2 + x_2^2 \le 1, x_2 \ge 0\} \,,$$

$$D_2 = \{(x_1, x_2): x_2 + 2x_1 \ge -2, x_2 - 2x_1 \ge -2\} \,,$$

and Ω_{nor} is a regular hexagon (Fig. 20) with the length of its sides equal to unity and center at origin; that is, Ω_{nor} is given by

$$x_2 - 0.866 \le 0 \,, \qquad -x_2 - 0.866 \le 0 \,,$$

$$-1.732x_1 - x_2 - 1.732 \le 0 \,, \qquad 1.732x_1 - x_2 - 1.732 \le 0 \,,$$

$$1.732x_1 + x_2 - 1.732 \le 0 \,, \qquad -1.732x_1 + x_2 - 1.732 \le 0 \,.$$

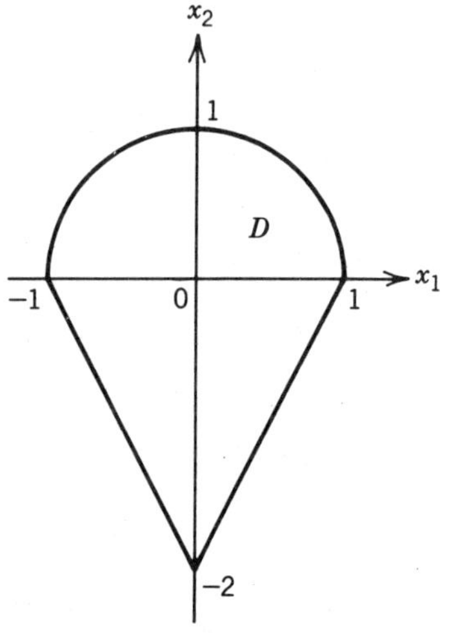

Fig. 19.

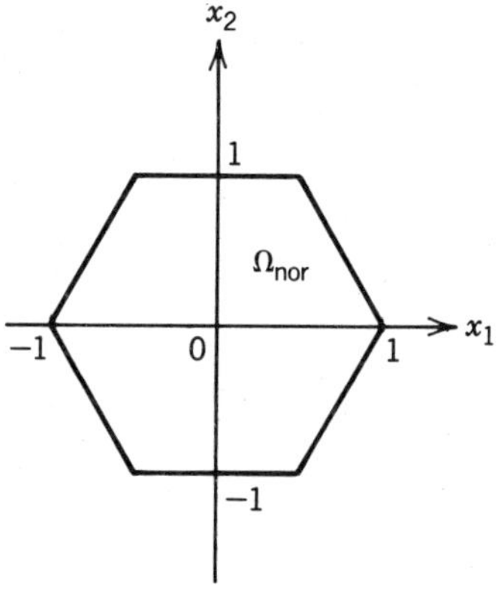

Fig. 20.

Applying steps 1–6 of the algorithm for finding Ω_{nor}, we have, respectively,

Step 1.

$$C_1 = \max_{x \in D} A_1 x = \max_{x \in D} x_2 = 1 ,$$

$$C_2 = \max_{x \in D} A_2 x = \max_{x \in D} (-x_2) = 2 ,$$

$$C_3 = \max A_3 x = \max_{x \in D} (-1.732x_1 - x_2)$$

$$= \max\{-1.732 \times (-1) - 1 \times 0, -1732 \times 0 - 1 \times (-2)\} = 2 ,$$

$$C_4 = \max_{x \in D} A_4 x = \max_{x \in D} \{1.732x_1 - x_2) = 2 ,$$

$$C_5 = \max_{x \in D} A_5 x = \max_{x \in D} \{1.732x_1 + x_2) .$$

To find C_5, we consider the following Lagrange function:

$$g(x, \lambda) = (1.732x_1 + x_2) - \lambda(x_1^2 + x_2^2 - 1) ,$$

from which we have

$$\frac{\partial g}{\partial x_1} = 1.732 - 2\lambda x_1 = 0 ,$$

$$\frac{\partial g}{\partial x_2} = 1 - 2\lambda x_2 = 0 ,$$

$$\frac{\partial g}{\partial \lambda} = x_1^2 + x_2^2 - 1 = 0 .$$

Solving this system of equations, we obtain $\lambda = 1$, $x_1 = 0.866$, $x_2 = 0.5$, and finally $C_5 = 2$. Similarly to C_5, we have

$$C_6 = \max_{x \in D} A_0 x = \max_{x \in D} (-1.732x_1 + x_2) = 2 .$$

Step 2. Assume $x^0 = (0, 0)$.

Step 3. $A_i x^0 - C_i < 0$ for all i, that is, for $i = 1, 2, 3, 4, 5,$.

Step 4. $\min_i \dfrac{b_i}{A_i x^0 C_i} = k^*(x^0) = \dfrac{-0.866}{-2} = 0.433.$

Step 5. Solve problem (A4.27), that is, $k^*(x^*) = \sup_{x^0 \in R^2} k^*(x^0)$.

Instead of problem (A4.27), we shall solve the equivalent problem [(A4.29)], that is,

$$k^*(x^*) = \sup_{x^0 \in S} k^*(x^0) ,$$

where

$$S = \{x \in R^2 \colon |x - y_0| \le rl) .$$

We have

$$r = \sup_{x \in \Omega_{nor}} |x| = 1 , \qquad l = \frac{1}{k^*(x^0)} = \frac{1}{0.433} = 2.31 .$$

Assuming $y_0 = (0, 0)$, we obtain

$$S = \{(x_1, x_2)\colon x_1^2 + x_2^2 \le 2.31^2 = 5.33\} .$$

The solution of $k^*(x^*)$ was approximated by applying the grid method to the square

$$\hat{S} = \{(x_1, x_2)\colon -2.4 \le x_1 \le 2.4,\ -2.4 \le x \le 2.4\} , \qquad \hat{S} \supset S .$$

From the symmetry of D about x_2, it follows that the optimal value of x_1 is zero.

Table 4 presents values of $k^*(x) = k^*(0, x_2)$ for different x_2 with grid

Table 4

x_2	$k(0, x_2)$	x_2	$k(0, x_2)$
−1.2	0.394	0.1	0.412
−1.1	0.412	0.2	0.394
−1.0	0.433	0.3	0.377
−0.9	0.456	0.4	0.361
−0.8	0.481	0.5	0.346
−0.7	0.509	0.6	0.333
−0.6	0.541	0.7	0.321
−0.5	0.577	0.8	0.309
−0.4	0.541	0.9	0.299
−0.3	0.509	1.0	0.289
−0.2	0.481	1.1	0.279
−0.1	0.456	1.2	0.771
0	0.433		

step equal to 0.1. The optimal $k^*(0, x_2)$ equals 0.577 and corresponds to $x_2 = -0.5$.

Step 6. The optimal hexagon can be presented as [see (A4.22)]

$$\Omega_{\text{nor}}: k^*(x^*) = \{x \in R^2: A_i(x - x^*)0.577 + b_i \leq 0\,,\ i = 1, \ldots, 6,$$
$$\text{where } x^* = (0, -0.5)\}\,.$$

The length of each of its sides equals $1/k^*(x^*) = 1/0.5777 = 1.733$, with the center at $x^* = (0, -0.5)$. Comparing the areas of $\Omega_{\text{nor}}^{x_0}: k^*(x^0)$ and the optimal $\Omega_{\text{nor}}: k^*(x^*)$, we find

$$\frac{\text{mes}\, \Omega_{\text{nor}}^{x_0}: k^*(x^0)}{\text{mes}\, \Omega_{\text{nor}}: k^*(x^*)} = 1.78\,.$$

B. Coverage with Quadratic Configurations

In this section we consider the problem of coverage of convex regions D with a hypersphere and a hyperellipsoid of minimal volume.† We shall show that in this case, we have to solve *two convex programming problems* [see (A4.31) and (A4.33)]. Consider first the coverage problem with a hypersphere.

As the first step, we choose some point $y \in R^n$. Then, solving the (first) convex programming problem,

$$\max_{x \in D} ((x - y), (x - y))\,, \tag{A4.31}$$

we find the minimal hypersphere with center y that covers D.

Let $r^2(y)$ be the solution of (A4.31). Lemma 6 proves that $r^2(y)$ is a convex function of y.

Lemma 6 $r^2(y)$ is a convex function of y.

Proof It is obvious that $\rho(x, y) = ((x - y), (x - y))$ is convex with respect to both x and y. So for every $x \in D$ we have

$$\begin{aligned}\rho[x, \alpha y_1 + (1 - \alpha)y_2] &\leq \alpha\rho(x, y_1) + (1 - \alpha)\rho(x, y_2)\\ &\leq \alpha r^2(y_1) + (1 - \alpha)r^2(y_2)\,.\end{aligned} \tag{A4.32}$$

† Due to Rockafellar, R. T., *Convex Analysis*, Princetion University Press, 1970. We can consider CoD instead of D, if D is not convex.

Maximizing (A4.32) with respect to x, we obtain

$$r^2[\alpha y_1 + (1-\alpha)y_2] \le \alpha r^2(y_1) + (1-\alpha) r^2(y_2)\,. \qquad \text{Q.E.D.}$$

In order to find the optimal covering hypersphere of minimal radius, we have to solve the (second) convex programming problem

$$R^2 = \inf_{y \in R^n} r^2(y)\,. \tag{A4.33}$$

The following lemma shows that it suffices to search for the optimal solution within the set D rather than in R^n.

Lemma 7 If D is closed, then problem (A4.33) is equivalent to

$$R^2 = \min_{y \in D} r^2(y)\,. \tag{A4.34}$$

Proof Since $r^2(y)$ is convex, it is also continuous. Taking now any hypersphere with radius r^* that covers D (such a hypersphere always exists because D is assumed to be bounded), we construct a new set

$$G = \{x \in R^n \colon \rho_D(x) \le r^*\}\,,$$

where $\rho_D(x)$ is the distance from x to D. Since D is a compact set, so is G, and it is readily seen that problem (A4.33) is equivalent to

$$\inf_{y \in G} r^2(y) \tag{A4.35}$$

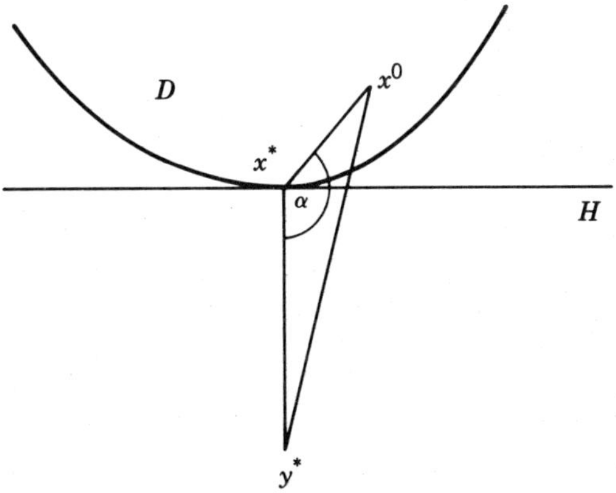

Fig. 21.

[as for the remaining points $r^2(y) > r^{*2}$, a radius of r^* we have already achieved]. Since G is compact and $r^2(y)$ is continuous, the infimum of (A4.35) is achieved at some point, say y^*, that belongs to G.

Assume that $y^* \not\in D$. Then by the closedness of D: (a) $\rho_D(y^*) = \varepsilon > 0$; (b) there exists a point $x^* \in D$ such that $\rho(x^*, y^*) = \varepsilon$, where $\rho(x^*, y^*)$ is the distance between x^* and y^*. Now we can draw a hyperplane H (see Fig. 21) through the point x^*, orthogonal to the line x^*y^*; this hyperplane obviously separates D and y^*. Take additionally any point $x^0 \in D$ and consider the triangle (x^0, x^*, y^*) (see Fig. 21). Since the hyperplane H separates D and y^*, the angle α (see Fig. 21) is obtuse and therefore $\rho(x^*, x^0) < \rho(y^*, x^0)\ \forall x_0 \in D$. But this contradicts the fact that y^* is the solution of (A4.33). Q.E.D.

Example 6 Let D be defined by (see Fig. 22)

$$x_1 + x_2 \le 1\,, \qquad x_2 - x_1 \le 1\,,$$

$$|x_1| \le 1\,, \qquad x_2 \ge -1\,.$$

Denote (see Fig. 22) by Ω_A the set of points whose distance to A is maximal among all other points of the region $D = ABCDE$. Similarly,

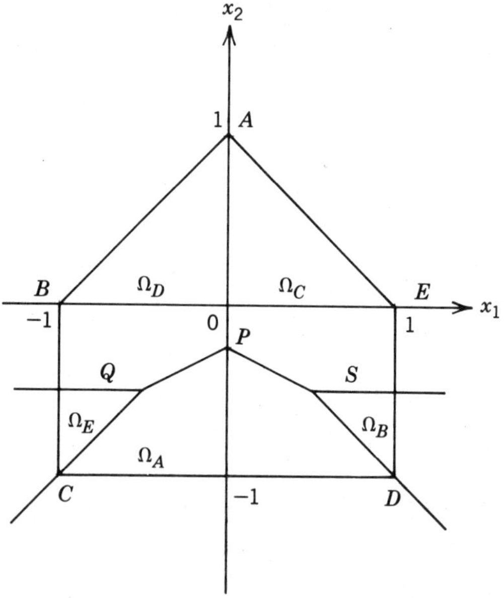

Fig. 22.

denote Ω_B, Ω_C, Ω_D, and Ω_E. Solving (A4.33) now separately for Ω_A, Ω_B, Ω_C, Ω_D, and Ω_E, we easily obtain

$$P = (0, -0.25) \quad \text{is the solution of (A4.33) for } \Omega_A ;$$

$$Q = (-0.5, 0.5) \quad \text{is the solution of (A4.33) for } \Omega_B, \Omega_C ;$$

$$S = (0.5, -0.5) \quad \text{is the solution of (A4.33) for } \Omega_D, \Omega_E .$$

Comparing the covering hyperspheres with centers at P, Q, and S, respectively, we readily find that the optimal one corresponds to that with center at P. It is also readily shown that $R = PA = 1.25$, and

$$\frac{\text{mes } D}{\pi R^2} = 0.611 .$$

Consider now the coverage problem with a hyperellipsoid. Nothing that a hyperellipsoid is essentially a flattened hypersphere, it is easily shown that under a proper transformation of the coordinates, all the results obtained for the hypersphere are valid for the hyperellipsoid. Indeed, consider, for example, the two-dimensional case. The new coordinates are $\tilde{x} = (\tilde{x}_1, \tilde{x}_2) = (x_1, (a/b)x_2)$, where a, b are, respectively, the major and minor axes of the ellipsoid. Another possibility is to solve, instead of (A4.31), the problem

$$\max_{x \in D} \{(x - y)A(x - y)\} ,$$

where A is a given positive-definite matrix of the hyperellipsoid, problems (A4.32)–(A4.35) remaining unchanged.

REFERENCES

Anderson, R. L. (1958). Recent Advances in Finding Best Operating Conditions, *J. Amer. Stat. Assoc.*, **48**, 789–798.

Antonov, I. V. and V. Ya. Katkovnik (1972). Generalization of the Concept of Statistical Gradient, *Automat. Remote Control*, No. 6, 26–33.

Archetti, F. (1975). A Sampling Technique for Global Optimization, in *Towards Global Optimization*, edited by L. C. W. Dixon and G. P. Szego, North-Holland, American Elsevier, New York.

Archetti, F. and F. Schoen (1984). A Survey on the Global Optimization Problem: General Theory and Computational Approaches, *Ann. Oper. Res.*, **1**, 87–110.

Archetti, F., B. Betro, and S. Steffe (1975). A Theoretical Framework for Global Optimization via Random Sampling, Quaderni Dipartimento Ricerca Operative Scienze Statische, Universitá di Pisa, A-25.

Avriel, M. (1976). *Nonlinear Programming. Analysis and Methods*. Prentice-Hall, Englewood Cliffs, NJ.

Avriel, M. and I. Zang (1974). Generalized Convex Functions with Applications to Nonlinear Programming, Chapter 2 in *Mathematical Programs for Activity Analysis*, North-Holland, Amsterdam.

Baba, N., T. Shoma, and Y. Sawaragi (1977). A Modified Convergence Theorem for a Random Optimization Algorithm, *Infor. Sci.*, **13**.

Boender, C. G. E., A. M. G. Rinnoy Kan, L. Stougie, and G. T. Timmer, (1982). A Stochastic Method for Global Optimization, *Math. Programm.*, **22**, 125–140.

Brooks, S. H. (1958). A Discussion of Random Methods for Seeking Maxima, *Oper. Res.*, **6**, 244–251.

Chichinadze, V. K. (1967). Random Search to Determine the Extremum of the Function of Several Variables, *Eng. Cybern.*, **1**, 115–123.

Dam, W. B., J. B. G. Frenk, and J. Telgen (1983). Randomly Generated Polytops for Testing Mathematical Programming Problems, *Math. Program.*, **26**, 172–181.

Deak, I. (1979). Comparison of Methods for Generating Uniformly Distributed Random Points in and on a Hypersphere, *Problems Control Infor. Theory*, **8**, No. 2, 105–113.

De Haan, L. (1981). Estimation of the Minimum of a Function Using Order Statistics, *JASA*, **76**, No. 374, 467–469.

Devroy, L. P. (1976). On the Convergence of Statistical Search, *Inst. Elec. Electron. Eng. Trans. Syst., Man, Cybern.*, **6**, 46–56.

Devroy, L. P. (1979). Recursive Estimation of the Mode of a Multivariate Density, *Canad. J. Stat.*, **7**, No. 2, 159–167.

Dixon, L. C. W. (1977). Global Optimization without Convexity. Technical Report N85, The Hatfield Polytechnical Numerical Optimization Center.

Dixon, L. C. W. and G. P. Szegö, Eds. (1975), *Toward Global Optimization*, North-Holland, Amsterdam.

Dixon, L. C. W. and G. P. Szegö, Eds. (1978), *Toward Global Optimization*, Vol. 2, North-Holland, Amsterdam.

Ermakov, S. M. and A. A. Zhiglyavskii (1984). On Random Search for a Global Extremum, *Theory Prob. Its Applic.*, **XXV**, No. 1, 136–141.

Ermoliev, Yu. M. (1969). On the Method of Generalized Stochastic Gradients and Quasi-Fejer-Sequences, *Cybernetics*, **5**, 208–220.

Ermoliev, Yu. M. (1976). *Stochastic Programming Methods*, Nauka, Moscow (in Russian).

Ermoliev, Yu. M. (1983). Stochastic Quasigradient Methods and Their Application to System Optimization, *Stochastic*, **9**, 1–36.

Gaviano, M. (1975). Some General Results on the Convergence of Random Search Algorithms in Minimization Problems, in *Towards Global Optimization*, edited by L. C. W. Dixon and G. P. Szegö, North-Holland, American Elsevier, New York.

Gelfand, S. B. and S. K. Mitter (1985). Analysis of Simulating Annealing for Optimization, in *Proceedings of the 24th Conference on Decision and Control*, Ft. Lauderdale, Florida, pp. 779–786.

Geman, S. and C. R. Hwang (1985). Diffusions for Global Optimization, to appear in *SIAM J. Control and Optimiz.*

Gidas, B. (1985a). Non-Stationary Markov Chains and Convergence of the Annealing Algorithm, *J. Stat. Phys.* **39**, 73–131.

Gidas, B. (1985b). Global Optimization Via the Langevin Equation, in *Proceedings of the 24th Conference on Decision and Control*, Ft. Lauderdale, Florida, pp. 774–777.

Gnedenko, B. V. (1943). Sur la Distribution du Terme Maximum d'une Serie Aleatoire, *Ann. Math.*, **44**, 423–453.

Gupal, A. M. (1977). Direct Method for Solution of Stochastic Programming Problem, *Automatic Remote Control*, No. 4, 502–506.

Gurin, L. S., Y. S. Demarsky, and A. D. Merkulov (1978). *Problems and Methods for Optimal Distribution of Resources*, Nauka, Moscow (in Russian).

Guseva, O. V. (1973). Convergence Rate of Generalized Gradient Method, *Kibernetika*, **6**, No. 6.

Hajek, B. (1985a). A Tutorial Survey of Theory and Application of Simulated Annealing, in *Proceedings of the 24th Conference on Decision and Control*, Ft. Lauderdale, Florida, pp. 755–760.

Hajek, B. (1985b). Cooling Schedules for Optimal Annealing, to appear in *Math. Oper. Res.*

Hicks, J. S. and R. F. Wheeling (1959). An Efficient Method for Generating Uniformly Distributed Points on the Surface of an *n*-dimensional, *Sphere Comm. ACM*, **3**, 17–19.

Katkovnik, V. Ya. (1976). *Linear Estimations and Stochastic Optimization Problems*, Nauka, Moscow (in Russian).

Katkovnik, V. Ya. and Yu. Kulchitsky (1972). Convergence of a Class of Random Search Algorithms, *Automat. Remote Control*, No. 8, 1321–1326.

Kirkpatrick, S., C. D. Gelatt, and M. P. Vecchi, (1983). Optimization by Simulated Annealing, *Science*, **220**, No. 4598, 671–680.

Kreimer J. and R. Y. Rubinstein (1985). Smoothed Functionals and Constrained Stochastic Approximation. Mimeograph Series No. 351 Technion, Haifa, Israel.

Lawrence, J. P. and K. Steiglitz (1972). Randomized Pattern Search, *IEEE Trans. Computers*, **C-21**, 382–385.

Matyas, J. (1965). Random Optimization, *Automat. Remote Control*, **26**, 246–253.

Meerkov, S. M. (1972). Deceleration in the Search for the Global Extremum of a Function, *Automat. Remote Control*, No. 12, 129–139.

Metropolis, N., A. W. Rosenbluth, M. N. Rosenbluth, A. Teller, and H. Teller (1953). Equations of State Calculations by Fast Computing Machines, *J. Chem. Phys.* **21**, 1087–1091.

Mitra, D. and F. Romeo (1986). Convergence and Finite-Time Behavior of Simulated Annealing, to appear in *Advances in Applied Probability*, Sept. 1986.

Muller, M. E. (1959). A Note on a Method for Generating Points Uniformly on *n*-dimensional Spheres, *Comm. ACM*, **2**, 19–20.

Nurminski, E. A. (1973a). Convergence Conditions of Stochastic Programming Algorithms, *Kibernetika*, **3**.

Nurminski, E. A. (1973b). Quasigradient Method for Solving Nonlinear Programming Problems, *Kibernetika*, **1**.

Patel, N. R. and R. L. Smith (1984). The Asymptotic Extreme Value Distribution of the Sample Minimum of a Concave Function under Linear Constraints, *Oper. Res.*, **31**, No. 4, 789–794.

Patel, N., R. L. Smith, and Z. Zabinsky (1985). Pure Applied Search in Monte Carlo Optimization. Manuscript, Dept. of Industrial and Operations Engineering, University of Michigan, Ann Arbor.

Pincus, M. (1968). A Closed Form Selection of Certain Programming Problems, *Oper. Res.*, **16**, 690–694.

Pincus, M. (1970). A Monte-Carlo Method for the Approximate Solution of Certain Types of Constrained Optimization Problems, *Oper. Res.*, **18**, 1225–1228.

Rastrigin, L. A. (1968). *The Stochastic Methods of Search*, Nauka, Moscow (in Russian).

Rastrigin, L. A. and R. Y. Rubinstein (1969). The Comparison of the Random Search and the Stochastic Approximation While Solving the Problem of Optimization, *Automat. Control Comp. Sci.*, **2**, No. 5, 23–29.

Rinnooy Kan, A. M. G. and G. T. Timmer (1983). *Stochastic Methods for Global Optimization*, Erasmus University, Rotterdam, Report 8317/0.

Rinnooy Kan, A. M. G. and G. T. Timmer (1985a). *Stochastic Global Optimization Methods, Part I: Clustering Methods*, Erasmus University, Rotterdam, Report 8539/A.

Rinnooy Kan, A. M. G. and G. T. Timmer (1985b). *Stochastic Global Optimization Methods, Part II: Multilevel Methods*, Erasmus University, Rotterdam, Report 8540/A.

Rubinstein, R. Y. (1969). Convergence of the Random Search Algorithm, *Automat. Control Comp. Sci.*, **3**, No. 1, 46–49.

Rubinstein, R. Y. (1981). *Simulation and the Monte Carlo Methods*, Wiley, New York.

Rubinstein, R. Y. (1982). Generating Random Vectors Uniformly Distributed Inside and on the Surface of Different Regions, *Euro. J. OR*, **10**, 205–209.

Rubinstein, R. Y. (1983). Smoothed Functionals in Stochastic Optimization, *Math. Oper. Res.*, **8**, No. 1, 26–33.

Rubinstein, R. Y. and A. Karnovsky (1978). Local and Integral Properties of a Search Algorithm of the Stochastic Approximation Type, *Stochastic Proces. Appl.*, **6**, 129–134.

Rubinstein, R. Y. and G. Samorodnitsky (1982). The Efficiency of the Random Search Method, *Math. Comp. Simulation*, **XXIV**, 257–268.

Rubinstein, R. Y. and G. Samorodnitsky (1986). Optimal Coverage of Complex Regions, Mimeograph Series N309, *J. Optimization Theory Applic.* (*JOTA*), (to be published).

Rubinstein, R. Y. and I. Weissman (1977). The Monte-Carlo Method for Global Optimization, *Cah. Cen. Etud. Rech. Opér.*, **21**, No. 2, 143–419.

Smith, R. L. (1984). Efficient Monte Carlo procedures for generating points uniformly distributed over bounded regions. *Oper. Res.*, **32**, No. 6, 1296–1308.

Solis, F. I. and I. B. Wets (1981). Minimization by Random Search Techniques, *Math. Oper. Res.* **6**, No. 1, 19–30.

Strongin, R. G. (1972). Simple Search Algorithm for Global Extremum of Function of Several Variables and Its Use in Functions Approximation Problem, *Radiofizika*, **7**, No. 15, 1077–1085.

Timmer, G. T. (1984). *Global Optimization: A Stochastic Approach*, Centrum voor Wiskunde en Informatica, Amsterdam.

von Neumann, J. (1963). *Various Techniques Used in Connection with Random Digits*, Collected works, Pergamon Press, New York, pp. 768–770.

Wasan, M. T. (1969). *Stochastic Approximation*, Cambridge University Press, New York.

Wilde, D. J. (1964). *Optimum Seeking Methods*, Prentice-Hall, Englewood Cliffs, NJ.

Wilks, S. S. (1962). *Mathematical Statistics*, Wiley, New York.

Yakowitz, S. J. and L. Fisher (1972). On Sequential Search for the Maximum of Unknown Function, *J. Math. Anal. Appl.*, **41**, 234–259.

Zielinski, R. (1970). A Monte Carlo Estimation of the Maximum of a Function, *Algorithms*, **VII**, No. 13, 5–7.

Yudin, D. B. (1966). Quantitative Analysis of Complex Systems, Part II, *Eng. Cybern.* No. 1, 1–23.

Index

Applied Probability and Statistics (Continued)

JUDGE, HILL, GRIFFITHS, LÜTKEPOHL and LEE • Introduction to the Theory and Practice of Econometrics
JUDGE, GRIFFITHS, HILL, LÜTKEPOHL and LEE • The Theory and Practice of Econometrics, *Second Edition*
KALBFLEISCH and PRENTICE • The Statistical Analysis of Failure Time Data
KISH • Survey Sampling
KUH, NEESE, and HOLLINGER • Structural Sensitivity in Econometric Models
KEENEY and RAIFFA • Decisions with Multiple Objectives
LAWLESS • Statistical Models and Methods for Lifetime Data
LEAMER • Specification Searches: Ad Hoc Inference with Nonexperimental Data
LEBART, MORINEAU, and WARWICK • Multivariate Descriptive Statistical Analysis: Correspondence Analysis and Related Techniques for Large Matrices
LINHART and ZUCCHINI • Model Selection
McNEIL • Interactive Data Analysis
MAINDONALD • Statistical Computation
MANN, SCHAFER and SINGPURWALLA • Methods for Statistical Analysis of Reliability and Life Data
MARTZ and WALLER • Bayesian Reliability Analysis
MIKÉ and STANLEY • Statistics in Medical Research: Methods and Issues with Applications in Cancer Research
MILLER • Beyond ANOVA, Basics of Applied Statistics
MILLER • Survival Analysis
MILLER, EFRON, BROWN, and MOSES • Biostatistics Casebook
MONTGOMERY and PECK • Introduction to Linear Regression Analysis
NELSON • Applied Life Data Analysis
OSBORNE • Finite Algorithms in Optimization and Data Analysis
OTNES and ENOCHSON • Applied Time Series Analysis: Volume I, Basic Techniques
OTNES and ENOCHSON • Digital Time Series Analysis
PANKRATZ • Forecasting with Univariate Box-Jenkins Models: Concepts and Cases
PIELOU • Interpretation of Ecological Data: A Primer on Classification and Ordination
POLLOCK • The Algebra of Econometrics
PRENTER • Splines and Variational Methods
RAO and MITRA • Generalized Inverse of Matrices and Its Applications
RIPLEY • Spatial Statistics
RUBINSTEIN • Monte Carlo Optimization, Simulation, and Sensitivity of Queueing Networks
SCHUSS • Theory and Applications of Stochastic Differential Equations
SEAL • Survival Probabilities: The Goal of Risk Theory
SEARLE • Linear Models
SEARLE • Matrix Algebra Useful for Statistics
SPRINGER • The Algebra of Random Variables
STEUER • Multiple Criteria Optimization
STOYAN • Comparison Methods for Queues and Other Stochastic Models
STOYAN, KENDALL, and MECKE • Stochastic Geometry and Its Applications
TIJMS • Stochastic Modeling and Analysis: A Computational Approach
TITTERINGTON, SMITH, and MAKOV • Statistical Analysis of Finite Mixture Distributions
UPTON • The Analysis of Cross-Tabulated Data
UPTON and FINGLETON • Spatial Data Analysis by Example, Volume I: Point Pattern and Quantitative Data
WEISBERG • Applied Linear Regression, *Second Edition*

(*continued from front*)